COPPER OXIDE SUPERCONDUCTORS

COPPER OXIDE SUPERCONDUCTORS

Charles P. Poole, Jr.
Timir Datta
Horacio A. Farach

with help from

M. M. Rigney
C. R. Sanders

Department of Physics and Astronomy
University of South Carolina
Columbia, South Carolina

WILEY

A Wiley-Interscience Publication
JOHN WILEY & SONS
New York • Chichester • Brisbane • Toronto • Singapore

Copyright © 1988 by John Wiley & Sons, Inc.

All rights reserved. Published simultaneously in Canada.

Reproduction or translation of any part of this work
beyond that permitted by Section 107 or 108 of the
1976 United States Copyright Act without the permission
of the copyright owner is unlawful. Requests for
permission or further information should be addressed to
the Permissions Department, John Wiley & Sons, Inc.

Library of Congress Cataloging in Publication Data:

Poole, Charles P.
  Copper oxide superconductors / Charles P. Poole, Jr., Timir Datta,
  and Horacio A. Farach: with help from M. M. Rigney and C. R. Sanders.
  p.   cm.

  "A Wiley-Interscience publication."
  Bibliography: p.
  Includes index.

  1. Copper oxide superconductors.   I. Datta, Timir.   II. Farach,
Horacio A.   III. Title.

QC611.98.C64P66   1988
539.6'23-dc 19   88-18569 CIP
ISBN 0-471-62342-3

Printed in the United States of America

10 9 8 7 6 5 4 3 2 1

PREFACE

The unprecedented worldwide effort in superconductivity research that has taken place over the past two years has produced an enormous amount of experimental data on the properties of the copper oxide type materials that exhibit superconductivity above the temperature of liquid nitrogen. The time is now ripe to bring together in one place the results of this research effort so that scientists working in this field can better acquire an overall perspective, and at the same time have available in one place a collection of detailed experimental data. This volume reviews the experimental aspects of the field of oxide superconductivity with transition temperatures from 30 K to above 120 K, from the time of its discovery by Bednorz and Müller in April 1986 until a few months after the award of the Nobel Prize to them in October 1987. During this period a consistent experimental description of many of the properties of the principal superconducting compounds such as BiSrCaCuO, LaSrCuO, TlBaCaCuO, and YBaCuO has emerged. At the same time there has been a continual debate on the extent to which the BCS theory and the electron–phonon interaction mechanism apply to the new materials, and new theoretical models are periodically proposed. We discuss these matters and, when appropriate, make comparisons with transition metal and other previously known superconductors. Many of the experimental results are summarized in figures and tables.

The field of high-temperature superconductivity is still evolving, and some ideas and explanations may be changed by the time these notes appear in print. Nevertheless, it is helpful to discuss them here to give insights into work now in progress, to give coherence to the present work, and to provide guidance for future work. It is hoped that in the not too distant future the field will settle down enough to permit a more definitive monograph to be written.

The literature has been covered almost to the end of 1987, and some 1988 work has been discussed. This has been an enormous task, and we apologize for any omissions in the citing and discussion of articles.

We wish to thank the following for giving us some advanced notice about their work: R. Barrio, B. Battlogg, L. A. Boatener, G. Burns, J. Drumheller, H. Enomoto, P. K. Gallagher, R. Goldfarb, J. E. Graebner, R. L. Greene, J. Heremans, T. C. Johnson, J. K. Karamas, M. Levy, J. W. Lynn, A. Malozemoff, K. A. Müller, T. Nishino, N. Nucker, J. C. Phillips, R. M. Silver, G. Shirane, J. Stankowski, B. Stridzker, S. Tanigawa, G. A. Thomas, and W. H. Weber. We appreciate comments on the manuscript from S. Alterowitz, C. L. Chien, D. K. Finnamore, J. Goodenough, J. R. Morton, and C. Uher, and helpful discussions with J. Budnick, M. H. Cohen, M. L. Cohen, R. Creswick, S. Deb, M. Fluss, A. Freeman, D. U. Gubser, A. M. Hermann, V. Z. Kresin, H. Ledbetter, W. E. Pickett, M. Tinkham, C. E. Violet, and S. A. Wolf. Support from the University of South Carolina, the Naval Research Laboratory, and the National Science Foundation Grant ISP 80 11451 is gratefully acknowledged.

Michael A. Poole helped to develop the computer data storage techniques that were used. Jesse S. Cook is thanked for editorial comments on the manuscript. C. Almasan, S. Atkas, J. Estrada, N. Hong, O. Lopez, M. Mesa, T. Mouzghi, and T. Usher are thanked for their interest in this project.

CHARLES P. POOLE, JR.
TIMIR DATTA
HORACIO A. FARACH

Columbia, South Carolina
July 1988

CONTENTS

COPPER OXIDE SUPERCONDUCTORS

I

INTRODUCTION

A. INTRODUCTORY REMARKS

For the past year there has been a dramatic intensification of activity in the search for a room-temperature superconductor. Until recently the record for the highest superconducting transition temperature T_c was 23.2 K achieved in 1973 for niobium–germanium thin films near the stoichiometric composition Nb_3Ge. In mid-1986 Bednorz and Müller (Bedno) reported transition temperatures in the 30-K range for metallic, oxygen-deficient compounds of the barium, lanthanum, copper oxide system, and the search for a room-temperature superconductor began in earnest. By the end of 1986 and the beginning of 1987 other lanthanum compounds were fabricated which went superconducting close to 40 K (Cavaz, Chuz1, Taka5, Tara1), and shortly after that the yttrium, barium system went superconducting at 85–95 K (Wuzzz, Zhaoz). Early in 1988 superconductivity reached the 120-K range with the discovery of the BiSrCaCuO (Chuz5, Maeda, Mich1) and TlBaCaCuO (Haze1, Sheng, Shen1) systems.

The enthusiasm reached a peak at the hastily organized symposium held during the March 1987 meeting of the American Physical Society in New York where dramatic new discoveries were reported, and on October 14 of that year the Nobel Prize in Physics was awarded to Johannes Georg Bednorz and Karl Alexander Müller for their discovery. During the months between these events many research articles have appeared and many more preprints were in circulation. The two superconducting systems LaSrCuO and YBaCuO have been fairly well characterized by this work, and a coherent overall picture of their properties has emerged. The Materials Research Society meeting held in early December 1987 in Boston confirmed this conclusion by its emphasis on the routine business of systematically determining the physical properties of the materials and under-

1

standing the nature of their superconductivity. Just when the world of superconductivity seemed to be calming down, triple-digit superconductivity became routine when two new systems, BiSrCaCuO and TlBaCaCuO, were added to the
copper oxide family. These new compounds achieve T_c values in excess of 120 K
without the presence of transition ions, which were formerly considered as necessary constituents. The time now seems opportune to assemble the results of the
first year of feverish activity into a review that will integrate the findings and
provide some perspective for the future.

This introductory chapter provides some of the history, comments on the development of the literature, and establishes the nomenclature to be used for the
various superconducting compounds. The second chapter summarizes some of
the earlier work on the transition metal, oxide, and heavy Fermion superconductors, to put the more recent discoveries into perspective. The third and fourth
chapters, respectively, provide an overview of experimental and theoretical
backgrounds relevant to the oxide superconductors. The fifth chapter explains
how to prepare and characterize samples; the sixth chapter presents the details
of the crystallographic structures and compares them to their perovskite prototypes. The seventh–tenth chapters describe in detail the properties of the oxide
superconductors that are, respectively, structural (e.g., defects, anisotropies, instabilities), magnetic (e.g., susceptibility, critical fields), spectroscopic (e.g.,
band gap, photoemission, infrared), and in the transport category (e.g., resistivity, Hall effect, tunneling). The book ends with a short chapter containing our
concluding remarks.

A brief appendix lists the main conversion factors for energy, pressure, magnetic field, and other factors that are used throughout the text. The appendix
also provides numerical values for important quantities, such as the unit cell
volumes of the LaSrCuO and YBaCuO compounds, and the BCS predictions for
such items as the energy gap ratio E_g/kT_C and the specific heat discontinuity
at T_c.

Throughout the book many of the important measurement results are presented in figures and tables. Quite often the figures have been chosen to be representative of particular phenomena. A number of the tables begin with data
from elemental and transition ion superconductors to put the newer oxide compound results in perspective. From time to time the text and tables compare the
results with the BCS theory and point out how well the older and recent superconductors conform to the BCS predictions.

A five-letter code is used to designate the references because it is short enough
not to take up too much space, and long enough to permit most first authors to
be recognized. This is convenient for continuously keeping up to date with the
fast-growing literature. When the first author's name contains less than five letters the blank spaces are filled up with the letter z because our data management
computer program becomes inefficient with blank spaces.

B. HISTORY

1. Early History

In 1908 H. Kamerlingh Onnes started the field of low-temperature physics by liquifying helium in his laboratory at Leiden. Three years later he found that below 4.15 K the dc resistance of mercury dropped to zero. With that finding the field of superconductivity was born. The next year Onnes discovered that the application of a sufficiently strong axial magnetic field restored the resistance to its normal value. In 1913 the element lead was found to be superconducting at 7.2 K, and 17 more years were to pass before this record was surpassed by the element niobium ($T_c = 9.2$ K).

A considerable amount of time passed before physicists became aware of the second distinguishing characteristic of a superconductor, namely, its perfect diamagnetism. In 1933 Meissner and Ochsenfeld found that when a sphere was cooled below its transition temperature in a magnetic field it excluded the magnetic flux.

The report of the Meissner effect stimulated the London brothers to propose their equation, which explained the Meissner effect and predicted a penetration depth Λ for how far a static external magnetic flux can penetrate into a superconductor (Londo). The next theoretical advance came in 1950 with the theory of Landau and Ginzburg, which described superconductivity in terms of an order parameter and provided a derivation for the London equations. Both of these theories were macroscopic in character.

In the same year the isotope effect, whereby the transition temperature decreases when the average isotopic mass increases, was predicted theoretically by H. Fröhlich (Frohl) and discovered experimentally by E. Maxwell and coworkers (Maxwe). This effect provided support for the electron–phonon interaction, the "phonon mechanism" of superconductivity.

Our present theoretical understanding of the nature of superconductivity is based on the BCS microscopic theory, which was proposed by J. Bardeen, L. Cooper, and J. R. Schrieffer in 1957 (Barde). This theory involves the formation of bound electron pairs that carry the supercurrent, and an energy gap that stabilizes the superconductive state. The Landau–Ginzburg and London results fit well into the BCS formalism. Much of the present theoretical debate involves how well, if at all, the BCS theory explains the properties of the new high-temperature superconductors.

The overall status of the field of superconductivity at the end of 1969 was summarized in the two-volume work *Superconductivity* (Parks). This is still the basic comprehensive reference for the properties of the older type superconductors, although some more recent developments, such as heavy Fermion systems, had not made their appearance at that time.

2. Discovery and Development

On April 17, 1986 the brief article "Possible High T_c Superconductivity in the Ba–La–Cu–O System," written by J. G. Bednorz and K. A. Müller (Bedno), was received by the editor of the Condensed Matter section of the West German physics journal *Zeitschrift für Physik,* and thus began the era of high-temperature superconductivity. When the article appeared in print later in the year it met with some initial skepticism because 19 years (1954–1973) of concentrated effort had been expended to raise the transition temperature of a superconductor from 18.1 to 23.2 K, and the previous record was suddenly surpassed by 50%. It is interesting to note that this same year (1986) saw a prediction of superconductivity at 25 K in the A-15 compound Nb_3Si (DewHu).

Other researchers soon became active, and the highest achieved transition temperature began a rapid rise. By the end of 1986 and the beginning of 1987 the La, Sr, Cu system had been fabricated, and it went superconducting close to 40 K at atmospheric pressure (Cavaz, Taka5, Tara1) and up to 52 K under high pressure (Chuzz). Soon thereafter the yttrium, barium system was discovered that went superconducting at 90–95 K (Wuzzz, Zhaoz). Another year passed before the bismuth and thallium systems raised transition temperatures to the triple-digit range (Chuz5, Maeda, Gaoz2, Haze1, Sheng, Shen1). From time to time there are still unconfirmed reports of even higher onsets of superconductivity. The phenomenon of high-temperature superconductivity is here to stay!

3. The Woodstock of Physics

The "Special Panel Discussion on Novel High Temperature Superconductivity" held on March 18th, 1987 at the meeting of the American Physical Society (APS) in New York introduced to the world the recently discovered phenomenon of high-temperature superconductivity (Stron). A last-minute addition to the program, this all-night session, which lasted from 7:30 PM to 3:15 AM the next morning, was attended by a crowd of about 3000 persons. Most of them, unable to fit in the 1200-seat meeting room, watched the proceedings on one of the many TV monitors hastily dispersed throughout the corridors and promenades in the vicinity of the room. There was great excitement in the air, and several scientists referred to it as the "Woodstock of Physics."

The proceedings began with reports from research groups around the world: K. Alex Müller of the IBM Zürich Research Center, Zhao Zhongxian from Beijing, Shoji Tanaka from Tokyo, Bertram Batlogg of AT & T, and C. W. (Paul) Chu of the University of Texas at Houston. There were initial presentations by two invited panels, one of experimentalists and one of theorists, together with discussions and the answering of questions from the audience. This was followed by a seemingly endless sequence of 5-minute contributed reports periodically interrupted by intervals of discussion and questions from the audience. The highest transition temperature reported at this session was 125 K by Constantan Politis of the Karlsruhe Nuclear Research Center in West Germany.

The excitement revolved around the imminent capability of routinely operating superconducting devices at liquid nitrogen temperature and perhaps eventually at room temperature. Because the ceramic materials are brittle, the best possibility seemed to be the use of superconducting films. K. Alex Müller reported the fabrication of such films 400 nm thick at IBM's Yorktown Heights Laboratory, and Aharon Kapitulnic of Stanford discussed measurements made with such films, but neither worker gave many details. Bertram Batlogg showed a thick ring just large enough to fit on a finger and a flexible sheet of the ceramic material, but, unfortunately, the flexibility of the latter was lost when it was sintered to render it superconducting.

A year after the Woodstock event the March 1988 Meeting of the American Physical Society convened in New Orleans, and 62 of the 433 sessions were devoted to superconductivity. A special session was held on the recently discovered bismuth and thallium compounds, with a leadoff talk by Allen Hermann, and it has been referred to as "Woodstock II" (Pool4). This session was attended by an overflow crowd of 750 scientists who listened to reports of the latest research results from 7:30 PM until after midnight.

C. LITERATURE

1. Development of the Literature

The initial report of high-temperature superconductivity by Bednorz and Müller lay dormant for several months before being taken seriously. Then there were some isolated confirmations followed by feverish activity. Preprints began to circulate, talks were presented at meetings, and communications and letters to the editor began to appear. Special topical conferences on the subject were organized, special issues and sections of journals were devoted to the subject, and full-length articles finally appeared in print.

At first there was great excitement because of the inability to predict what would be discovered next, and then articles began to follow the pattern of logical extensions and refinements of previous work.

The literature has been converging to a self-consistent experimental description of the oxide superconductors and their properties. Thus the time is ripe for a review of the present status of the field to summarize what is now known, to organize and tabulate various properties, and perhaps to stimulate future research.

Another justification for this review is the massive volume of literature that is appearing. In the beginning, a large percentage of the articles were concentrated in several important rapid communication journals and the proceedings of topical conferences. Now that the work is becoming more routine, it is appearing in a larger number of journals.

During 1987 several short reviews of work at particular laboratories such as Argonne (Capo2, Jorg2), Berkeley (Cohen), IBM Almaden and Thomas J. Watson Research Centers (Engl1, Maloz), IBM Zürich (Mull1), KfK Karlsruhe (Wuhlz), the Naval Research Laboratory (Osofs), and the Dutch scene (Mydos) have been presented at conferences and/or published.

2. Updating the Literature

A source of references to the recent literature is *High T_c Update*, distributed by the Ames Laboratory of Iowa State University. Volume 1, No. 15 of *High T_c Update* mentions that the Russian literature is more extensive than would be expected from an examination of the English language sources. For example, 62 reports were published in a special supplement of Pis'ma Zh. Eksp. Teor. Fiz. (*46*, Suppl., 1987), representing two months' work in the USSR. The long delay in the availability of the translations prevented us from covering this portion of the literature.

The cutoff date for the literature covered in this review was between November and December of 1987, although some earlier articles were not included. Nevertheless, some articles published after this date have been mentioned, particularly those for which we had preprints. We apologize for any articles published prior to the cutoff date that we have overlooked.

3. General Articles

The phenomenon of superconductivity intrigues the general public as well as the scientific community at large. As a result many popular and semipopular articles have appeared in a variety of publications ranging from the *New York Times* (front page, March 19, 1987) and *Time* magazine (May 11, 1987), which are read by the general public, to more technical reports in, for example, *CERN Courier* (Larb2), *Nature, New Scientist* (Sutto), *Physics Today* (Khura, Khur1), *Research News Science* (Pool4, Robin), and the *Scientific American* (Haze2). No effort has been made to keep track of this type of publication.

D. PACE OF CHANGE

This pace of change, of improvement in superconductors, exceeds that of earlier eras, as the data listed on Table I-1 and plotted in Fig. I-1 demonstrate. In Section B we mentioned the discovery of the first superconductor, mercury, and the subsequent finding that the elements Pb and, eventually, Nb had much higher T_c values. Soon alloys began to be studied, and the so called A-15 compounds Nb_3Sn, Nb_3Ga, and Nb_3Ge appeared over a decade and a half with successively higher T_c values. Finally, the initial report of superconducting behavior in the 30–35-K range inaugurated the new era of high-temperature ceramic superconductivity. Less than a year later the $YBa_2Cu_3O_{7-x}$ class of superconductors was discovered with T_c of 85–95 K, above the temperature (77 K) of liquid nitrogen. In January and February of 1988 the BiSrCaCuO and TlBaCaCuO systems with transition temperatures in the 110–120-K range were reported.

From Table I-1 it can be seen that for for 56 years niobium and its compounds had dominated the field of superconductivity. In addition to providing the highest T_c values, the best magnet materials are NbTi with $H_{c2} = 100$ kg and Nb_3Sn with $H_{c2} = 220$ kg at 4.2 K. This period, from 1930 to 1986, might be called the

obtain higher critical fields is to raise T_c from the assumed 92 K to a higher value because, for example, $T_c = 110$ K could double H_{C2} at 77 K.

The critical current of YBa* is less favorable relative to standard magnet materials (Newho, Schwa), partly because 77 K is so close to T_c. An upper limit on the critical current density is given by the Ginzburg–Landau expression

$$J_{Cmax} = \left[\frac{2}{3} \left(1 - \frac{T}{T_c} \right) \right]^{3/2} \left[\frac{10 H_C}{4 \pi \Lambda} \right] \qquad (I\text{-}1)$$

where for YBa* the thermodynamic field $H_C \approx 1$ T, the penetration depth $\Lambda \approx 0.2$ μm, and $T_c \approx 92$ K. This gives $J_C \approx 3 \times 10^8$ A/cm^2 at 0 K and $J_C \approx 1.2 \times 10^7$ at 77 K, respectively. The former 0 K value $J_{Cmax} = 10 H_C / 4 \pi \Lambda$ is called the depairing current density.

Achievable critical currents are typically a factor of 10 less than the limiting values calculated from Eq. (I-1), and these values are listed in column 5 of Table I-2. The table compares critical fields and critical currents of YBa* thin films at 4.2 and 77 K with the corresponding values for Nb–Ti and Nb$_3$Sn at 4.2 K. We see from the table that YBa* is superior to the standard magnet materials in its critical fields for both 4.2 and 77 K operation, but that its critical current is better only at 4.2 K and not at 77 K. Limits of resistivity between 4×10^{-16} and 7×10^{-23} Ωcm have been reported (Kedve, Skoln, Wells, Yehzz; cf. Section X-B-3).

The ability of copper oxide superconductors to achieve high fields may be limited by the strains that the material can endure due to stresses associated with containing the fields. In addition, wire configurations of YBa* may not give as high fields as the single crystals that have been studied (Jinzz, Jinz1, Jinz3). Much more development work is needed.

For low-field applications refrigeration costs can dominate and make YBa* magnets cheaper to operate. For high-field applications material costs are more important and can render Nb$_3$Sn superior (Malo1).

Superconducting quantum interference devices called SQUIDs have been fabricated from YBa* films (Koch1, Koch2). Both ac and dc Josephson junctions have been operated above 77 K (Tsai1, Zimme; see also Colc1, Kawab).

TABLE I-2. Typical Upper Critical Fields H_{c2} and Critical Currents J_c for Three Superconductors

Material	T_c (K)	Temp (K)	H_{c2} (T)	10% $J_c$max (A/cm^2), calc, $H = 0$	J_c (A/cm^2) for $H = 0$	J_c (A/cm^2) for $H = 4$T
Nb–Ti	9	4.2	10	—	2×10^5	9×10^4
Nb$_3$Sn	18	4.2	20	—	10^7	3×10^5
YBaCuO	92	4.2	120–200	3×10^7	3×10^6	1.7×10^6
YBaCuO	92	77	30	1.2×10^6	10^5	—

F. NOMENCLATURE

Two general classes of oxide superconductors are the lanthanum, strontium copper oxides $(La_{1-x}Sr_x)_{2-y}CuO_{4-\delta}$, which we will refer to as LaSrCuO or LaSr for short, and the yttrium, barium copper oxides $Y_{1-x}Ba_{2-y}Cu_3O_{7-\delta}$, which will be called YBaCuO or YBa. The compound $(La_{1-x}Ba_x)_{2-y}CuO_{4-\delta}$, which is less common than LaSr, will be called LaBaCuO or LaBa. These superconductor types differ (a) in their crystallographic structures, as described in Chapter V, (b) in the mole ratio of Cu to other cations, which is $1:2$ for LaSr and $1:1$ for YBa, and (c) in their transition temperatures, which are ≈ 35 K for LaSr and ≈ 90 K for YBa.

The structure of the LaSrCuO superconductors is based on that of the prototype La_2CuO_4 compound with two La atoms for each Cu, as discussed in Section VI-E. Therefore we will refer to this as the 21 structure. The YBaCuO superconductors have a structure based on that of the prototype $YBa_2Cu_3O_8$ with Y, Ba, Cu in the mole ratio $1:2:3$, as noted in Section VI-F, so this will be called the 123 structure.

A great deal of work has been done finding the optimum conditions and compositions for the fabrication of superconducting samples. As a result, $(La_{0.925}Sr_{0.075})_2CuO_4$, $(La_{0.925}Sr_{0.075})_2CuO_4$, and $YBa_2Cu_3O_7$ are by far the most widely used. We shall refer to these as LaSr*, LaBa* and YBa*, respectively.

For samples $(La_{1-x}Sr_x)_2CuO_4$ and $(La_{1-x}Ba_x)_2CuO_4$ which contain other values of x we will sometimes write LaSr(x), and LaBa(x), respectively. The chemical formula will be written out explicitly for other cases.

Frequently we will not distinguish between compounds that are written with an unknown δ and with $\delta = 0$. Thus both $YBa_2Cu_3O_{7-\delta}$ and $YBa_2Cu_3O_7$ will be called YBa*. This is done because many of the compounds reported as, for example, $(La_{1-x}Sr_x)_2CuO_4$ or $YBa_2Cu_3O_7$, have not actually had their oxygen content determined quantitatively.

Sometimes YBaCuO compounds with a rare earth substituted for the yttrium will be designated by writing, for example, GdBaCuO, HoBa, or DyBa*. No abbreviations will be used for rare-earth-substituted LaSrCuO compounds.

The more recently discovered bismuth and thallium compounds with the formulas $Bi_2Sr_2Ca_nCu_{n+1}O_{6+2n}$ and $Tl_2Ba_2Ca_nCu_{n+1}O_{6+2n}$ will be referred to generically as BiSrCaCuO and TlBaCaCuO, respectively. Another commonly employed notation for these compounds is $(2\ 2\ n\ n+1)$, according to which $Bi_2Sr_2Ca_2Cu_3O_{10}$ is called (2223) and $Tl_2Ba_2Ca_3Cu_4O_{12}$ is referred to as the (2234) compound. In this spirit the $YBa_2Cu_3O_7$ compound is sometimes called (123).

II

TRANSITION METAL, OXIDE, AND HEAVY FERMION SYSTEMS

A. INTRODUCTION

The new superconductors are all oxides that contain the crucial transition element copper and most of them contain other transition elements such as yttrium, lanthanum, or another rare earth. Consequently, it will be appropriate to survey the properties of the older type superconductors that contain transition elements. This group has produced some of the highest transition temperatures in the past. In addition, some of the properties of transition metal superconductors are likely to provide valuable clues for understanding the nature of the newer superconductors.

The status of research on the superconducting transition elements and their compounds is surveyed through 1981 in the book by Vonsovsky, Izyumov, and Kurmaev (Vonso) and in the earlier article by Gladstone, Jensen, and Schrieffer (Glads). Several of the figures and some of the data in the tables of this chapter are from these works.

B. ELEMENTS

The majority of the elements that are superconducting are members of the various transition series, namely, the first transition series from scandium to zinc that has an incomplete $3d^n$ electron shell, the second transition series from yttrium to cadmium with $4d^n$ electrons, the third such series from lutecium to mercury with $5d^n$ electrons, the rare earths from lanthanum to lutetium that have an incomplete $4f^n$ electron shell, and the actinides from actinium to lawrencium with $5f^n$ electrons. Figure II-1 shows which elements are superconducting

11

Fig. II-1. Periodic table showing the elements that are superconducting. Elements that only become superconducting in thin films, under pressure, or after irradiation are so indicated. Regular superconductors have their transition temperature T_c in the upper right corner of the box and from top to bottom the Debye temperature θ_D, the electronic specific heat factor γ, the electron–phonon coupling constant λ, and the density of states $N(E_F)$ at the Fermi level. This format and the units used are shown. The number N_e of (valence) electrons outside of closed shells, 1–12, is shown for each column under the main body of the table. The Sn data are for white tin.

Legend for each box:

SYMBOL	T_C		units
	DEBYE TEMP		KELVINS / KELVINS
	ELECT. SP HEAT		mJ/MOLE K
	E - PH COUPL		DIMENSIONLESS
	DEN OF ST $N(E_F)$		STATES/ATOM eV

Element data:

Element	T_c	θ_D	γ	λ	$N(E_F)$	Notes
Li						FILM
Be	0.03					
Sc	0.01	470	10.9			
Ti	0.4	415	3.3	0.54	1.4	
V	5.4	383	9.8	1.0	2.1	
Cr						FILM
Y						PRES
Zr	0.6	290	2.8	0.22	0.8	
Nb	9.3	276	7.8	0.85	2.0	
Mo	0.9	460	1.8	0.35	0.6	
Tc	7.8	411	6.3			
Ru	0.5	580	2.8	0.47	0.9	
La	4.9	6.3				(α) (β)
Hf	0.1	252	2.2	0.14	0.8	
Ta	4.4	258	6.2	0.75	1.7	
W	0.02	383	0.9	0.25	0.5	
Re	1.7	415	2.4	0.37	0.74	
Os	0.7	500	2.4	0.44	0.68	
Ir	0.1	425	3.2	0.4	0.35	
Pd						IRRAD
Zn	0.9	316	0.7			
Cd	0.5	210	0.67			
Hg	4.2	75	1.8			
Al	1.2	423	1.4			
Ga	1.1	317	0.60			
In	3.4	108	1.7			
Tl	2.4	88	1.5	0.8		
Si						FILM PRES
Ge						FILM PRES
Sn	3.7	196	1.8			(W)
Pb	7.2	102	3.1	1.55		
P						PRES
As						PRES
Sb						PRES
Bi						FILM PRES
Se						PRES
Te						PRES
Ba						PRES
Cs						FILM PRES
Ce						PRES
Th	1.4	165	4.3			
Pa	1.4					
U						PRES
Eu						FILM
Am	1.0					(β)
Lu	0.1					

Column N_e values (shown under the table): top row 1 2 3 4 5 6 7 8 9 10 11 12 3 4 5 6 7 8.

12

in the bulk state, and also some that only become superconducting in thin films, under pressure, or after irradiation. This figure gives the transition temperature T_c, the Debye temperature θ_D, the Sommerfeld constant or normal state electronic specific heat constant γ from the expression $C_n = \gamma T$, the electron–phonon coupling constant λ (cf. Section IV-B-1), and the density of states $N(E_F)$ at the Fermi level (cf. Sections IV-G and IX-C) for the various superconductors. The columns of the periodic table are labeled with the number of (valence) electrons N_e outside of closed shells. Table II-1 lists various properties of some of the transition elements. Figure II-2, which illustrates how T_c depends on N_e, has two peaks, one near $N_e = 5$ and the other near $N_e = 7$ (Matt2). Graphs of the specific heat constant γ, the magnetic susceptibility $\chi = M/B$ and the inverse Debye temperature squared $1/\theta_D^2$ exhibit the same dependence on N_e, with the $N_e = 7$ peak somewhat suppressed in the Debye case (Glads, Vonso).

Among the elements niobium has the highest transition temperature, and perhaps not coincidently it also is a constituent of higher T_c compounds like Nb_3Ge. Niobium has not appeared prominently in the newer oxide superconductors.

Of the transition elements most commonly found in the newer ceramic type superconductors lanthanum is superconducting with a moderately high T_c (4.88 K for the α or fcc form and 6.3 for the β or hcp form), yttrium becomes superconducting only under pressure ($T_c \approx 2$ K for $110 \leq P \leq 160$ kbar) and copper is not known to superconduct. Studies of the transition temperature of copper alloys as a function of the copper content have provided an extrapolated value of $T_c = 6 \times 10^{-10}$ K for Cu, which is extremely low. The nontransition elements oxygen and strontium in these compounds do not superconduct, barium only does so under pressure ($T_c = 1$–5.4 K for pressures from 55 to 190 kbar), bismuth likewise superconducts only under pressure, and thallium is a superconductor with $T_c = 2.4$ K. Thus the superconducting properties of the elements are not always indicative of the properties of their compounds, although niobium seems to be an exception, as was mentioned above.

C. ALLOYS AND COMPOUNDS

Transition elements combine with a number of other elements to form superconducting materials that sometimes have higher transition temperatures than any of their constituents. These materials may be classified into alloys with the subdivisions solid solutions (with random atomic ordering) and intermetallic compounds or intermetallides (ordered crystallographically), and chemical compounds with the subdivisions ordinary compounds, semiconductors, layered compounds, and polymers. The intermetallides and ordinary compounds provide the highest transition temperatures, with solid solutions and layered compounds also moderately high.

These materials tend to be stoichiometric, and T_c is often sensitive to it. For example, the gradual approach of Nb_3Ge to stoichiometry raised its measured T_c

TABLE II-1. Superconducting Properties of Transition Metals[a,b]

Element	N_e	Crystal Structure	T_c (K)	H_c (mT)	θ_D (K)	γ (mJ/mole K²)	$\chi \cdot 10^6$ (cm³/mole)	λ	$\mu*$	$\partial T_c/\partial P$ (K/GPa)	P (GPa)	α	E_g/kT_c	$N(E_F)$ (No. States/atom eV)
Ti	4	hcp	0.40	5.6	415	3.3	155	0.54		0.6	0–1.4			~1.4
V	5	bcc	5.4	141	383	9.82	300	1.0		6.3	0–2.5		3.4	~2.1
Y	3	hcp	2.7(16 GPa)				183			37	11–16			
Zr	4	hcp	0.61	4.7	290	2.77	129	0.22	0.17	15	0–2.0	0		~0.83
Nb	5	bcc	9.2	206	276	7.80	212	0.85		−2.0	0–2.5		3.6	~2.0
Mo	6	bcc	0.92	9.6	460	1.83	89	0.35	0.09	−1.4	0–2.5	0.37	3.4	0.65
Tc	7	hcp	7.80	141	411	6.28	270			−12.5	0–1.5		3.6	
Ru	8	hcp	0.49	6.9	580	2.8	39	0.47	0.15	−2.3	0–1.8	0	3.5	0.91
La	3	hcp	4.9	80	152					190	0–2.3		3.7	
La	3	fcc	6.0	110	140					110			3.7	
Hf	4	hcp	0.13	1.3	252	2.21	70	0.14	0.10	−2.6	0–1.0			0.83
Ta	5	bcc	4.47	83	258	6.15	162	0.75		−2.6			3.5	~1.7
W	6	bcc	0.015	0.12	383	0.90	53	0.25						~0.5
Re	7	hcp	1.7	20	415	2.35	68	0.37	0.10	−2.3	0–1.8	0.38	3.3	0.76
Os	8	hcp	0.66	7	500	2.35	13	0.44	0.12	−1.8		0.21		0.70
Ir	9	fcc	0.11	1.6	425	3.19	24	0.35						0.70

[a] Includes number of valence electrons N_e, transition temperature T_c, critical field H_c, Debye temperature θ_D, electronic specific heat factor γ, magnetic susceptibility χ, electron–phonon coupling constant λ, coulomb pseudopotential $\mu*$, pressure derivative of T_c, pressure range P, isotopic shift exponent ($\alpha = \frac{1}{2}$ for BCS), energy gap ratio ($E_g/kT_c = 3.53$ for BCS), and density of states at the Fermi level $N(E_F)$.
[b] Compiled mostly from Table 4.1, Vonso, p. 180.

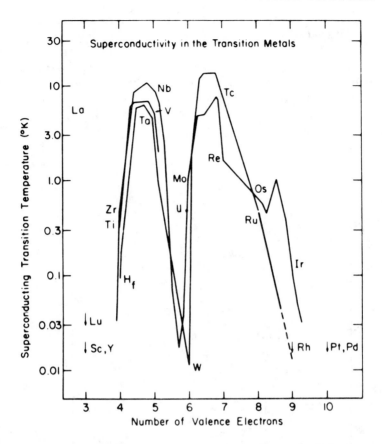

Fig. II-2. The dependence of T_c on the number of valence electrons N_e in elements and solid solutions formed by neighboring transition metals. (From Glads, p. 736; see also Hamil and Vonso, pp. 184, 239.)

from 6 to 17 K and finally to 23.2 K. In contrast, there are cases such as Cr_3Os, Cr_3Ir, Mo_3Ir, Mo_3Pt, and V_3Ir in which T_c is less composition dependent and the highest value does not occur at the stoichiometric composition. This latter case is quite common among the newer superconductors.

Systematic studies of mixed alloys of neighboring transition elements produce a graph similar to Fig. II-2 with intermediate points filled in and the same two maxima. Matthias interpreted these results in terms of the presence of favorable and unfavorable regions of N_e (Matt1). Amorphous alloys only exhibit one maximum for each series. Other properties such as the electronic specific heat factor γ, the magnetic susceptibility χ, the Debye temperature θ_D, and the electron–phonon coupling constant λ have dependencies on electron concentration quite similar to the T_c versus N_e graph of Fig. II-2.

The highest transition temperatures of the older superconductors were obtained with what are called the A-15 intermetallic compounds A_3B of which Nb_3Ge is the prototype. There are two peaks in the T_c versus N_e plot, as shown in Fig. II-3. These compounds are cubic (Pm3n, O_h^3) with the two B atoms in the unit cell at the body center $(\frac{1}{2}\frac{1}{2}\frac{1}{2})$ and apical (000) positions and the six A atoms paired on each face at the sites $[\pm(\frac{1}{4},0,\frac{1}{2}; \frac{1}{2},\frac{1}{4},0; 0,\frac{1}{2},\frac{1}{4})]$, forming chains with spacings of half the lattice constant a. The A atom is always a transition element,

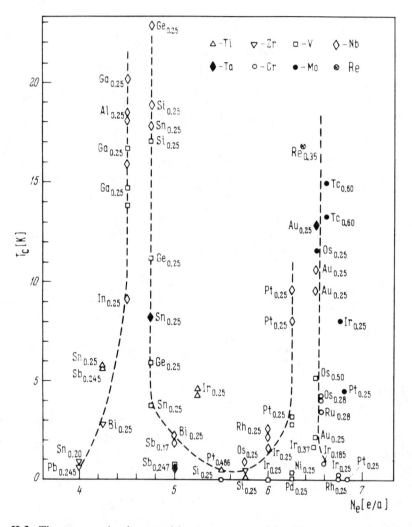

Fig. II-3. The superconducting transition temperature of $A_{0.75}B_{0.25}$ compounds with the A-15 structure as a function of the electron concentration N_e. The A components are designated by the symbols at the top and the B components are specified at the experimental points. (From Vonso, p. 269.)

and sometimes the B atom is also. The superconducting energy-gap data vary over a wide range, from $E_g = 0.2$ to 4.8 for Nb_3Sn and from 1.0 to 3.8 for V_3Si, which suggests a considerable anisotropy of the gap. Some A-15 compounds transform from cubic to tetragonal at T_c. There is no isotope effect and it is difficult to predict the variation of T_c with pressure. The questions of isotopic substitution and pressure in oxide materials will be discussed in Sections VII-G and VII-H, respectively.

The NaCl-type superconductors have metallic atoms A and nonmetallic atoms B arranged on two interpenetrating fcc lattices with each atom in the center of an octahedron of the other type. These superconductors are compositionally stoichiometric, but are not structurally so, since some have substantial concentrations of vacancies in the lattice, as indicated in Table II-2. Nonstoichiometric NaCl type representatives of this class such as $V_{0.763}O_{1.0}$ also exist. Ordinarily the vacancies are random, but sometimes they are ordered and produce a larger unit cell in, for example, $Ti_{1.0}O_{0.7}$. The vacancies can be ordered on both sublattices, as is the case for $Nb_{0.75}O_{0.75}$. Table II-2 lists several NaCl-type compounds with ordered vacancies.

TABLE II-2. Vacancy Content of Stoichiometric Compounds with NaCl Structure[a,b]

Compound	T_c (K)	Vacancy (%)	Ordered Compound	Number of Sublattices
VC	0.03	8.5	$V_{1.0}C_{0.84}$	1
VO	<0.3	11–15	$V_{0.763}O_{1.0}$	1
ZrC	<0.3	3.5		
HfC	<1.2	4		
NbO	1.4	25	$Nb_{0.75}O_{0.75}$	2
YS	1.9	10		
TiO	2.0	15	$Ti_{1.0}O_{0.7}$	1
			$Ti_{0.85}O_{0.85}$	2
ZrS	3.3	20		
TiC	3.4	2		
TiN	5.5	4		
TaN	6.5	2		
VN	8.5	1	$V_{1.0}N_{0.75}$	1
TaC	10.4	0.5	$Ta_{1.0}C_{0.76}$	1
ZrN	10.7	3.5		
NbC	12	0.5–3	$Nb_{1.0}C_{0.75}$	1
NbN	17.3	1.3		

[a] Also shown are vacancy ordering of stoichiometric compounds on two sublattices and of nonstoichiometric compounds on one sublattice.
[b] Data from Vonso, pp. 394–395.

D. TERNARY COMPOUNDS CONTAINING GROUP VI ATOMS

A number of superconducting ternary compounds contain one of the group-VI elements S, Se, Te, or O. An example is the compound $Li_xTi_{1.1}S_2$, where x lies in the range $0.1 < x < 0.3$ with T_c in the range 10–13 K. This compound is unstable from 77 to 600 K, and T_c is sensitive to this instability. There are many superconducting ternary molybdenum sulfides and selenides called Chevrel phases (Fisch) with the general formula $A_xMo_6B_8$, where B is a chalcogenide S, Se, or Te and A can be practically any element. The parameter x assumes various values such as $x = 1$ (e.g., YMo_6S_8, $LaMo_6S_8$), $x = 1.2$ (e.g., $V_{1.2}Mo_6Se_8$), $x = 1.6$ (e.g., $Pb_{1.6}Mo_6S_8$), and $x = 2$ (e.g., $Cu_2Mo_6Se_8$). Substituting oxygen for sulphur in $Cu_{1.8}Mo_6S_8$ raises T_c (Wrigh). Superconductivity and magnetic order are known to coexist in Chevrel phase compounds.

E. OXIDES

A large percentage of the nonalloy type compounds discussed in the book by Vonsovsky et al. contain an element of row VI in the periodic table, namely O, S, Se, or Te, with oxygen the least represented among the group (Vonso). In contrast, the newer superconductors are all oxides. Since the Debye temperature is inversely proportional to the lightest atom in the compound, oxides have the advantage of a high Debye temperature (Gallo). Thus the presence of a group VI element is a commonality that links many of the older and the newer superconductors.

One oxide superconductor is the well-known ferroelectric perovskite $SrTiO_3$ which has a very low transition temperature (0.03–0.35 K). Niobium-doped $SrTiO_3$ with its small carrier concentration $n_e \approx 2 \times 10^{20}$ and a large electron phonon coupling has $T_c = 0.7$ K (Barat, Binni, Chuz1, Thorn). Another oxide example is the spinel $LiTi_2O_4$ with a moderately high T_c of 13.7 K (John2). The system $Li_xTi_{3-x}O_4$ is superconducting in the range $0.8 \leq x \leq 1.33$. It is interesting to note that the stoichiometric compound with $x = 1$ is near the composition where the metal-to-insulator transition occurs. A recent band structure calculation of $LiTi_2O_4$ (Satpa) is consistent with resonance valence bond superconductivity (cf. Section III-G-6) and a large electron–phonon coupling constant ($\lambda \approx 1.8$). Only three more of the 200 known spinels superconduct, and they all contain copper, namely, $CuRh_2Se_4$ with $T_c = 3.5$ K, CuV_2S_4 with $T_c = 4.5$ K, and $CuRh_2S_4$ with $T_c = 4.8$ K.

In their pioneering article Bednorz and Müller (Bedno) called attention to the discovery over a decade ago of superconductivity in the mixed valence compound $BaPb_{1-x}Bi_xO_3$ by Sleight et al. (Gilbe, Sleig, Slei1, Suzu1, Thorn). They pointed out that the stoichiometric form of this compound presumably had the composition $Ba_2Bi^{3+}Bi^{5+}O_6$. It is clear that admixing with $BaPb^{4+}O_3$ together with oxygen deficiency could change the Bi^{5+}/Bi^{3+} ratio. The highest T_c for homogeneous oxygen-deficient mixed crystals of this system was 13 K, and this came with the comparatively low carrier concentration of 2–4×10^{21} (Thanh). The intensity of the strong vibrational breathing mode (cf. Fig. VIII-6) near 100 cm^{-1} was found to be proportional to T_c (Bedno, Masak). These works led Bednorz and Müller to reason that "Within the BCS system, one may find still higher T_c's in the perovskite type or related metallic oxides, if the electron–phonon interactions and the carrier densities at the Fermi level can be enhanced further." It was their determination to prove the validity of this conjecture that led to the biggest breakthrough in physics of the present decade. Their choice of materials to examine was influenced by the 1984 article of Michel and Raveau (Miche) on mixed valent Cu^{2+}, Cu^{3+} lanthanum–copper oxides containing alkaline earths.

A novel family of superconducting oxides close to the composition $Sr_2 Bi_2Cu_2O_{7+\delta}$ with $T_c \approx 22$ K was isolated in mid-1987 (Mich1). It is a semimetal from 100 to 300 K with $\rho \approx 2.5 \times 10^4 \ \mu\Omega$ cm, a value 10 times larger than that found in the LaSrCuO and YBaCuO systems. The authors suggested that the structure may be related to perovskite. This compound was the predecessor of the BiSrCaCuO superconductor with $T_c > 100$ K (Maeda, Chuz5).

Many workers, starting with Bednorz and Müller, have pointed out that the newer superconductors tend to have a structural resemblance to perovskite; we will say more about this in Section VI-D.

F. VALENCE ELECTRONS AND DENSITY OF STATES

We have shown that some transition metal properties, such as the electronic specific heat factor γ, the susceptibility χ, and the Debye temperature θ_D, correlate with the number N_e of valence electrons. The melting points, in contrast, show the opposite correlation since the highest values occur for $N_e = 6$, where T_c is the lowest.

The chemical bonding of the transition metals is mainly ionic, but there can also be covalent contributions. The amount of covalency is particularly strong in the two metals molybdenum and tungsten, which have five valence electrons, and this has been used as an explanation for the low transition temperatures of these two elements. It was suggested (Stern) that the formation of covalent bonds with antiparallel spins "condensed" in coordinate or real space may compete with the formation of Cooper pairs with opposite spins and momenta "condensed" in momentum space.

Another important electronic parameter of a transition metal is its density of states $N(E_F)$ at the Fermi level, and Table II-1 lists $N(E_F)$ for the various transition elements. In some cases the value in the table is an average of several measurements with a large amount of scatter. The d electrons dominate this density of states, with small contributions coming from the remaining valence electrons. For example, in niobium the percentage contributions to $N(E_F)$ from the s, p, d, and f electrons are 3, 14, 81, and 2%, respectively.

When a transition metal is subjected to high pressure the density of states at the Fermi level changes, and this may be detected by the change in the value of the conduction electron heat capacity factor γ. The derivative dT_c/dP is sometimes positive and sometimes negative. The T_c of some transition metals is raised dramatically in thin films.

G. COMMONALITIES OF OLD AND NEW SUPERCONDUCTING COMPOUNDS

The newer superconductors are mainly quaternary compounds such as $(La_{1-x} Sr_x)_2CuO_4$ with a divalent cation such as Sr occupying some of the La sites, and those based on the compound $YBa_2Cu_3O_{7-\delta}$ where a rare-earth ion can replace Y. More recent newer compounds with five types of atoms are BiSrCaCuO and TlBaCaCuO. Thus the newer ones tend to have more types of atoms than the older ones. Another difference, of course, is the tendency for the newer ones to be less stoichiometric, especially with respect to their oxygen content. There are also likenesses between them. A great deal has been said about the layered characteristics of the newer oxide materials. Layered-type superconductors with transition temperatures in the reasonably high range from 4 to 7 K have been known for some time (Vonso). In fact Ginzburg and Kirzhnits devote three whole chapters of their 1977 book (Ginzb) to the possibility of high-temperature superconductivity in layered crystals, systems with one-dimensional anisotropy and sandwich-type compounds. The closeness to an instability condition is another common characteristic that many of the older superconductors share with the newer ones. The crucial role of transition element ions and of a group-VI atom like oxygen or sulphur is common for each group.

H. HEAVY FERMION SYSTEMS

For the 5 or 6 years prior to 1987 much of the superconductivity research that was carried out involved the study of heavy electron, or more generally heavy Fermion, superconductors in which the effective conduction electron mass $m*$ is typically 100 electron masses. The first such superconductor, $CeCu_2Si_2$, was discovered in 1979 (Stegl), and some time passed before the phenomenon was confirmed by the discovery of the further examples UBe_{13} (Ottz1) and UPt_3 (Stew1) in 1983 and 1984, respectively. Since then quite a few additional cases have been found. Many of the investigators who are now active in the field of oxide superconductivity obtained their experience with the heavy Fermion types.

We will not attempt an overall survey of the field here, but rather we refer the reader to several recent reviews (Coles, Mapl1, Ottzz, Stewa) and articles (Choiz, Geshk, Monie, Rodri, Svozi). We will be content to describe the phenomenon and mention some of its interesting features.

The properties of heavy fermion superconductors are due to the presence of transition ions with partially filled f electron shells. For example, cerium is a rare earth with 4f electrons and uranium is an actinide element with 5f electrons, and the electrons in these partially filled f shells are responsible for the formation of the superconducting state. In many other compounds these same f electrons are responsible for producing magnetically ordered systems.

Much of the evidence for the high effective mass comes from experimental observations in the normal state. For example, the dependence of the specific heat of UBe_{13} on the temperature is large and flat above T_c, as expected for heavy mass electrons. The heavy Fermion superconductors have anisotropic properties which manifest themselves in such measurements as ultrasonic attenuation, thermal conductivity, and NMR relaxation. Lower critical fields H_{C1} are several mT at 0 K, while upper critical fields H_{C2} at 0 K approach 1 or 2 tesla. The London penetration depth Λ_L is several thousand angstroms, consistent with the large effective mass $m*$ which enters as a square root factor into the classical expression

$$\Lambda_L = (m*c^2/4\pi n_S e^2)^{1/2} \qquad \text{(II-1)}$$

In ordinary superconductors, magnetic impurities suppress the superconducting state because of the pair-breaking effect of the magnetic moments (Abrik). In anisotropic superconductors, on the other hand, any kind of impurity is pair-breaking. In particular, small amounts of nonmagnetic impurities replacing, for example, the uranium, beryllium, or platinum produce a pronounced lowering of T_c. The suggestion has been made that the heavy Fermion materials involve an unconventional type of superconductivity, one not involving the usual electron–phonon interaction, a claim also made about the newer oxide superconductors.

The heavy Fermion superconductors have very large densities of states $N(E_F)$ at the Fermi level, whereas the oxide types have low values. This is demonstrated by the very large values of the electronic specific heat coefficient γ (several J/mole K^2) in the former and the small values (several mJ/mole K^2) in the latter compounds. The two types may share the common feature of a magnetic mechanism being responsible for the formation of the Cooper pairs (Maple).

There have also been reports of high effective masses in the oxide superconductors. For example, $m*/m \approx 12$ (Mats2) in LaSrCuO, and $m*/m \approx 5$ (Gottw), $m*/m = 9$ (Salam), and $m*/m \approx 10^2$ in YBaCuO (Kres2).

III

BACKGROUND AND OVERVIEW

A. INTRODUCTION

The literature covering the newer oxide superconductors contains many articles devoted exclusively to describing their structural properties, and other articles that provide explanations of these properties involving such factors as copper–oxygen layering schemes, oxygen deficiency, copper valence states, "charge pair" coupling mechanisms, and dimensionality. There are also discussions of how the newer oxide superconductors compare with the previous heavy Fermion, oxide, transition metal, alloy, A-15, and other types. Comparisions are made with respect to energy gap, transition temperature, critical fields, critical current, coherence length, penetration depth, and other superconducting properties.

This chapter provides an overview of the oxide superconductors which addresses some of these questions and at the same time provides an introduction to their salient properties. The properties themselves will be more systematically examined in subsequent chapters.

B. OXIDE-TYPE SUPERCONDUCTORS

It will be appropriate to begin with a short summary of the various oxide materials. There is a sort of hierarchy of these types with gradually increasing transition temperatures:

(a) There are several previously known low-temperature oxide superconductors such as the perovskite $SrTiO_3$ with its very low $T_c \approx 0.3$ and the spinel $LiTi_2O_4$ with its comparatively high T_c of 13.7 K (cf. Section II-E).

(b) A number of researchers (e.g., Maeno, Matth, and Sleig) have discussed the $BaPb_{1-x}Bi_xO_3$ system with its phases of perovskite-related structures (Coxz1, Coxz2, Matt6) which superconduct over the range of x from 0.05 to 0.3 and compared it with LaSrCuO types. In these BaPbBiO compounds Bi can have different valence states, such as Bi^{3+}, Bi^{4+}, and Bi^{5+} (cf. Section II-E).

(c) LaSrCuO with transition temperatures in the range 35–40 K was the initial high-temperature superconductor type. It has the 21 structure and the general formula $(La_{1-x}M_x)_{2-y}CuO_{4-\delta}$, where ordinarily M = Sr or Ba, x is small, and $\delta = 0$. More recently 90-K-range superconductors with this 21 structure have been made in which Y replaces La and x is much larger.

(d) YBaCuO with the 123 structure and the general formula $Y_{1-x}Ba_{2-y}Cu_3O_{7-\delta}$ has its transition temperature in the 90-K range, which is above the boiling point of liquid nitrogen (77 K). Of particular importance has been $YBa_2Cu_3O_7$ with $x = y = 0$, $\delta \approx 0$, although sometimes the yttrium and barium contents are much closer to equal (e.g., Mats1).

(e) The BiSrCaCuO and TlBaCaCuO superconductors, which are quite close to each other in structure, have transition temperatures up to the 120-K range. Some of the early articles on these compounds appeared when this manuscript was being prepared for publication.

C. TRANSITION TEMPERATURE

An important question that arises is how to operationally define the transition temperature T_c. Many phase transitions have a finite width, and when this is the case a typical approach is to define T_c in terms of the point of most rapid change-over from the old to the new phase. Critical exponents are evaluated in this region near T_c. Ordinarily much less account is taken of the more gradual changes that take place at the onset or during the final approach to the new equilibrium state. In contrast to this superconductivity, researchers often place great significance on the onset or highest temperature where superconducting properties begin to manifest themselves, and on the point of zero resistance, which can be significantly lower in temperature.

1. Resistivity Curve Definition

Many authors talk in terms of the onset, 5%, 10%, midpoint, 90%, 95%, and zero resistance points, and Fig. III-1 shows some of these on an experimental resistivity curve. These values are determined by selecting the onset or 0% point at the place where the experimental curve appears to start dropping below the extrapolated high-temperature behavior shown dotted on the figure. The T_c values that we quote or list in tables are ordinarily midpoint values for which $\rho(T)/\rho_0$ is $\frac{1}{2}$. Quite often the 0% point is not easy to select unambiguously. Most of the reports of very high transition temperatures have involved citing onset values.

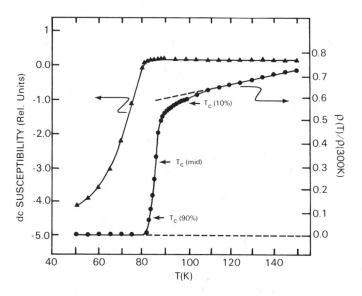

Fig. III-1. Temperature dependence of the magnetization (zero field cooled) and electrical resistivity between 50 and 150 K for HoBa$_2$Cu$_3$O$_7$ (Kuzzz).

2. Derivative Curve Definition

If one obtains digitized susceptibility or resistivity data of the type shown on Fig. III-2, it is an easy matter to calculate the first and second derivatives with respect to the temperature to obtain the plots shown on Figs. III-3 and III-4, respectively (Almas, Babic, Datta, Datt2, Gurv1, Pool3, Pureu). The point at which the first derivative reaches its maximum value could be selected as defining T_c since it is the inflection point or point of most rapid change on the original curve. The width ΔT between the points where the first derivative curve is half of its maximum value, as shown on the figure, is a good quantitative measure of the width of the transition. The second derivative shown on Fig. III-4 permits the transition temperature to be more precisely measured as the temperature where the curve crosses zero, and in addition it gives the peak-to-peak width ΔT_{pp} of the distribution. The width ratio $\Delta T/\Delta T_{pp}$ and the asymmetry parameter $[(A - B)/(A + B)]$ obtained from Fig. III-4 are sensitive measures of the mechanism or model used to explain the shape of the original experimental curve of Fig. III-2.

If the initial susceptibility or resistivity versus temperature curve is antisymmetric about its midpoint, the first derivative curve will be symmetric and the second derivative curve will be antisymmetric. When this is the case, the inflection point of χ versus T occurs at the midpoint and the three curves give the same value of T_c. When the initial curve is unsymmetrical, the first derivative curve is also, the midpoint and inflection points occur at different field positions, and

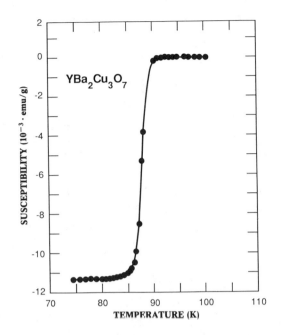

Fig. III-2. Zero-field cooled susceptibility $\chi(T)$ as a function of temperature of a YBa$_2$Cu$_3$O$_7$ specimen for a field of 0.1 mT. (Provided by C. Almasan.)

the plots of Figs. III-2–III-4 do not all give the same value of T_c. The greater the asymmetry, the more these values will differ.

D. RESISTIVITY AND SUSCEPTIBILITY CRITERIA

There are two classic ways to define superconductivity, one involving the property of zero resistance and the other based on perfect diamagnetism, and in an ideal homogeneous superconductor both measurements should provide the same transition temperature. Many articles have reported resistivity and magnetic susceptibility data on the same sample, and in most cases the resistivity versus temperature curve is sharper than its susceptibility counterpart (e.g., Akimi, Grant, Kuzzz, Maple, Sampa, Takag, Yoshi). In addition, sometimes the onset temperature of the resistivity curve occurs at a higher temperature than is the case for the susceptibility (e.g., Grant, Odazz, Sampa), and sometimes the two onsets are close together (e.g., Maple, Takag, Yoshi). Ordinarily T_c determined from the resistivity midpoint is at a higher temperature than its susceptibility counterpart. Figure III-5 shows resistivity, specific heat, Meissner magnetization, and ac susceptibility data obtained on the same YBa$_*$ sample (Juno1). The resistivity curve on this figure is shifted to a relatively higher temperature than is usually the case.

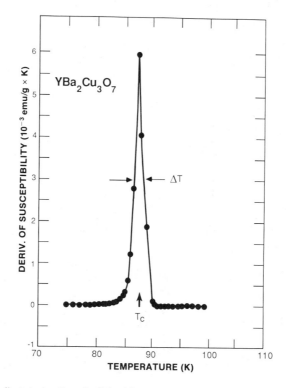

Fig. III-3. The first derivative of $\chi(T)$ with respect to temperature for the data shown in Fig. III-2. Notice the pronounced peak at $T_c = 87.65 \pm 0.15$ K with the width $\Delta T = 0.95 \pm 0.15$ K. (Provided by C. Almasan and J. Estrada.)

Susceptibility measurements determine the magnetic state of the entire sample, and also give a better indication of the extent to which the sample has transformed to the superconducting state. Resistivity measurements merely show when continuous superconducting paths are in place. The dc susceptibility or magnetism measurements provide a better experimental indicator of the overall superconducting state, while the resistivity measurement is a better practical guide for application purposes. We should also note that magnetization is a thermodynamic state variable, whereas resistivity is not.

E. EFFECT OF GRANULARITY

It is fairly well established that these ceramic superconductors have a granular structure with grain sizes that are usually of the order of 1 μm (Farre, Finne, Larba, Ourma, Ourm1). In London's model (Londo) spherical grains of radius R have a susceptibility that depends on the ratio of R to the penetration depth Λ

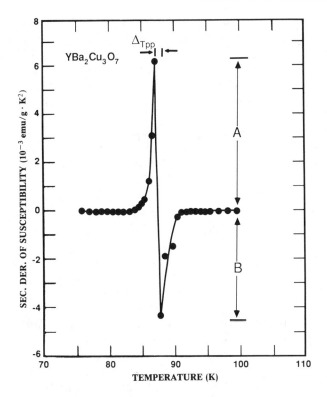

Fig. III-4. Second derivative of $\chi(T)$ with temperature. T_c is the temperature at which the second derivative crosses the base line, and the peak-to-peak transition width is $\Delta T_{pp} = 0.5$ K. (Provided by C. Almasan and J. Estrada.)

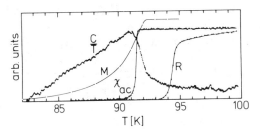

Fig. III-5. Temperature dependence of the resistivity R, ac susceptibility χ_{ac}, Meissner magnetization M (field cooled in 1.7 mT), and specific heat/temperature ratio for a YBa$_*$ sample. The resistivity transition is at a higher relative temperature than is usually the case (Juno1).

which is defined in Section III-G-4, and Λ is strongly temperature dependent near T_C, as will be explained in that Section. Another model considers the grains to have cores of normal metallic material with a surface coating of superconducting material (Ginle, Ginl1, Ventu). It has been suggested that single crystallites may have localized superconducting regions of reduced dimensionality coupled by tunneling and an overall polycrystalline nature with a connectivity that allows total screening (Larb1). In addition, individual grains can have a distribution of sizes and the possibility of a range of transition temperatures which broadens the normal-to-superconductor transformation. All of these factors affect the temperature dependence or the shapes of susceptibility and resistivity versus temperature curves.

Susceptibility measurements tell us the extent to which the grains exclude or expel the applied magnetic flux and resistivity measurements monitor the extent to which the superconducting grains form completed paths between the electrodes.

In an idealized case, the (volume) susceptibility of an individual grain will change from a very small value above T_c to the value $-1/4\pi$ far below T_c as a result of its transformation to the superconducting state. The macroscopically measured susceptibility at an intermediate temperature depends upon the extent to which the grains have undergone this transformation. In practice the situation is more complex owing to the difference between the zero-field and Meissner-effect flux expulsion cases caused by voids, nonsuperconducting inclusions, trapped flux, and weak links which are described in Section X-D-2.

The macroscopic resistivity is also dependent on the granularity of the samples. Each grain of superconducting material contributes to lowering the resistivity only if it links a connected path or contributes to a partial path across the sample. Current that reaches a grain will be shunted or channeled through the superconducting portions of it. The formation of partial zero-resistivity paths will cause the normalized resistivity curve $\rho(T)/\rho_0$ to drop faster initially than its susceptibility counterpart, as shown on Fig. III-1. As more and more grains superconduct, nearby groups and clusters can form that short-circuit the nearby normal material, and there is a possibility that a filamentary type of current-channeling mechanism may cause the resistivity to continue dropping faster than the susceptibility, as indicated on Fig. III-1. Eventually complete superconducting paths will form between the measuring electrodes, and the state of zero resistance may be achieved long before most of the grains become superconducting.

Mathematically, granular resistivity can be treated as a percolation-type problem, and the arrival at the zero-resistance state might occur near the percolation threshold. Some authors (Gebal) attributed sharp resistivity drops to the formation of percolative paths (Larba). In a homogeneous bulk superconductor, on the other hand, the zero-resistance current paths can be along the surface. Finally, granularity, particulary the grain boundries, controls many macroscopic properties of the materials. Some examples of such properties are oxygen diffusion, elasticity, chemical reactivity, and critical current density. Some of these will be discussed later in the appropriate sections.

F. VERY HIGH TRANSITION TEMPERATURES

The LaSrCuO or 21-type superconductors have transition temperatures that are typically 35–40 K, the YBaCuO or 123 types have T_c values of 90–95 K, and the BiSrCaCuO and TlBaCaCuO systems have been reported from 110 to 120 K. In addition to these accepted values many authors have reported such high onsets as ≈ 110 K for $Y_2Ba_3Cu_5O_{25-\delta}$ (Liizz, Zhong), 120 K for the YBaCuO system (Akimi, Raozz), 159 K in fluorine-doped YBaCuO after temperature cycling (Bharg), T_c's in YBaCuO as high as 160 K (Caizz) and 173 K in $(Y_{0.8}Ba_{0.2})$ $Cu_2O_{8-\delta}$ (Kirs3), a 240-K onset based on reverse Josephson behavior (Chenz, Garwi), and low resistive fluctuations as high as 234 K (Bour1). A resistance decrease in $EuBa_2Cu_3O_{7-\delta}$ by 4 orders of magnitude at 230 K disappeared after several thermal cyclings and has not been reproduced (Huang). A resistance loss while cooling $YBa_2Cu_3F_2O_{6-\delta}$ at ≈ 168 K was regained on warming at ≈ 148 K (Ovshi) with a minimal accompanying Meissner anomaly. Exposure to nitrogen gas was also reported to increase T_c of YB* to 130 K (Matt3). A number of additional nonreproducible zero-resistive claims have been reported (Ayyub, Bharg, Bour3, Gebal, Kirs1, Paulo, Swinb, Weng1).

A few samples exhibited nonreproducible zero-resistance anomalies as high as 260 K produced by temperature-dependent current contacts. This could explain some of the earlier reported anomalies (Torra); others might arise from filamentary conductivity or perhaps a high T_c minority fraction, less than 1% of the sample (Capli, Raubz). The existence of isolated regions with transition temperatures of $4T_c$ has been postulated (Phill). The rounding of the resistivity curve around and above T_c can be a manifestation of thermodynamic fluctuations rather than a signature of the onset of a higher T_c phase (Freit).

Perhaps it should be mentioned that in the early 1970s there were a number of unconfirmed reports of superconductivity above 77 K, even as high as 30°C (Edels), but these are no longer believed. We mentioned the two criteria for the establishment of superconductivity: (a) zero resistivity and (b) the Meissner effect. Two more constraints can be added, namely (c) stability and (d) reproducibility (Chuz4), and these would rule out the sporadic recent reports of very high transition temperatures, as well as the various unconfirmed reports from the 1970s.

G. CHARACTERISTIC LENGTH PARAMETERS

There are several characteristic lengths that take part in determining the properties of superconductors, and in the order of increasing magnitude they are lattice parameter (4–13 Å), average distance between conduction electrons (≈ 10 Å), coherence length (14–32 Å), penetration depth (1400–3300 Å), grain size (0.05–2 μm), and sample size (1 mm–1 cm). Typical ranges of values are given in parentheses for each. For comparison it might be mentioned that a superconducting magnet wire is sometimes 5 μm in diameter (Cheng, Grego).

1. Unit-Cell Dimensions

The new oxide superconductors have crystallographic structures which are tetragonal, or orthorhombic close to tetragonal, with two of the lattice parameters $a \approx b \approx 3.8$ Å the same. The third lattice parameter c varies with the compound, with the values 13.2 Å for LaSrCuO, 11.7 Å for YBaCuO, and ≈ 30 Å for BiSrCaCuO and TlBaCaCuO. Details are presented in Sections VI-E,F,H,I, respectively.

2. Conduction Electron Separation

The average distance between conduction electrons in a conductor may be taken as the cube root of the reciprocal of the electron density $n_C \approx 10^{21}$ electrons/cm³, corresponding to an electron–electron separation of $(1/n_C)^{1/3} \approx 10$ Å.

3. Coherence Length

There are several ways to define the coherence length ξ of a superconductor: (1) it is the spacial dimension over which the magnetic field B of the London equation must be averaged to provide the associated current density J; (2) it is the distance over which the energy gap cannot change appreciably in a spacially varying magnetic field; (3) it is the minimum width of an intermediate layer between normal and superconducting material; and (4) it is related to the upper critical field H_{C2} through the expression $\Phi_0 \approx 2\pi \xi^2 H_{C2}$. Since we are dealing with Type II superconductors the coherence length is shorter than the intrinsic coherence length ξ_0 given by the Pippard expression $\xi_0 = 0.39 h v_F / \pi E_g$ in the BCS theory (Tinkh). Some articles mention the temperature-independent Pippard coherence length and others give the Ginzburg–Landau coherence length which agrees with the Pippard expression for pure Type I materials well away from T_c, and diverges as $(T_c - T)^{-1/2}$ at the approach to T_c.

In impure or "dirty" superconductors the coherence length is reduced by electron scattering, and it becomes dependent on the mean free path ℓ through the expression (Tinkh)

$$\frac{1}{\xi} = \frac{1}{\xi_0} + \frac{1}{\ell} \qquad \text{(III-1)}$$

which gives $\xi \approx \ell$ in the extreme dirty limit $\ell \ll \xi_0$. The approximation of $(\xi_0 \ell)^{1/2}$ for ξ has also been used in the dirty limit (Rosei).

A number of workers have measured coherence lengths, and their results are summarized in Table III-1. Some authors report lengths calculated from onset and midpoint transition temperatures, the former being perhaps 70% of the latter (Mura1), and for these cases only the midpoint values are listed. The coherence lengths in the table range from about 1.3 to 3.4 nm for both the LaSrCuO and the YBaCuO compounds. These lengths are close to or exceed the lattice constants in the c direction of 1.32 nm for LaSrCuO and 1.17 nm for YBaCuO.

We see from the table that coherence lengths of the 123 compound are aniso-

TABLE III-1. Coherence Lengths (ξ), Penetration Depths (Λ), and Their Ratios (Ginzburg–Landau Parameter) $\kappa = (\Lambda/\xi)$ for Oxide Superconductors.[a]

Material	T_c (K)	ξ (nm)	Λ (nm)	κ (Λ/ξ)	Ref.
LaSr[b]	37		250		Aeppl
	36	3	330	110	Finne
		1.8			Kobay
	34		200		Koss1
	38	1.3	210	160	Orla2
				~ 75	Renke
				~ 40	Takag
	37		230		Wappℓ
LaSr(0.096)	38	2.0			Nakao
LaSr(0.1)		2.6			Kobay
	35	1.3	210	160	Orla3
	39	2.0	100	50	Uchid
LaSr(0.3)	35	3.2			Mural
YBa[b]				180	Bezi1
			22.5		Felic
		1.5	120	80	Gottw
				70	Grant
	95	2.7($\parallel$)			Hikit
	95	0.6($\perp$)			Hikit
	84		130		Koss1
	89	1.7			Orla1
	89	3.4($\parallel$)	26($\perp$)	7.6($\parallel$)	Wortl[b]
	89	0.7($\perp$)	125($\parallel$)	37($\perp$)	Wortl[b]
			400		Zuoz1
$YBa_2Cu_3O_{6.9}$	92.5	2.2	140	65	Cava1
$Y_{0.4}Ba_{0.6}CuO_{3-\delta}$	89	1.4			Mura1
BiSrCaCuO		4.2($\parallel$)			Hida1[c]
		0.1($\perp$)			Hida1[c]

[a] The notation used is: $(La_{1-x}M_x)_2CuO_{4-\delta}$ = LaM(x); LaM (0.075) = LaM$_*$; $YBa_2Cu_3O_{7-\delta}$ = YBa$_*$. Several values of the coherence length $\xi_\parallel = \xi_{ab}$ in the Cu-O planes and $\xi_\perp = \xi_c$ perpendicular to these planes are given.

[b] The Hikit direction convention was used for ξ.

[c] Y. Hidaka et al., J.J.A.P. *27*, L538 (1988)

tropic. The interplanar coherence length $\xi_c = \xi_\perp$ along the c axis, perpendicular to the planes, has the value $\xi_\perp = 0.65$ nm. This is about $\frac{1}{5}$ of the value $\xi_{ab} = \xi_\parallel = 3.1$ nm of the in-plane coherence length parallel to, or in, the a,b plane (Hikit, Wort1). The former value of $\xi_\perp$ is between the average spacing 0.39 nm and twice the average spacing 0.78 nm between Cu–O layers, suggesting that the superconductivity may have a three-dimensional character. The in-plane coherence length $\xi_\parallel$, on the other hand, is more than 11 unit-cell spacings (≈ 0.386 nm each) along the a and b directions. This indicates that the Cooper pairs are delocalized in Cu–O planes over average distances of 11 lattice spacings. Most reported coherence-length values are averages over anisotropies. (Hikita et al. used the opposite notation for designating parallel and perpendicular coherence lengths (Hikit).

Table III-2 lists coherence lengths of several elements for comparison purposes; they are all very much greater than typical copper oxide compound values because the elements are Type I superconductors. We also see from the table that there is a tendency for the coherence length to decrease with increasing T_c.

4. Penetration Depth

The penetration depth Λ is the distance into a superconductor where an external dc magnetic field has decreased to 0.368 ($1/e$) of its value outside (in the limit of zero or low-vortex concentration). We see from the data listed in Table III-1 that the penetration depths reported for the oxide superconductors are in the range from 1200 to 4000 Å, 2 orders of magnitude greater the corresponding coherence lengths, as is expected because they are Type II superconductors. Table III-2 shows that the Type I elements have much smaller penetration depths.

The penetration depth varies with the temperature in the manner shown on Fig. III-6. It becomes extremely large as T_c is approached from below, and it is fairly constant near absolute zero.

TABLE III-2. Comparison of Coherence Length (ξ), Penetration Depth (Λ), and Their Ratio ($\kappa = \Lambda/\xi$) of Elemental and Oxide Superconductors.[a]

Material	T_c (K)	ξ (nm)	Λ (nm)	κ (Λ/ξ)
Cd	0.56	760	110	0.14
Al	1.2	1400	50	0.04
In	3.4	270	45	0.17
Sn	3.7	150	60	0.40
Pb	7.2	80	45	0.56
Nb	9.26	38	39	1.02
LaSrCuO	~35	~2	~220	~100
YBaCuO	~90	~2	~150	~90

[a]The values of the latter are rough estimates from Table III-1. This table gives averages of several reported values (see Heube, p. 10; Meser, p. 174).

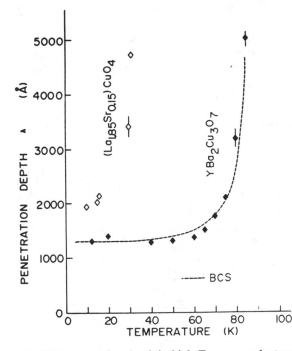

Fig. III-6. Magnetic-field penetration depth in high-T_c superconductors determined by muon spin relaxation measurements (Koss2).

5. Ginzburg–Landau Parameter

The Ginzburg–Landau parameter κ is the ratio of the penetration depth Λ to the coherence length ξ, and representative values of κ for a number of superconducting materials are given in Tables III-1 and III-2. Superconductors are classified as Type I or II depending on whether the parameter κ is less than $1/\sqrt{2}$ or greater than $1/\sqrt{2}$, respectively, and we see from the tables that the copper oxide superconductors are Type II.

6. Grain Size

The ceramic oxide superconductors have a granular structure, with typical sizes between 0.3 and 3 μm. Many authors have used electron microscopy, metallography, and other means to estimate the linear dimensions of their grains, and some reported values are 0.01 μm (Tsue1), ≈ 0.5 μm (YBa, Ourma, Ourm1), ≈ 1.0 μm, (LaSr, Larba), 1–2 μm (Farre), 1.3–2.0 μm (LaSr, Finne), and 33–170 μm (LaSr, Suena). Three-dimensional sizes have also been determined, such as $10 \times 10 \times 35$ μm and $10 \times 10 \times 125$ μm (YBa, Suena), rods 10×2–4 μm (Ventu), and 2–4-μm spheres with islands 100–500 μm (Farre). There was a report of superconducting clusters in LaSrCuO 1 μm in diameter, much smaller than the grain size of 10–20 μm (Larba), and of microstructure subdivisions in LaSrCuO on a scale of 0.05 μm (Larb1). Some authors have suggested that superconducting phases exist on the surface or on grain boundaries of normally conducting material (Takab, Kirt3). Small orthorhombic domains of size ≈ 10 μm might be in YB* single crystals (Hemle).

7. Length Parameters and Critical Fields

The oxide superconductors are Type II ($\kappa > 1/\sqrt{2}$) so they have two critical fields, as shown on Fig. III-7, a lower critical field H_{C1} below which they are ideally perfect diamagnets, and an upper one H_{C2}, above which they lose their superconductivity and become normal. Between these two critical fields they are in a mixed or vortex state with both normal and superconducting regions present. In this state the superconducting material is threaded parallel to the applied magnetic field with filaments called vortices of normal material which are penetrated by the magnetic field. Each vortex contains exactly the same amount of flux equal to the fluxoid or flux quantum Φ_0

$$\Phi_0 = hc/2e = 2.0679 \times 10^{-7} \text{ G cm}^2$$
$$= 20.7 \text{ G } (\mu\text{m})^2 \tag{III-2}$$

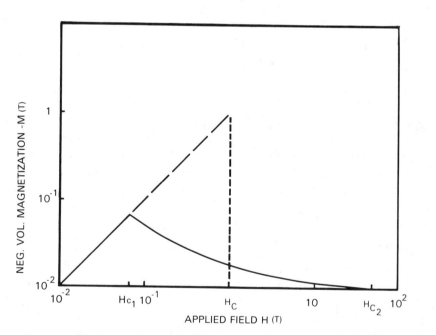

Fig. III-7. Plot of magnetization M versus magnetic field H. A log–log scale is used because the ratio $H_{c2}/H_{c1} \sim 10^3$ for these oxide materials is so large. This graph was made for the values of H_{c1} and H_{c2} typical for high-T_c superconductors.

where the factor of 2 appears in the denominator because the charge of a Cooper pair is $2e$. In Section VIII-C we discuss the critical fields of oxide superconductors in more detail.

The critical fields H_{C1} and H_{C2} have the following relationships with the fluxoid, Φ_0, the penetration depth Λ, and the coherence length ξ introduced in the previous section (Forga, Grant, Takag, Tinkh)

$$H_{C1} = \Phi_0 \ell n \ \kappa/2\pi\Lambda^2$$
$$H_{C2} = \Phi_0/2\pi \ \xi^2 \tag{III-3}$$

where $\kappa = \Lambda/\xi$. The thermodynamic critical field H_C is related to the lower H_{C1} and upper H_{C2} critical fields through the approximate expression

$$(H_{C1}H_{C2})^{1/2} = H_C \ (\ell n \ \kappa)^{1/2} \tag{III-4}$$

These expressions will be used in Chapter VIII.

Vortices have central cores of normal material with diameters approximately equal to the coherence length surrounded by thick outer cylindrical layers of mixed material with diameters approximately equal to the penetration depth. They tend to form a two-dimensional hexagonal lattice structure which was observed on the surfaces of $YBa_2Cu_3O_{7-\delta}$ by the Bitter pattern technique (Gamme). The density of the vortices was found to be linear in the applied field, and the slope of this line provides an estimate of the flux quantum $\Phi_0 = hc/2e$. The vortex pattern has been analyzed theoretically (e.g., Klein).

H. DIMENSIONALITY

Many authors have discussed the significance of the CuO_2 planes and the two-fold dimensionality of the overall structure. Low-dimension band structure calculations (e.g., Chang, Labbe, Mohan) have been carried out which tend to support a two-dimensional picture of the superconductivity. Two 2D Cu2-O and two 1D Cu1-O bands were found in $YBa_2Cu_3O_{7-\delta}$ (Freem).

In the LaSrCuO case substitutions for the copper ions have a much more drastic effect on the superconductivity than substitutions for lanthanum or an alkali metal ion (cf. Section VII-F). This stresses the importance of the –Cu–O–Cu–O– linkages. These superconductors have a strong anisotropic diamagnetism with principal directions perpendicular to and within the planes. Several authors have used a two-dimensional band model and discussed the two-dimensional character of the Fermi surface (Leezz, Oguch, Tajim, Uchi1, Xuzzz, see also Shapi, Shap1).

We mentioned in Section III-G-3 that the factor of 5 anisotropy in the coherence length $\xi_\perp = 0.65$ nm, $\xi_\parallel = 3.1$ nm of $YBa_2Cu_3O_{8-\delta}$ correlates well with the extensive delocalization of Cooper pairs in the Cu–O planes, together with some three-dimensional character of the superconductivity (Wort1).

The presence of two CuO_2 planes with the copper in a distorted square planar coordination was suggested as the key to the superconductivity of the YBaCuO system (Siegr). Higher transition temperatures have been predicted for bigger assemblies of CuO_2 layers coupled by Ba layers (Horzz).

I. FLUCTUATIONS AND NOISE

Fluctuation effects are important both above and below T_c (Tinkh). Qualitatively, fluctuations may cause small vestiges of normal state behavior to linger below T_c and superconducting phases to persist above the transition (e.g., for $T \approx 2 T_c$). Magnetic and structural fluctuations in polycrystalline YBa*, studied as a function of the reduced temperature $t = (T - T_c)/T_c$, were interpreted as being three dimensional (Freit). This is suggestive of an appreciable coupling between the low-dimensional superconducting channels, namely the Cu–O planes and chains. Critical field calculations (Lobbz) based on the Ginzberg–Landau theory modify the mean field predictions within 0.1 K of T_c. In this approximation the correlation length was shown to diverge as $(T - T_c)^{-\beta}$ with $\beta \approx \frac{2}{3}$. An examination of quantum fluctuations and quasi-particle tunneling in superconducting granular films at $T = 0$ led to the conclusion that the normal state resistivity has to be less than or equal to $h/4e^2$ for superconductivity to be present (Chakr). A neutron scattering experiment designed to measure magnetic fluctuations found them to be vanishingly small (Bruck), thereby making unlikely the possibility of magnetic couplings as a source for superconductivity.

The noise power spectrum $S(f)$ of YBa* and ErBa* measured between 77 K and room temperature reached a peak at a temperature where dR/dT also peaked, and it vanished in the superconducting phase. The frequency dependence $S(f) \approx f^{-\alpha}$ was mentioned (Testa). Peculiar noise behavior in Nb/YBa* point contact junction current as a function of bias has also been reported (Kuzni).

J. VALENCE

The conjecture "It is the ability of $YBa_2Cu_3O_{8-\delta}$ to sustain a large range of mixed valency, as demonstrated by its sensitivity and quenching conditions, that is responsible for its high and variable transition temperature, and the even higher temperatures that might be achieved by exploring other materials" (Grant) underlines the importance of this section. Average copper valences obtained in a crystallographic study of YBaCuO (Hewa1) over a range of temperature from 20 to 750°C showed that one Cu^{3+} and two Cu^{2+} atoms of $YBa_2Cu_3O_7$ are distributed over both the CuO planes and the chains, with somewhat more Cu^{3+} in the chains. Others (e.g., Xiaoz) maintain that Cu^{2+} is in the planes and Cu^{3+} in the chains. The $(La_{1-x}Sr_x)_{2-y}CuO_{4-\delta}$ materials have a much a smaller range of mixed valency, since ordinarily $x \ll 1$ and $\delta \ll 1$, and all of the copper is in the

$+2$ state for $x = \delta = 0$. There is evidence that this mixed valence situation may be necessary, but it is certainly not sufficient for superconductivity (Torr1).

The charge states Cu^+ and Cu^{2+} are quite common in chemical compounds and crystals while Cu^{3+} is rather rare. The other cations such as Sr^{2+}, Ba^{2+}, Y^{3+} and La^{3+} that are present in oxide superconductors have definite valence states, as does O^{2-}. To estimate the copper charge states in a material we first calculate the average copper charge Q by assuming that the other cations and oxygen have their usual charges, and then we divide this charge between the two copper ions that bracket this value (Rigne). Thus we have, for example:

$$(La_{0.9}Sr_{0.1})_2CuO_4: Q = 2.20; \quad Cu^{2+} = 80\%; \quad Cu^{3+} = 20\%$$
$$YBa_2Cu_3O_7: Q = 2.333; \quad Cu^{2+} = 67\%; \quad Cu^{3+} = 33\% \tag{III-5}$$

This provides an upper limit on the amount of Cu^{2+} and a lower limit on the amount of Cu^{3+} (or Cu^+) that is present. Some authors (Pauli) have suggested that divalent copper can participate in the charge interchange

$$2 Cu^{2+} \longrightarrow Cu^+ + Cu^{3+} \tag{III-6}$$

and if we assume that, all of the divalent copper disproportionates in this way when we obtain upper limits on the amount of Cu^+ and Cu^{3+} that can be present. For the example given above we have, in this approximation:

$$(La_{0.9}Sr_{0.1})_2CuO_4: Q = 2.20; \quad Cu^+ = 40\%; \quad Cu^{3+} = 60\%$$
$$YBa_2Cu_3O_7: Q = 2.333; \quad Cu^+ = 33\%; \quad Cu^{3+} = 67\% \tag{III-7}$$

Tables III-3 and III-4 list the average copper charge state Q and the percentages of each copper ion in these two limiting cases for a large number of LaSrCuO and YBaCuO superconductors, respectively (Rigne). The results are listed in the order of increasing Q for each superconductor type.

Proposed theoretical models for the superconductivity of YBaCuO are sensitive to oxygen stoichiometry and copper valence in the following manner (Stoff): The resonance valence bond model described in Section IV-D (Ander) postulates singlet pairs of electrons on adjacent Cu^{2+} ions which are immobile and yield an insulating state unless there are enough holes in the form of Cu^{3+}. The excitonic model relies on a virtual transfer of charge between Cu and O for binding the Cooper pairs. Band structure calculations put less emphasis on valence states and more on the positions of various Cu–O bonding and antibonding bands relative to E_F, which in turn is controlled by the doping effect of oxygen vacancies.

Some explanations for oxide superconductivity are based on the preference of copper for the Cu^{2+} state (Stein, Tokur, Tranq), and others assume the breakup of Cu^{2+} into Cu^+ and Cu^{3+} states (Ihara, Ihar2, Pauli, Zhang). Of course other possibilities also exist. For example it has been proposed that charge fluctuations might occur between $Cu^{2+}O^{2-}$ and Cu^+O^- in the CuO_2 planes of

TABLE III-3. Calculated Values of Average Charge Q on Cu and Possible Cu Valence States for $(La_{1-x}Sr_x)_{2-y}CuO_{4-\delta}$ Type Superconductors with Oxygen Deficiency δ Explicitly Known.[a]

	x	y	δ	Q	Cu^{2+} Present			Cu^{2+} Absent		Ref.
					% Cu^{1+}	% Cu^{2+}	% Cu^{3+}	% Cu^{1+}	% Cu^{3+}	
La–Sr	0.35	0.3	1.1	1.295	70.5	29.5	0	85.3	14.7	Tera1, Oyana
	0.42	0.1	0.7	1.698	30.2	69.8	0	65.1	34.9	Tera1
	0.14	0	0.17	1.93	7	93	0	53.5	46.5	Shafe
	0.13	0	0.15	1.95	5	95	0	52.5	47.5	Shafe
	0.11	0	0.1	2.02	0	98	2	49	51	Shafe
	0.03	0	0.01	2.04	0	96	4	48	52	Shafe
	0.06	0	0.03	2.06	0	94	6	47	53	Shafe
	0.09	0	0.06	2.06	0	94	6	47	53	Shafe
	0.04	0	0.01	2.06	0	94	6	47	53	Shafe
	0.08	0	0.04	2.07	0	93	7	46.5	53.5	Ohish
	0.08	0	0	2.15	0	85	15	42.5	57.5	Shafe

[a]The table is sorted by the average charge. Both the upper and lower valence states of copper are given under the assumption that only two valence states are present for the two limiting cases, namely with Cu^{2+} and without Cu^{2+}, as discussed in the text (prepared by M. M. Rigney).

$YBa_2Cu_3O_{7-\delta}$ (Horzz). Charge transfer can also take place between Cu^+ and Cu^{2+} or between Cu^{2+} and Cu^{3+} (Fuzzz, Littl, Sarma, Sarm1, Sreed). These various approaches are alluded to in many articles, and will be mentioned from time to time in this review.

Many authors give the contents of the positive ions present in the materials that they study, but most do not report a value for the oxygen deficiency factor δ in the chemical formulas $(La_{1-x}Sr_x)_{2-y}CuO_{4-\delta}$ and $Y_{1-x}Ba_{2-y}Cu_3O_{7-\delta}$. Sometimes they mention working with, for example, $(La_{0.9}Sr_{0.1})_2CuO_4$ or $YBa_2Cu_3O_7$, without having precisely determined the amount of oxygen that is present. In these cases the subscripts 4 and 7, respectively, could be nominal values. Fortunately there are many cases in which researchers either provide an estimate or quote a definite value for δ.

Experimental determinations of copper valence states are rather sparse. X-ray photoelectron spectroscopy (XPS) of $YBa_2Cu_3O_7$ (Ihar2) produced a Cu $2p_{1/2}$ line shown in Section IX-E-2 which was decomposed into three peaks of intensity ratio $1:2:1$ that were assigned, respectively, to monovalent, bivalent, and trivalent copper. Both XPS and ultraviolet (UPS) photoemission experiments of $YBa_2Cu_3O_{7-\delta}$ with $\delta \approx 0$ show a small density of states at the Fermi level, Cu 2p spectra with only a Cu^{2+} contribution and an indication of large O^{2-} vacancies coordinated with Ba^{2+} sites (Stein) or 2p holes localized mainly on the oxygens on the Cu-O chains (Yarmo). Carriers doped into LaSrCuO or YBaCuO are claimed to be oxygen p-like holes interacting with Cu d^9 states (Fuji1).

Another way to measure the amount of Cu^+ and Cu^{3+} present with Cu^{2+} is through iodometry by dissolving the samples with HCl + KI (Matac). The Cu^+ was verified by potassium dichromate titration of Fe^{2+}. The total oxygen content was found by graphite reduction at high temperature followed by gas chromatographic detection. Others (e.g., Onell) found no evidence for monovalent copper.

Some believe that the substitution of Sr for La is accompanied by the oxidation reaction $Cu^{2+} \rightarrow Cu^{3+}$; others (Kraka, Tranq) find no support for this. Pressure enhances the Cu^{3+}/Cu^{2+} ratio in LaSrCuO (Wuzzz). X-ray absorption edge spectra suggest that the copper is predominantly monovalent in $(La_{0.65}Sr_{0.35})_{1.7}CuO_{2.9}$ (Oyana) and $(La_{0.65}Sr_{0.35})_{1.7}CuO_3$ (Ihara). In contrast, photoemission studies have indicated that most of the copper in $(La_{0.75}Sr_{0.25})_2CuO_{4-\delta}$ is divalent with the remainder monovalent despite the formal charge $Q = 2.2$ in this nonsuperconductor (Fujim). The hole concentration in $(La_{1-x}Sr_x)_2CuO_{4-\delta}$ was found to equal the Sr concentration to about $x = 0.075$, and for higher x the hole [i.e., $(Cu-O)^+$ group] concentration decreases and oxygen vacancies are formed (Shafe). X-Ray absorption of the same compound indicates that $Q > 2$ (Jeonz), and another X-ray absorption edge study shows that the main effect of Sr doping of La_2CuO_4 is a slight enhancement of the Cu^{3+} peak (Alpzz). Still another similar work on $(La_{1-x}M_x)_2CuO_4$ with M = Sr, Ba and $x = 0-0.3$ shows copper as divalent for all concentrations with the number of (O-2p) empty states near E_F increasing with doping (Tranq).

TABLE III-4. Calculated Values of Average Charge Q on Cu and Possible Cu Valence States for $R_{1-x}Ba_{2-y}Cu_3O_{7-\delta}$ Type Superconductors with Oxygen Deficiency δ Explicitly Known.[a]

	x	y	δ	Q	Cu²⁺ Present			Cu²⁺ Absent		Ref.
					% Cu^{1+}	% Cu^{2+}	% Cu^{3+}	% Cu^{1+}	% Cu^{3+}	
Y-Ba	0	0	0.1	1.6	40	60	0	70	30	Johns
	0.1	−0.1	1	1.7	30	70	0	65	35	Borde
	0	0	0.7	1.87	13.3	86.7	0	56.7	43.3	VanBe
	0	0	0.66	1.89	10.7	89.3	0	55.3	44.7	Ongz1
	0	0	0.62	1.92	8	92	0	54	46	Kuboz
	0	0	0.59	1.94	6	94	0	53	47	Ongz1
	0	0	0.57	1.95	4.7	95.3	0	52.3	47.7	Kuboz
	0	0	0.51	1.99	0.7	99.3	0	50.3	49.7	Ongz1
	0	0	0.5	2	0	100	0	50	50	Hazen,Hemle,Poli1
	−0.2	0.2	0.4	2	0	100	0	50	50	Nakam
	0	0	0.47	2.02	0	98	2	49	51	Kuboz
	0	0	0.44	2.04	0	96	4	48	52	Ongz1
	−0.2	0.2	0.34	2.04	0	96	4	48	52	Syono
	0.06	0.8	1.3	2.06	0	94	6	47	53	Koch2
	0	0	0.35	2.1	0	90	10	45	55	Kuboz
	0	0	0.33	2.113	0	88.7	11.3	44.3	55.7	Ongz1
	0	0	0.31	2.127	0	87.3	12.7	43.7	56.3	Fujim
	0	0	0.28	2.147	0	85.3	14.7	42.7	57.3	Tara3
	0	0	0.2	2.2	0	80	20	40	60	Kuboz,Stoff

	0	0	0.19	2.21	0	79.3	20.7	39.7	60.3	Benoz
	0	0	0.15	2.23	0	76.7	23.3	38.3	61.7	Bonn1,Nevit
	0	0	0.11	2.26	0	74	26	37	63	Ongz1
	0	0	0.1	2.7	0	73.3	26.7	36.7	63.3	Jorge,Orens
	0	0	0.07	2.287	0	71.3	28.7	35.7	64.3	Kuboz
	0	0	0	2.333	0	66.7	33.3	33.3	66.7	[b]
	0	0	-0.05	2.37	0	63.3	36.7	31.7	68.3	Mawds
	0	0	-0.1	2.4	0	60	40	30	70	Kuboz
	-0.1	0	-0.4	2.53	0	46.7	53.3	23.3	76.7	Beye1
Dy–Ba	0	0	0.72	1.85	1.44	85.3	0	57.3	42.7	Tara3
Er–Ba	0	0	0.84	1.77	22.7	77.3	0	61.3	38.7	Tara3
Eu–Ba	0	0	-0.1	2.4	0	60	40	30	70	Eibsc,Tara3
Gd–Ba	0	0	0.47	2.02	0	98	2	49	51	Tara3
Ho–Ba	0	-0.16	0.87	1.75	24.7	75.3	0	62.3	37.7	Tara3
Nd–Ba	0	0	0.16	2.44	0	56	44	28	72	Tara3
Sm–Ba	0	0	-0.11	2.41	0	59.3	40.7	29.7	70.3	Tara3
Tm–Ba	0	0	0.93	1.71	28.7	71.3	0	64.3	35.7	Tara3
Yb–Ba	0	0	0.29	2.14	0	86	14	43	57	Tara3

[a] The table is sorted by the cations and then within the cations it is sorted by the average charge. Both the upper and lower valence states of copper are given under the assumption that only two valence states are present for the two limiting cases, namely with Cu^{2+} and without Cu^{2+}, as discussed in the text (prepared by M. M. Rigney).

[b] Data are from Ihar2, Ongz1, Siegr, Stein, Ventu, and Yarmo.

K. CONCLUDING REMARKS

This completes the background and overview of the experimental aspects of the oxide superconductors. In the next chapter we will introduce some of the theory that has been used to explain their superconducting behavior. We then proceed to a more systematic and detailed survey of their properties.

IV

THEORY

A. INTRODUCTION

This article is a review of the oxide superconductivity literature through most of 1987 designed primarily for the experimentalist. Thus of necessity these few sections on theory are written more to help the experimentalist gain a better perspective of their results than to provide a systematic account of the theory literature (DiSal).

We now introduce the reader to the very successful BCS theory of conventional superconductivity. The next few sections discuss one general approach for explaining copper oxide superconductivity, namely, the weak coupling BCS model with electron–electron coupling arising from the exchange of one or more types of Boson excitation, such as the phonons of conventional superconductivity, excitons, or plasmons. Following this we will examine a second general approach, which consists of explaining the superconductivity by a different theory based on, for example, the magnetic interaction or Hubbard model. In the next section the resonant model will be described, in the following one the nonresonant antiferromagnetic model will be considered, and this will be followed by a survey of some other models. After this we will discuss some results of band structure calculations, and in the final section we will make some concluding remarks. Recent theoretical articles which are not particularly relevant to the oxide-type superconductors will not be mentioned.

B. BARDEEN–COOPER–SCHRIEFFER THEORY

In this section we will very briefly describe the nature of the theory introduced by Bardeen, Cooper, and Schrieffer (BCS) in 1957 (Barde), and we will mention some of the ways in which experimental data can be compared to it. This theory explains well the properties of most low-temperature superconductors.

1. Description of BCS Theory

The BCS theory is based upon the existance of a net attractive interaction V between the electrons in a narrow energy range near the Fermi surface. This produces a ground state separated from the upper excited states by a gap in energy. The attractive interaction forms two-electron singlet bound states in momentum space, called Cooper pairs. The BCS theory presupposes the formation of Cooper pairs via an electron–electron attractive potential $V(\omega)$. The simplest approximation is to assume that $V(\omega) = V$ for electron energies within the range $E_F \pm E_D$ of the Fermi energy E_F and $V = 0$ beyond this range. Here E_D is a limiting energy which, in the case of a phonon mechanism, is equal to the Debye energy $k_B \theta_D$ characteristic of the lattice vibrations, where θ_D is the Debye temperature. A transcendental integral equation is solved to give the following expression for the energy gap E_g and the transition temperature T_c

$$T_c = 1.134 \, \theta_D \exp[-1/\lambda] \tag{IV-1}$$

$$E_g = 4 \, E_D \exp[-1/\lambda] \tag{IV-2}$$

where $\lambda = V N(E_F)$ is the dimensionless electron–electron interaction called the electron–phonon coupling constant, and $N(E_F)$ is the density of states at the Fermi level. Table IV-1 tabulates values of λ for some of the old and new superconductors. We see from the table that λ tends to increase as T_c increases. Dividing Eq. IV-2 by Eq. IV-1 provides the universal dimensionless ratio

$$E_g/k_B \, T_c = 3.528 \tag{IV-3}$$

which is independent of the form of $V(\omega)$.

In the next-higher-order approximation the repulsive screened Coulomb potential V_C is taken into account, and we can define the dimensionless Coulomb constant $\mu = \langle V_C \rangle \, N(E_F)$ in terms of $N(E_F)$ and the averaged Coulomb interaction $\langle V_C \rangle$. The Coulomb interaction pseudopotential μ_* given by

$$\mu_* = \frac{\mu}{1 + \mu \ell n \, (E_F/E_D)} \tag{IV-4}$$

enters into the expression for T_c

$$T_c = 1.14 \, \theta_D \exp\left[- \frac{1}{\lambda - \mu_*} \right] \tag{IV-5}$$

where Eq. (IV-3) is still valid. Table IV-1 lists values of μ_* for some of the old and new superconductors. The so-called canonical value of $\mu_* = 0.13$ is some-

TABLE IV-1. Electron Phonon Interaction Constant λ and Coulomb Interaction Pseudopotential μ_* for Some Older and Newer Superconductors.[a]

Material	T_c (K)	λ	μ_*	Ref.
Ru	0.49	0.47	0.15	Table II-1
Zr	0.61	0.22	0.17	Table II-1
Os	0.66	0.44	0.12	Table II-1
Mo	0.92	0.35	0.09	Table II-1
Re	1.7	0.37	0.1	Table II-1
V_3Ge	6.1	0.7		Vonso, p. 303
Pb	7.2	1.55	0.6	Ginzb, p. 171
Pb–Bi alloy	9	2.13		Ginzb, p. 171
Nb	9.3	0.85		Ginzb, p. 171
NbC	11.1	0.61		Ginzb, p. 169
TaC	11.4	0.62		Ginzb, p. 169
$Ba(Pb,Bi)O_3$	12	1.3		Schl1
V_3Si	17.1	1.12		Vonso, p. 303
Nb_3Sn	18.1	1.67		Ginzb, p. 171
Nb_3Ge	23.2	1.80		Vonso, p. 303
LaSr$_*$	18	1.2,2.4	0.12	Kita1
LaSr(0.1)	27	1–2		Schl1
LaBaCuO	30–40	2.5	0.1	Pick1
LaBa ($0.05 \leq x \leq 0.15$)	30–40	≥ 2		Weber
LaSr ($0.05 \leq x \leq 0.15$)	30–40	≥ 2		Weber
LaSr$_*$	35	~ 2.0	0.2	Ramme
YBa$_*$		≤ 0.3		Gurv1
	85	2.5	0.1	Kirt3
	92	2.5		Mawds
	91	~ 5.0	0.2	Ramme

[a]The following notation is used: $(La_{1-x}M_x)_2CuO_4$ = LaM(x); LaSr(0.075) = LaSr$_*$; $YBa_2Cu_3O_7$ = YBa$_*$.

times used (Weber). A more sophisticated approach involving electron–phonon coupling of strength $\alpha(\omega)$ consists of averaging over the phonon density of states $D(\omega)$ to give (Elias, Elia1, Wuzz5)

$$\lambda = 2 \int_0^\infty \frac{\alpha^2(\omega)\, D(\omega)\, d\omega}{\omega} \tag{IV-6}$$

The integrand $\alpha^2(\omega)D(\omega)$ is called the Eliashberg function and it has been measured in $(La_{0.92}Sr_{0.08})_2CuO_{4-\delta}$ (Ramir, Sule1), fit to thermopower data (Mawds), and discussed in the recent literature (e.g., Ashau, Marsi, Ramme). The following semiempirical formula was proposed for the transition temperature arising from a phonon mechanism (McMil):

$$T_c = \frac{\theta_D}{1.45} \exp\left[-\frac{1.04(1 + \lambda)}{\lambda - \mu_*(1 + 0.62\lambda)} \right] \tag{IV-7}$$

and a number of related expressions may be found in the literature (e.g., Ruval).

It has been pointed out that quantitative estimates of the characteristic values of the Ginzburg–Landau and BCS parameters of high-temperature superconductors are consistent with a Type II dirty limit conventional superconductor (Salam).

In the next section we will discuss some of the important predictions of the BCS theory. This will be helpful for putting into perspective the experimental results.

2. Predictions of BCS Theory

The BCS theory makes the following predictions which have been compared with experiments for a number of the oxide superconductors:

(a) The energy gap E_g is proportional to the transition temperature through the expression (IV-3), $E_g/kT_c \approx 3.53$ (except for the gapless superconductors).

(b) The London equation is a consequence of the BCS theory and hence one expects the state of perfect diamagnetism to exist below T_c with $\chi = -1/4\pi$.

(c) The transition temperature is related to the Debye temperature θ_D, the electron–phonon coupling constant λ and the density of states $N(E_F)$ at the Fermi surface, as mentioned above.

(d) There is a discontinuity in the electronic contribution to the specific heat at the transition temperature given by

$$\frac{C_S - C_n}{C_n} = 1.43 \qquad\qquad \text{(IV-8)}$$

where the subscripts s and n denote the superconducting and normal states, respectively, and $C_n = \gamma T_c$ (cf. Section IX-F).

(e) Magnetic flux is quantized with the value $\Phi_0 = hc/2e$ given by Eq. (III-2).

(f) For phonon coupling the transition temperature depends on the average isotopic mass M through the expression

$$T_c \approx M^{-\alpha} \qquad\qquad \text{(IV-9)}$$

and in a simple model $\alpha = \frac{1}{2}$ (cf. Section VII-G).

Tables II-1, III-1, III-2, and IV-1 list values of various BCS factors and parameters for a number of superconducting elements and compounds. We see from these tables that there is a tendency for λ, $N(E_F)$ and E_g to increase with T_c, as expected.

3. Applicability of BCS Theory

It has been suggested (Garoc) that the BCS theory cannot apply to the present oxide superconductors, while others support it (e.g., Boyce, Messm, Phill, Wang1, Wuzz1). In conventional superconductors the mechanism for pair formation is the electron–phonon interaction, but other mechanisms are possible involving, for example, polarons (Aleks, Littl, Scal1) or excitons (Gallo, Gutfr, Litt1). The isotope effect discussed in Section VII-G is an indicator of the electron–phonon mechanism. The results of flux quantization (Gough) and tunneling (Section X-D) experiments have established that the superconducting condensate is produced by carriers with two units of charge (2e). Two major theoretical questions about the new superconductors are whether the paired electrons are Cooper-type pairs coupled together in momentum space, as in the BCS theory (Barde, Schri), and whether the mechanism for pair formation is mediated by a phonon or nonphonon process. Many articles comment on these questions, and undoubtedly more discussion will be forthcoming. These interactions will be discussed further in the next few sections. The very brief outline of the theory given above was for the case of electron–phonon coupling.

C. COUPLING MECHANISMS

We mentioned above that the BCS theory presupposes some kind of mechanism that couples electrons to form Cooper pairs. In the next few subsections we will discuss the phonon, exciton, and plasmon mechanisms which various workers have suggested as being operative in the oxide superconductors. Phonons were dominant in the lower-temperature superconductors, and they might also contribute strongly here, but the final verdict on this question awaits future work.

1. Phonons

In the last section we mentioned that conventional superconductors form Cooper pairs through the electron–phonon interaction λ. Most elements are weakly coupled, which means that λ is significantly less than 1, as the data in Tables II-1 and IV-1 indicate. These elements tend to have values close to the BCS ones for the test equations (IV-3) and (IV-8).

The strongly coupled elements Hg and Pb deviate from these BCS expectations. For example Pb with $\lambda = 1.55$ has significantly higher values 4.3 and 2.7 than the corresponding BCS ones 3.53 and 1.43 for Eqs. (IV-3) and (IV-8), respectively. Niobium with the moderate coupling $\lambda = 0.85$ has intermediate values 3.6 and 1.9. The common magnet material Nb_3Sn with $T_c = 18.1$ K has the large value $\lambda = 1.67$, and the amorphous Pb–Bi alloy with $T_c = 9$ K has $\lambda = 2.13$. The oxide spinel $LiTi_2O_4$ was recently assigned the value $\lambda = 1.8$ (Satpa).

If the newer copper oxide superconductors are phonon mediated types, they are probably strong coupling cases with relatively large values of λ (Aleks,

Bourn, Hangs). Band-structure calculations (Papac, Picke) for the LaSrCuO system give $\lambda \approx 2.5$ at the upper end of the range of experimental values 0.5–2.4 listed in Table IV-1. The table gives $\lambda = 2.5$ measured for YBaCuO (Mawds).

It has been shown (Zache) that high transition temperatures may be accommodated in the electron–phonon interaction mechanisms by the use of highly anisotropic potentials. A mean field approach showed that electron–phonon intramolecular vibrations depress T_c while intermolecular ones enhance it (Laizz). Nickel-substituted perovskites and other systems with nonlinear potential wells may be candidates for high T_c materials. Interlayer coupling effects on the phonon–electron coupling have also been studied (Inoue). The phonon mechanism has been judged by some as important for high-temperature superconductivity (e.g., Brunz, Herrz, Matti, Picke, Pick1, Phill, Weber, Wuzz1, Yuzzz). In this context a high T_c is achieved near a metal-to-insulator transition.

Fermi surface nesting occurs when two sheets of the Fermi surface are parallel to each other, separated by the wave vector q, so that the energy denominator in the expression $1/(E_k - E_{k+q})$ approaches zero for a wide range of k values. This causes a strong response of the system to a perturbation, and tends to induce a charge-density or spin-density wave. This closeness to an instability can enhance T_c (Gaoz3, Matt5, Pick1, Take1).

The possible contribution of charge-density wave instabilities to the electron–phonon matrix elements in increasing T_c has been reported (Matt4, Okabe). The role of interband transfer and exchange effects has also been examined (Ohka2). Since the Cu–O–Cu bond angle of 180° introduces a strong phonon-mediated repulsion, a higher T_c could result from a smaller bond angle.

The implications of strong electron–phonon coupling have been investigated (Ashau, Bulae, Hangs, Mats2). It appears that $\lambda \gg 1$ can make $E_g/kT_c < 3.54$. It was argued that in two dimensions strong phonon coupling leads to the ordering of oxygen vacancies (Matt5). Using the electron–phonon spectral density of LaSr* and a large mass enhancement gives $\Delta C/\gamma(0)T_c = 2.8$ and $\gamma(0)T_c^2/H_c^2(0) = 0.124$, values similar to those of Pb (Schos). It has been noted that although strong coupling (i.e., large λ) increases E_g/kT_c, its effect on other dimensionless parameters is in the opposite direction (Mars3).

A combined diagrammatic and Monte Carlo technique was employed to investigate the effects of localized phonon impurities (Schut, Schu3). It was concluded that such impurities enhance T_c.

2. Excitons

For polymers and other systems of low dimensionality it has long been suggested that Cooper pairs might form through the exciton mechanism. The same proposal has also been made (Beill, Coll1, Daole, Gallo, Gutf1, Hsuzz, Kamar, Littl, Norto) for the Cu–O system. By exciton we mean any type of electronic excitation in place of the usual vibrational or phonon ones. In other words, the attraction between the paired electrons can be obtained from the virtual movement of electrons or by the exchange of excitons. Exciton coupling has been reformulated to avoid some of the initial difficulties (Nakaj). Exciton enhanced superconductivity has been reported in YBa* (Ching). Eliashberg theory calculations (Marsi–Mars3) support the possibility of exciton or phonon–exciton mechanisms being operative.

Let us consider a simple one-dimensional model system (Gutfr, Littl). Consider that a conduction electron moving along the main chain or spine induces a movement of charge in the side chains which reaches its maximum a short time after the electron passes. This induced charge could be attractive to a second electron passing the same point a short time later, and a pairing might occur between the electrons. Thus a bound electron paired state might be established which is mediated by the virtual electronic excitation of the side chains, and the coupling will cause the paired electrons to remain near each other as they move together along the spine.

Equations (IV-1), (IV-5), and (IV-7) show that at several levels of approximation the transition temperature is proportional to the phonon excitation energy E_D, and replacing this by a larger electronic excitation energy should increase T_c accordingly. This is the reasoning behind the quest for exciton-mediated superconductivity.

The ordering of oxygen vacancies and cations could give rise to phonon softening and excitonic coupling (Ganzz). The exciton plus phonon model (Alle2) has been invoked to show that for the oxide superconductors T_c is a monotonically increasing function of carrier density, as is the electron–phonon coupling constant. A combined phonon–exciton mechanism could explain the observed E_g/kT_c ratios and the isotope shifts of LaSrCuO (Marsi). A model was proposed involving superconducting planes with $T_c \approx 40$ K and superconducting chains with $T_c \approx 180$ K to produce intermediate T_c values (Bard1). They point out that the usual BCS model might not be appropriate for $T_c > 100$K.

The absence of sharp absorption peaks from the mid-IR through the UV spectral regions (0.1–3.5 eV) has been advanced as evidence against the presence of exciton-mediated pairing (Bozo1). Others (e.g., Mars2) suggest that the presence of excitons is indicated by experiments.

3. Plasmons

Plasmon and other bosonic mechanisms have also been discussed (Chan1, Ihara, Ruval, Zhaoz) and Kresin (Kresi, Kres1, Kres3) concluded that the high T_c arises from the combination of strongly coupled phonons and plasmons (quantum of plasma oscillation). There is IR–Raman evidence for plasmons in YBa* (Perko) and LaSr* (Schle), and a mechanism was proposed for LaSr (Leezz) involving an attractive interaction between plasmons in a band composed mainly of Cu $d_{x^2-y^2}$ and oxygen 2p orbitals and electrons in a lower band formed from copper d_z and oxygen 2p orbitals. Pair breaking arising from inelastic electron–phonon and electron–electron scattering suppresses T_c relative to the energy gap, leading to E_g/kT_c values in excess of the BCS one of 3.5 (cf. Section IX-B).

4. Polarons and Bipolarons

The effect of polarons (electron plus induced lattice polarization) on producing high T_c's has been discussed (Aleks, Kuram, Littl, Robas, Scal1). Photoconductivity data indicate that an ensemble of polarons and excitons plays a substantial role in the mechanism of high-temperature superconductivity (Mazum, Mazu1).

Retardation effects on the longitudinal optical (LO) phonon exchange between two Fröhlich polarons were shown to be attractive. This is analogous to the van der Waals interaction between neutral atoms which involves the retardation effect (or phase delay) due to the finite velocity of the virtual photons that are exchanged. The value $T_c \approx 200$ K with $E_g/kT_c \approx 1.3$ may be attainable for a LO frequency of 2×10^{14} Hz. This model (Yizzz) predicts that as T_c increases the ratio E_g/kT_c decreases and approaches 1.47 as $T_c \to \infty$.

The possible role of bipolarons has also been examined (Alexa). In one approach involving an extended Hubbard model, the singlet superconducting state is unstable toward disordering if the intersite interaction is attractive, and it is unstable toward a charge-ordered superconducting state if the intersite interaction is repulsive (Hirsc, Mazum). Another such work (Wysok) determined the phase diagram of $(La_{1-x}M_x)_2CuO_4$ as a function of x, and also provided the expression

$$\frac{E_g}{kT_c} = \frac{\ln[(2-x)/x]}{1-x} \qquad \text{(IV-10)}$$

In this model the superconducting state is due to localized or real state pairs, and the low-temperature specific heat is believed to vary as a power of T.

5. Other Mechanisms

A proposed band model for $YBa_2Cu_3O_7$ involves heavy d-like holes interacting with light p-like electrons. It has been suggested that the d-hole behavior might be described as a plasmon from the point of view of a collective mode or as an exciton from the perspective of an individual mode. The plasmons and phonons act as attractors between electrons to form Cooper pairs, and T_c is enhanced (Ihar2). Several authors have considered the effects of intralayer interactions on the superconducting transition (Chang, Tesan), and this can make the superconductors doped Mott insulators (Cyrot). The interaction involving the spin density wave with the lattice through the second harmonic involves the generation of low- and high-frequency modes, and may provide a way to enhance T_c (Fento, Fent1).

Electron–electron or all electronic interactions have also been considered (Green, Newns, Newn1, Malet, Penne, Shafe), and an electronic mechanism that leads to d-wave pairing has been proposed (Leez1). A librational model advanced for explaining electron pair coupling predicts a reduced or absent isotopic shift (Hardy).

D. RESONANT VALENCE BONDS

In the previous sections we discussed the original BCS model with the phonon coupling mechanism that is operative in conventional superconductors. We also discussed generalizations and extensions of that model by considering interactions mediated not by phonons, but by excitons, plasmons, and other bosonic particles.

In this section we will outline a magnetic mechanism by discussing the resonant valence bond (RVB) approach championed by Anderson and Pauling (Ander, Ande1–Ande4, Baska, Kivel, Kive2, Pauli, Robas). Anderson noted that the newer superconductors all occur near a metal-to-insulator transition (perhaps a Mott transition) into an odd electron insulator state with peculiar magnetic properties. It has been pointed out (Wilso) that most copper oxides, including $YBa_2Cu_2O_5$, are Mott insulators. A Mott transition involves the conversion of a metal to a so-called Mott insulator when the interatomic spacing exceeds a certain critical value.

Anderson's 1973 article (Ande4) originally proposed the RVB state for a triangular lattice, and he has extended it to the two-dimensional square lattice (Hirs1) appropriate for the Cu–O planes of the new superconductors. The relevant Hamiltonian H is a second-order Heisenberg type

$$H = -J \Sigma \, b_{ij}^+ b_{ij} \qquad \text{(IV-11)}$$

with

$$J = 4t^2/U \qquad \text{(IV-12)}$$

and $U > t$. The Mott–Hubbard parameters are t, representing the electron hopping matrix and U, representing the electron–electron correlation energy. The second quantization operator b_{ij}^+ creates and its counterpart b_{ij} destroys a valence bonded pair (ij), respectively. At half-band filling the Wannier function is singly occupied and each unit cell with one Cu and two oxygens is electrically neutral. The RVB is related to what are known as the fractional quantum Hall effect Laughlin states, and the state Φ_{RVB} is given by

$$\Phi_{RVB} = \int d\theta \, \exp\left(-\frac{iN}{2\theta}\right) \pi_K \frac{\Phi_0}{(1 + a_{K2})} [1 + e^{i\theta} a_K b_{K+}] \qquad \text{(IV-13)}$$

The RVB state can be thought of as a Bose condensate of Cooper pairs of neutral solitons (Kivel, Kive2). Physically each atom is bonded to one neighboring atom by a single electron pair, and the ground state is a linear coherent superposition of all the states that can be formed by such a dimerization of the lattice (Thoul). A soliton is a solitary wave which preserves its shape even through interactions.

Anderson interprets this state as a quantum spin liquid, that is, as an interacting rather than noninteracting spin system (spin gas) with short-range correlations. The RVB state is a peculiar coherent state in the sense that it is a true quantum liquid with excitations that require both local and nonlocal operators involving the entire background liquid. He argues that low dimensionality (2D) and magnetic frustration favor this insulating magnetic phase. Frustration is the inability to achieve complete nearest-neighbor antiferromagnetic pairing, a pairing that is geometrically excluded on a planar triangular lattice.

The RVB ground state is not the conventional antiferromagnetic or Néel type, when one or both of following conditions are satisfied: (a) the next-nearest-neighbor coupling is antiferromagnetic (recall that conventional antiferromagnets have parallel next-nearest neighbors); or (b) there are virtual phonon interactions short of being strong enough to excite a spin-Peierls instability (Ander). However, experimentally the copper oxide system seems to exhibit long-range antiferromagnetic order, as opposed to the resonant dimerization of RVB. A Peierls instability involves the formation of an energy gap at the Fermi surface of a linear metal due to its instability with respect to a static lattice deformation of wave vector $2k_F$.

The question of competition between paramagnetism and RVB antiferromagnetism has been investigated (Grosz) by a variational Monte Carlo method applied to the stability of Gutzweiler wave functions. In the paramagnetic case d-wave pairing is favored (Cyrot, Leez1), but not s-wave pairing. There is a gap in energy for any charged excitation of the RVB bond state, with the Fermi energy in the gap for the stoichiometric compound, and hence this half-filled compound is an insulator. Doping removes the "half-filled" criterion (Ander) and is supposed to convert the system to a metal. For less than half filling, (i.e., by removing electron pairs), singly charged vacancies can be formed in pairs (Thoul). These vacancies will be Bosons (spin $= 0$) with a charge e, that is, the system is a Bose condensate of charged solitons. One kind of lowest excitation is a spin excitation which is a neutral soliton or spin $\frac{1}{2}$, charge 0 particle (Kivel).

The RVB also involves a Bose condensate of Cooper pairs of neutral solitons with a gap in energy $E_g = 2\Delta$ (Kivel), where E_g is the energy needed to break a bond. But Anderson et al. believe that the spectrum is gapless for the creation of neutral solitons. The spin-zero-charged Bosons in the Bose condensate may be mobile enough to carry a current. In this model the mechanism of superconductivity is predominately electronic and magnetic, although weak phonon interactions may favor the RVB state (Pauli).

Anderson reports that the RVB model predicts or is compatible with several measurable quantities, some of which are:

(a) The absence of gaps.
(b) Insulating and/or antiferromagnetism of undoped or weakly doped Cu–O compounds.
(c) The absence of an isotope effect, low-temperature specific heat, and elastic properties dominated by electronic energies.
(d) Anomalous temperature dependence of the normal state resistivity and the carrier density versus Hall–Seebeck coefficients.
(e) Effective mass $m* \approx m_e/\delta \approx 10\, m_e$, where δ is the doping fraction.

Initially there was a question concerning the unit of fluxoid involving singly charged Bosons in the doped state. However, since these Bosons must go around a current-carrying loop in pairs the effective charge for flux quantization is 2e, as usual (Thoul).

Calculations of the ratio E_g/kT_c for the RVB state (Nauen) seem to indicate a small value compared with the BCS prediction of 3.53. In this respect the RVB prediction $\Delta < kT_c$ is at variance with the gaps $\Delta > kT_c$ that have been reported (See Section IX-B, Table IX-1).

Anderson et al. identified the mysterious "twitch" transition in La_2CuO_4 with the mean-field resonating valence-bond transition of the Heisenberg model (Ande1, Baska). The topological order of the RVB state has been examined, and calculations have been carried out based on this model (e.g., Kivel, Fukuy).

Pauling agrees that the mechanism of superconductivity is predominantly electronic and magnetic, although weak phonon interactions may favor the RVB state. Sufficiently strong doping transforms preexisting magnetic singlet pairs into charged superconducting pairs. Lines of alternating Cu–O atoms interacting with layers of La and other cations give rise to the superconductivity. There is some resonance of $2Cu^{2+}$ to $Cu^+ + Cu^{3+}$. The possibility of unsynchronized resonance when some oxygens are missing stabilizes the conducting state. Other substances have favorable individual structural features, but only the copper oxide superconductors incorporate all of them.

The superconducting behavior of the oxides has also been described in terms of charge transfer resonance (Varma). In this approach both longitudinal and transverse resonances are included.

In the following section we will treat several other proposed mechanisms for "high T_c" superconductivity, and some of these will involve antiferromagnetic interactions.

E. ANTIFERROMAGNETIC MODELS

Emery maintained that it is important to deal with the largest energy in the problem before treating the others (Emery). His model of La_2CuO_4 has an antiferromagnetic insulating limit. The holes reside on O-2p states and not on Cu-3d states because the short-range repulsion is better screened by oxygen than it is by copper. The pairing is produced by strong coupling to local spins on the copper sites. This mechanism does not involve the much-discussed Cu^{3+} state (Wilso). Monte Carlo simulations on a statistically small lattice (Hirsc) seem to agree with these observations, but this work has been questioned (Mazum, Scala). The simultaneous presence of the antiferromagnetic and superconducting states has been calculated using a related model (Param).

The Emery model provides large estimates of T_c despite the possibility that intramolecular vibrations can renormalize the parameters t and U and restrict T_c (cf. Eq. IV-12). Geometric broken symmetry considerations (Mazum) suggest that the spin-Peierls state might be unique to 1D systems, which is in conflict with the assignment of a spin-Peierls state to the 2D Hubbard antiferromagnetic model (Hirsc). Another criticism is that the highest T_c values are observed in orthorhombic YBaCuO, while the models consider the square planar coordination of tetragonal symmetry. Thus these models may not be sufficiently realistic.

Antiferromagnetic models are attractive because some experimental data indicate the presence of antiferromagnetism in some of these oxides (cf. Section VIII-D-4), the isotope effect results do not support strong phonon coupling, and there is an observed gap in contrast to the expectations of the resonance models (Barri).

F. OTHER MODELS

In this section we will survey some of the theoretical literature that is different from both the phonon BCS and the RVB models. These two approaches, however, are not totally disjoint because some of the proposed models involve multiple mechanisms which may include BCS or RVB. Also many authors compare their results with these two popular models. In addition there is some theorizing which is not easy to categorize.

Many authors have analyzed the electronic properties of the Hubbard Hamiltonian (Bhatt, Emery, Emer1, Kosty, Param, Rucke). The Hamiltonian in the present case is similar to RVB but the proposed ground states are not of the RVB type. Both strongly and weakly correlated cases have been examined (Leder, Riecz). The strong correlation is essentially the Mott–Hubbard case, and in the weak coupling limit the Fermi surface nesting is dominant.

The temperature dependence of the normal state resistivity and the Hall mobility have been calculated from the Brinkman–Rice formula (Sokol). In both cases a $T^{1/2}$ dependence is obtained, which differs from the near linear behavior temperature dependence of the resistivity that is usually observed with LaSrCuO and YBaCuO.

There has been some discussion of chain and one-dimensional correlation models. Superconductivity was claimed to be located in the Cu–O planes with T_c raised by the charge-density wave (CDW) instability of the neighboring Cu–O chains (Okabe). One linear chain model (Gagli) exhibits a change from highly ionic to highly metallic behavior as electron states are emptied or holes are added. These holes appear to correlate strongly in space by occupying next-nearest-neighbor sites. It has been argued (Barri) that the real space correlation length is of the order of the lattice parameter, and that the density of states at the Fermi surface is not relevant to the interaction process. In another work it was mentioned that there is no definitive experimental evidence for lower dimensional instabilities such as spin- or charge-density waves (Barri), and the effects of Jahn–Teller instability have been discussed (Aokiz). The more recent discovery of the bismuth and thallium superconductors with their lack of chains has caused many to lose interest in chain model explanations.

The superconducting electron pairs have been assigned to Cu sites (Calla) induced by the electronic polarization of the oxygens on the transverse chains. The disproportionation of Cu^{2+} was reported to be absent in a negative U local pairing model (Wilso).

Many unusual properties in the critical behavior, fluctuations, and the magnetic field interaction in thin films and microcrystals of elemental superconductors such as Al, Ga, Gd, In, Nb, Pb, Sn, and other superconducting materials have been attributed to twin planes (Khlys). It was argued theoretically that twinning can assist conventional superconductivity and may be responsible for the high T_c of the orthorhombic copper oxide superconductors such as YBaCuO. Again the discovery of the tetragonal bismuth and thallium compounds in which twinning is not widespread has diminished interest in twinning explanations.

In the spin-bag model (Schr2) the ground state of the undoped two-dimensional system involves spins that are antiferromagnetically correlated over many unit cells, and form a spin-density wave (SDW) with an electronic pseudogap Δ_{SDW}. As the system is doped the SDW is suppressed by the holes which self-trap into spin bags. The energy is lowered when two holes share the same bag and give rise to a superconducting gap Δ_{SC},

$$\Delta_{SC} \approx \Delta_{SDW} \exp(-t/\alpha U) \tag{IV-14}$$

where $\alpha \approx 1$, and t and U are defined in Section IV-D (cf. Eq. IV-12).

Fractons or the quantized excitations of fractal structures (Aharo) have been invoked as the pairing interaction mechanism. The fracton density of states and the cutoff frequency are much higher in a fractal structure. This as well as the granular effects due to Josephson contacts (Chakr) were argued to be capable of enhancing T_c (Buttn).

Ceramic superconductivity has been discussed in the Fermi liquid formalism (Ohkaw). The large (≈ 2) Wilson ratio is capable of producing magnetic ordering at ≈ 250 K and "$d\gamma$" symmetry Cooper pairs at $T_c \approx 100$ K via the superexchange mechanism. It was suggested (Crisa) that Anderson localization effects in the strong coupling theory should be examined.

For the case of nonlinear attractive interactions, the lowest excited states may involve solitons instead of particles or plane waves. This is particularly important for low space dimensionality (d). Like any bosons, these solitons can undergo a Bose condensation and give rise to superconductivity.

The low-temperature specific heat of such a system is given by (Leez8)

$$C_P \approx T^{\frac{1}{2}d+n} \tag{IV-15}$$

where $n > 1$. Therefore, for two dimensions a nearly linear specific heat is predicted. This model does not address the microscopic mechanism for the attraction, but it does provide for superconductivity independent of the mechanism of the attraction.

G. BAND STRUCTURE

Much of our present understanding of the properties of metals and semiconductors is due to band-structure calculations. Using mainframe computers these calculations can be carried out very efficiently and precisely to provide information on such quantities as the density of states, the energy gap, the conduction electron velocity, and the energies of various bands.

The LaSrCuO and YBaCuO compounds are rather complicated, and they require all of the talents of a skilled band-structure theorist with access to a very large computer or supercomputer. However, the essential features of the electronic structure can be understood by considering the "kernel" of the copper oxide superconductors, which is the Cu–O cluster (Eschr) in the form of a CuO_6 octahedron. This may be expected to give the top of the band or the large k features of the electronic structure (Ricez). Assuming the formal charge state of copper to be Cu^{2+}, one obtains a d^9 configuration which corresponds to a hole in the full d^{10} shell. This hole hybridizes with O^{2-} to form Cu3d–O2p orbitals which are the antibonding states e_g. The difference in the Cu–O bond length perpendicular to the Cu–O planes (2.4 Å) compared with the in-plane Cu–O bond length (1.9 Å) of YBa* removes the degeneracy between the x^2-y^2 and $3z^2-r^2$ orbitals of the e_g band. This gives rise to a half-filled (undoped) band produced by the antibonding hybridization of the oxygen $2p_x$ and $2p_y$ orbitals with the x^2-y^2 one from the copper. The planar Cu–O arrangement will result in a two-dimensional band structure. For the half-filled case the Fermi surface will be the square boundary of the Fermi sphere given by the lines joining the reciprocal space k points (π/a) (1,0) and (π/a) (0, 1) (Ricez).

These simple considerations have been adequately justified by the results of band-structure calculations of $LaCuO_3$ (Takeg), La_2CuO_4 (Kasow, Kraka, Labbe, Matt5, Ruval, Take1, Temme, Zhao2), the oxide superconductors LaSrCuO (Free1, Fuku1, Fuzzz, Kraka, Terak), YBaCuO (Bulle, Ching, Fried, Herm1, Matt4, Ortiz, Temme, Yuzz1, Zhao2), and both (Freem, Matt6). Two-dimensional (Freem, Labbe, Lynnz) and perovskite analogues like $YCuO_3$ (Papac, Papa1) have also been studied. The symmetry classification and the difficulties with the usual s-wave, extended-wave, and d-wave characterizations of the states in these materials have been discussed (Sigri). Some of these calculations were compared with experimental data such as photoemission measurements (Nucke, Redin, Reihl, Temme).

The band structure of $(La_{1-x}M_x)_{2-y}CuO_{4-d}$ is confined to the Cu–O layer and acts like a two-dimensional electronic system (Xuzzz). Linear augmented plane wave (APW) calculations indicated significant O–2p orbital character at E_F, half-filled σ-antibonding bands, Fermi surface nesting (cf. Section IV-C-1), incipient charge density wave instabilities, low carrier densities, and a low density of states at E_F in $La_{2-x}(Ba,Sr)_xCuO_4$ and $YBa_2Cu_3O_7$, as in $BaPb_{1-x}Bi_xO_3$ (Matth). The calculations suggest that the enhanced T_c's are due to the strong interaction between the conduction band states near E_F and the high-frequency oxygen bond-stretching phonon modes.

An APW calculation of LaSrCuO showed the electronic structure dominated by the layered in-plane Cu-3d,O-2p interactions (Yuzzz). A strong Fermi surface instability via a soft electron–phonon mode leads to the observed orthorhombic phase. Adding Ba or Sr suppresses the instability and stabilizes the tetragonal phase, with a large electron–phonon interaction inducing a high T_c.

The La bands have been determined in a two-dimensional tight binding approximation (Fuku1). Disorder effects due to the layered structure of the Cu^{2+} states in this system appear to be less severe than those in the cubic perovskites, namely $BaPb_{1-x}Bi_xO_3$. Both T_c and the electron–phonon coupling constant λ were reported to be nonmonotonic functions of x with a critical value $x \approx 0.1$.

The electronic structures of semiconducting $YBa_2Cu_3O_6$ and superconducting $YBa_2Cu_3O_7$ differ by the presence of a dominant 1D contribution in O_7 which is absent in the O_6 compound (Yuzz1). This was thought to be a contributing factor in producing the larger T_c of the 123 structure. There is also the influence (Free1) of the van Hove saddle point singularity (Ohka2, Ricez) on the density of states (DOS) and on such properties as T_c, the specific heat, and the magnetic susceptibility, which depend upon the DOS. This DOS at the Fermi level is reported to be 1.1–1.6 states/eV Cu atom (Bulle); the DOS in YBr* was calculated to be small compared with that in LaSr* (Yuzz1). The calculated larger Stoner factor and the possible nesting of the Fermi surface along 110 could produce antiferromagnetic ordering in $YBa_2Cu_3O_6$. In $Bi_2Sr_2CaCu_2O_8$ the DOS was calculated to be 1.1 (Hyber) and 1.44 (Krak1) states/eV Cu atom. (A van Hove singularity is a vanishing of the gradient of the energy with respect to the wave vector k in the denominator of the integral used to calculate the density of states. The integration provides a finite but large density of states.)

A first-principles self-consistent pseudofunction calculation (Herm1) indicates that the variation in T_c with oxygen content observed in $YBa_2Cu_3O_{7-\delta}$ is reconcilable with reasonable values of the BCS parameters. The highest T_c values should be achieved with $YBa_2Cu_3O_7$ or with highly ordered oxygen vacancies.

A local density, functional type band-structure calculation carried out for $Bi_2Sr_2CaCuO_8$ found a pair of nearly half-filled two-dimensional Cu–O 3d,2p bands similar to those in the previous Cu–O planar superconductors, as well as a pair of slightly filled Bi 6p bands that provide additional carriers in the Bi–O planes (Hyber). Another density-functional calculation revealed Bi–O bands that cross the Fermi level, and found the two-dimensional character of the bands was even greater than that of previous cuprates (Krak2).

H. CONCLUDING REMARKS

This and the previous chapter complete the brief survey of the experimental and theoretical aspects of oxide superconductors. We have tried not to be opinionated about the different theoretical models that have been proposed because it is too soon to pass judgment on them. Many aspects of the BCS theory, however, do seem to apply.

We will now proceed to a more systematic and detailed treatment of the properties of these materials. Along the way we will see the extent to which they conform to the BCS predictions, and how well they agree with some of the other theoretical and experimental ideas presented in these two chapters.

V

PREPARATION AND
CHARACTERIZATION OF SAMPLES

A. INTRODUCTION

Copper oxide superconductors with a purity sufficient to exhibit zero resistivity or to demonstrate levitation (Early) are not difficult to synthesize. We believe that this is at least partially responsible for the explosive worldwide growth in these materials. Nevertheless, it should be emphasized that the preparation of these samples does involve some risks since the procedures are carried out at quite high temperatures, often in oxygen atmospheres. In addition, some of the chemicals are toxic, and in the case of thallium compounds the degree of toxicity is extremely high so ingestion, inhalation, and contact with the skin must be prevented.

The superconducting properties of the copper oxide compounds are quite sensitive to the method of preparation and annealing. Multiphase samples containing fractions with T_c above liquid nitrogen temperature (Monec) can be synthesized using rather crude techniques, but really high-grade single-phase specimens require careful attention to such factors as temperature control, oxygen content of the surrounding gas, annealing cycles, grain sizes, and pelletizing procedures. The ratio of cations in the final sample is important, but even more critical and more difficult to control is the oxygen content. However, in the case of the Bi- and Tl-based compounds, the superconducting properties are less sensitive to the oxygen content.

Figure V-1 illustrates how preparation conditions can influence superconducting properties. It shows how the calcination temperature, the annealing time, and the quenching conditions affect the resistivity drop at T_c of a BiSrCa-CuO pellet, a related copper-enriched specimen, and an aluminum-doped coun-

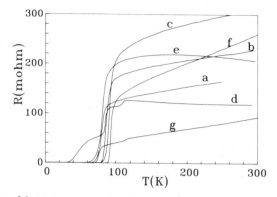

Fig. V-1. Effects of heat treatments on the resistivity transition of $BiSrCaCuO_{7-\delta}$ (*a*) calcined at 860°C, (*b*) calcined at 885°C, (*c*) calcined at 901°C, (*d*) aluminum-doped sample calcined at 875°C, prolonged annealing, (*e*) copper-rich sample calcined at 860°C, (*f*) aluminum-doped sample calcined at 885°C, slow quenching and (*g*) calcined at 885°C, prolonged annealing, and slow quenching (Chuz5).

terpart (Chuz5). These samples were all calcined and annealed in the same temperature range and air-quenched to room temperature.

Polycrystalline samples are the easiest to prepare, and much of the early work was carried out with them. Of greater significance is work carried out with thin films and single crystals, and these require more specialized preparation techniques. More and more of the recent work has been done with such samples.

Many authors have provided sample preparation information, and others have detailed heat treatments and oxygen control. Some representative techniques will be discussed.

The beginning of this chapter will treat methods of preparing bulk superconducting samples in general, and then samples of special types such as thin films and single crystals. The remainder of the chapter will discuss ways of checking the composition and quality of the samples. The thermodynamic or subsolidus phase diagram of the ternary Y–Ba–Cu oxide system illustrated in Fig. V-2 contains several stable stoichiometric compounds such as the end-point oxides Y_2O_3, BaO, and CuO at the apices, the binary oxides stable at 950°, (Ba_3CuO_4), Ba_2CuO_3, $BaCuO_2$, $Y_2Cu_2O_5$, $Y_4Ba_3O_9$, Y_2BaO_4, and ($Y_2Ba_4O_7$), along the edges, and ternary oxides such as ($YBa_3Cu_2O_7$), the semiconducting green phase Y_2BaCuO_5, and the superconducting black solid $YBa_2Cu_3O_{7-\delta}$ in the interior (Beye2, Bour3, Capo1, Eagl1, Frase, Hosoy, Jone1, Kaise, Kurth, Kuzzz, Leez3, Lian1, Mali1, Schni, Schn1, Schu1, Takay, Torra, Wagne). Compounds in parentheses are not on the figure, but are reported by other workers. The existence of a narrow range of solid solution was reported (Panso), and then argued against (Wagne) by the same group.

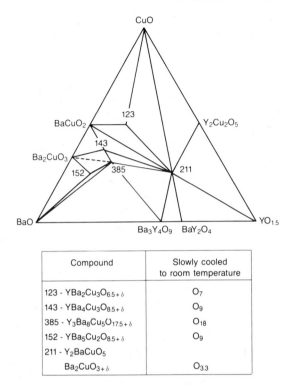

Compound	Slowly cooled to room temperature
123 - $YBa_2Cu_3O_{6.5+\delta}$	O_7
143 - $YBa_4Cu_3O_{8.5+\delta}$	O_9
385 - $Y_3Ba_8Cu_5O_{17.5+\delta}$	O_{18}
152 - $YBa_5Cu_2O_{8.5+\delta}$	O_9
211 - Y_2BaCuO_5	
$Ba_2CuO_{3+\delta}$	$O_{3.3}$

Fig. V-2. Ternary phase diagram of the Y_2O_3–BaO–CuO system at 950°C. The green phase [Y_2BaCuO_5, (211)] the superconducting phase [$YBa_2Cu_3O_{7-\delta}$, (123)], and three other compounds are shown in the interior of the diagram (DeLee).

B. METHODS OF PREPARATION

In this section three methods of preparation will be described, namely, the solid state, the coprecipitation, and the sol–gel techniques (Hatfi). The widely used solid-state technique permits off-the-shelf chemicals to be directly calcined into superconductors, and it requires little familiarity with the subtle physicochemical processes involved in the transformation of a mixture of compounds into a superconductor. The coprecipitation technique mixes the constituents on an atomic scale and forms fine powders, but it requires careful control of the pH and some familiarity with analytical chemistry. The sol–gel procedure requires more competence in analytical procedures.

In the solid-state reaction technique one starts with oxygen-rich compounds of the desired components such as oxides, nitrates, or carbonates of Ba, Bi, La, Sr, Tl, Y, or other elements. Sometimes nitrates are formed first by dissolving oxides in nitric acid and decomposing the solution at 500°C before calcination (e.g., Davis, Holla, Kelle). These compounds are mixed in the desired atomic ratios and ground to a fine powder to facilitate the calcination process. Then these room-temperature-stable salts are reacted by calcining for an extended period (≈ 20 hr) at elevated temperatures ($\approx 900°C$). This process may be repeated several times, with pulverizing and mixing of the partially calcined material at each step. As the reaction proceeds, the color of the charge changes. The process usually ends with a final oxygen anneal followed by a slow cool down to room temperature of the powder, or pellets made from the powder, by sintering in a cold or hot press. Sintering is not essential for the chemical process, but for transport and other measurements it is convenient to have the material pelletized. A number of researchers have provided information on this solid-state reaction approach (e.g., Allge, Finez, Galla, Garla, Gopal, Gubse, Hajk1, Hatan, Herrm, Hika1, Hirab, Jayar, Maen1, Mood1, Mood2, Neume, Poepp, Polle, Qadri, Rhyne, Ruzic, Saito, Sait1, Sawa1, Shamo, Takit, Tothz, Wuzz3).

Some of the earlier works on foils, thick films, wires, or coatings employed a suspension of the calcined powder in a suitable organic binder, and the desired product was obtained by conventional industrial processes such as extruding, spraying, or coating.

In the second or coprecipitation process the starting materials for calcination are produced by precipitating them together from solution (e.g., Asela, Bedno, Leez7, Wang2). This has the advantage of mixing the constituents on an atomic scale. In addition the precipitates may form fine powders whose uniformity can be controlled, which can eliminate some of the labor. Once the precipitate has been dried, calcining can begin as in the solid-state reaction procedure. A disadvantage of this method, at least as far as the average physicist or materials scientist is concerned, is that it requires considerable skill in chemical procedures.

Another procedure for obtaining the start-up powder is the sol–gel technique in which an aqueous solution containing the proper ratios of Ba, Cu, and Y nitrates is emulsified in an organic phase and the resulting droplets are gelled by the addition of a high-molecular-weight primary amine which extracts the nitric acid. This process was initially applied to the La materials, but has been perfected for YBaCuO as well (Cimaz, Hatfi).

When using commercial chemical supplies to facilitate the calcination process a dry or wet (acetone) pregrinding with an agate mortar and pestle or a ball mill is recommended. Gravimetric amounts of the powdered precursor materials are thoroughly mixed and placed in a platinum or ceramic crucible. Care must be taken to ensure the compatibility of the ceramic crucible with the chemicals to obviate reaction and corrosion problems.

Complete recipes for the YBa* material have been described (e.g., Gran2). Typically, the mixture of unreacted oxides is calcined in air or oxygen around 900°C for 15 hr. During this time the YBaCuO mixture changes color from the green Y_2BaCuO_5 phase to the dark gray $YBa_2Cu_3O_{7-\delta}$ compound. Then the charge is taken out, crushed, and scanned with X rays to determine its purity. If warranted by the powder pattern X-ray scan, the calcination process is repeated. Often, at this stage the material is very oxygen poor, and electrically it is semiconducting or even nonconducting. After pelletizing at $>10^5$ psi the pellet is sintered for several hours at $\approx 900°C$ in flowing oxygen and then slowly cooled at $\approx 3°C/min$ down to room temperature. Slow cooling from the elevated temperature is important for producing the low-temperature orthorhombic superconductor phase. The tetragonal nonsuperconducting phase may be obtained by quenching. The pellet may be used as is or it may be cut into suitable sizes by sand blasting, with a diamond saw, or with an arc. After vigorous machining another oxygen anneal (450°C, 1 hr, slow cool down) is often required to preserve the superconducting properties.

An example of preparing a Bi-based superconductor involves mixing gravimetric amounts of high-purity Bi_2O_3, $SrCO_3$, $CaCO_3$, and CuO powders, calcining them in air at 750–890°C, regrinding them, and then repeating these procedures several times. Then pellets of the calcined product were sintered at the same temperature and quenched to room temperature (Chuz5). Figure V-1 shows the effect of sample treatment on the resistance versus temperature curve.

WARNING: As was mentioned above, thallium is a toxic material and proper precautions must be taken when working with it. It is useful to start by preparing the high-quality precursor compound $BaCu_3O_4$ or $Ba_2Cu_3O_5$ by reacting the oxides in air at 925°C for 24 hr. Then appropriate amounts of Tl_2O_3 are added, powdered, and pelletized. The pellet is then heated to 880–910°C for a few minutes in flowing oxygen, and at the onset of melting it is quenched to room temperature (Shen1).

Allen Hermann has suggested consulting the following references for information on thallium poisoning and antidotes thereto: H. Heydlanf, Euro. J. Pharmacol. 6, 340 (1969), which discusses thallium poisoning and describes the antidote ferric cyanoferrate, and Int. J. Pharmacol. 10, 1 (1974), which discusses cases of thallium intoxication treated with Prussian Blue.

C. ADDITIONAL COMMENTS ON PREPARATION

This section will treat some additional methods which have been employed for the preparation of samples.

In one experiment coprecipitated nitrates of La, Sr, Cu, and Na carbonate were calcined for 2 hr at 825°C, pressed into pellets, and then subjected to shock compression of ≈ 20 GPa at an estimated peak temperature of $\approx 1000°C$ (Graha). The best superconductivity was observed after 1 hr of air exposure at 1100°C. Shock compression fabrication has also been reported (Murrz, Murr1) for YBa* and other rare-earth derivatives. This process produced "monoliths," distinct from the usual composites.

Another technique involved the formation of a precursor alloy of Eu, Ba, Cu or Yb, Ba, Cu by rapid solidification, with the superconducting materials obtained subsequently by oxidation (Halda). A novel method involved preparing the superconductors from molten Ba–Cu oxides and solid rare-earth-containing materials. In principle this process may be better controlled and complicated shapes can be molded or cast (Herma).

Pulsed current densities of 300–400 $\text{Å}/\text{cm}^2$ with rise times of 0.6 μsec at room temperature were used to convert the weakly semiconducting phase of YBaCuO to the stable metallic phase (Djure, Djur1).

A claim was made that thermal cycling from cryogenic temperatures to 240 K raised the T_c of YBa* and YBaCuO-F (with some F substituting for O) to 159 K. Cycling above 140 K lowered T_c. This cycling process could possibly change the density of twins and thereby enhance T_c.

A freeze-drying technique was reported as producing sintered materials homogeneous in composition and small in porosity (Stras). The low-temperature firing of oxalates ($T < 780°C$) has also been reported as producing a homogeneous material of small grain size (Manth).

Both Bi and Pb act as fluxes during the sintering process (Kilco). Bismuth substitution appears to reduce the normal state resistivity by about an order of magnitude without affecting the superconducting properties.

A convenient method of separating the superconducting particles from a powdered mixture using magnetic levitation has been reported (Barso). This may be used to select the superconducting fraction after each calcination process.

D. FILMS

The new ceramic oxide superconductors presently lack mechanical properties such as ductility which are needed for high-current applications like magnet wire fabrication (Jinzz–Jinz3) and power transmission. To circumvent some of these deficiencies for microelectronic applications one can prepare thin films on suitable substrates. Some devices such as Josephson junctions require thin superconducting films. Many workers have discussed the preparation and properties of LaSrCuO- (e.g., Adach, Delim, Kawas, Koinu, Matsu, Nagat, Naito, Tera1) and YBaCuO- (e.g., Burbi, Charz, Evett, Gurvi, Hause, Hongz, Inamz, Kwozz, Kwoz1, Manki, Scheu, Somek, Wuzz4) type films.

Almost every conceivable thin-film deposition technique such as electron beam evaporation, molecular beam epitaxy, sputtering, magnetron, laser ablation, screening, and spraying has been tried with the copper oxide system. Some of these techniques require expensive, elaborate apparatus, although descriptions of simple thin-film deposition systems are also available (e.g., see Koin1). Some representative examples of deposition procedures will be discussed.

Epitaxial films of $YBa_2Cu_3O_{7-\delta}$ on (100) $SrTiO_3$ were produced using three separate electron beam sources (e.g., Chaud, Chau1, Laibo). The deposition was done in 10^{-4}–10^{-3} torr O_2 with a substrate temperature of 400°C. The deposited films were atomically amorphous with a broad X-ray peak. The epitaxial ordering was achieved upon annealing in O_2 at 900°C with the orthorhombic c axis essentially perpendicular to the plane.

High-quality superconducting films were obtained using a multiple electron beam to evaporate metallic sources in a flow of molecular oxygen at 4–5×10^{-6} torr (Hammo, Ohzzz). The deposition rate was 10 Å/sec. To anneal the deposited film in oxygen it was heated for 3–6 hr in a flow of oxygen at $650°C$, raised to $750°C$ for 1 hr, then to $850°C$ for 1 hr, and finally slowly cooled down in the furnace.

Superconducting films were prepared using a double ion beam sputtering arrangement (Madak). The target beam was Ar at 40 mA, and the substrate beam was Ar or an Ar–O_2 mixture at 10–500 eV and 2 mA. The base pressure was 5×10^{-7} torr and, with the gas, 4×10^{-4} torr. The best substrate materials such as ZrO_2–9% Y_2O_3 did not appreciably interact, diffuse, or change the deposited films. The films were $\approx 1~\mu m$ thick and were rendered superconducting by oxygen annealing. Zero resistance was attained at 88 K. The superconducting properties depended upon the ion beam energy, substrate temperature, annealing conditions, composition, and the extent of poisoning from the substrate.

Films of dysprosium barium copper oxide were grown (Webbz) by molecular beam epitaxy (MBE) using a Varian 360 MBE system, and the nucleation process was monitored by reflection high-energy electron diffraction (RHEED). The copper was incompletely oxidized in metallic microcrystals growing in a sea of amorphous Ba and Dy. After deposition superconducting films were obtained by high-temperature oxygen annealing.

Films of $Y_{1.1}Ba_{1.5}Cu_3O_{6.4}$ approximately 3300 Å thick with a surface roughness of 500 Å were prepared (Dijkk, Inamz, Wuzz4). These films were deposited on $SrTiO_3$, sapphire, and vitron carbon by evaporation from a single bulk pellet of YBaCuO 1 cm diameter and 0.2 cm thick at a pressure of 5×10^{-7} torr. The evaporation was produced by several thousand pulses of laser irradiation (3–6 Hz, ≈ 30 nsec width, 1 J/pulse, 2 J/cm^2). For best results the substrate was heated to $450°C$. As deposited thin films were well bonded to the substrate and they appeared shiny dark brown and were electrically insulating. The films were oxygen annealed at $900°C$ for 1 hr and then slowly cooled over a period of several hours. Standard four-probe resistivity measurements indicated the onset of superconductivity around 95 K and, for a (100) $SrTiO_3$ substrate, with zero resistivity achieved near 85 K. The laser ablation technique was also employed for LaSr* (Moorj) and YBa* (Nara1).

Films were obtained from sandwiched multilayers by depositing Y_2O_3, BaO, and Cu in layers (Nasta, Tsaur) on ZrO_2, MgO, and sapphire substrates at $200°C$ and 10^{-5} torr. Oxygen treatment for 1–2 hr at $\approx 850°C$ permitted the layers to diffuse, homogenize, and oxygenate, and thereby form the superconducting compound (Baozz). Films on Ni have also been reported in which superconductivity was obtained by a diffusion process involving the Cu substrate, Y_2O_3, and $BaCO_3$ composite (Tachi).

Some 5000-Å thick films of YBaCuO have been deposited using an ultrahigh vacuum dc–magnetron getter-sputter deposition system. The deposition rate was 0.2 Å/sec, the substrate temperature was $1050°C$, and the target-to-substrate distance was 12 cm. The scattering was done in an Ar–O_2 atmosphere.

The X-ray and electron microscope examinations indicated some variation among the substrates arranged on the heater. Inhomogeneities were observed even within the film made on a single substrate. As deposited the films were oxygen deficient, and annealing produced suitable compositions. The reversible oxygen incorporation was monitored by the systematic splitting of the strongest X-ray peaks. The oxygen diffusion coefficient at 600°C was 10^{-15} m^2/sec and the activation energies for desorption and absorption were 1.1 and 1.7 eV, respectively. The highest onset temperature was 99 K with complete superconduction at 40 K. Exposure to water inhibited the superconductor (Barns, Kishi, Yanzz). A device structure with a Y_2O_3 barrier has also been studied (Blami).

Another work showed that films produced by dc magnetron sputtering are copper deficient if the substrate-to-target distance is large or if the substrate is at an elevated temperature (Leez5).

Superconducting YBaCuO thin films with a large surface area (≈ 5 cm $\times$ 5 cm) were grown on Al_2O_3, sapphire, and MgO up to a 500°C substrate temperature by magnetron and diode techniques. Rutherford back scattering (RBS) indicated a uniform composition across magnetron-deposited film areas with diameters up to 5 cm, and the diode film composition homogeneity was even better, but over a smaller area (≈ 2.5 cm diameter). The as-deposited films were annealed in oxygen at different temperatures and exposure times. Prolonged high-temperature annealing (> 850°C) increased the impurity phase. The highest T_c films had a wide range of composition, with the maximum T_c film copper rich. On the basis of an in-situ resistivity study of YBa* thin films a rapid heating to about 900°C in flowing helium followed by slow cool down in flowing oxygen was recommended (David).

The post-deposition anneal cycle was avoided by producing the films in a high-pressure reactive evaporation process involving rapid thermal annealing (Lathr). Smooth films were obtained on zirconia and $SrTiO_3$ substrates. Screen printing of oxide superconducting films is also possible (Budha, Fuzz1), and simple spray deposition has been reported (Gupta). Films have also been made by coating and spinning off the solutions. Aqueous and aqueous–alcoholic mixed solutions of the metal nitrates (Coop2), metal acetates in dilute acetic acid (Rice1), and sol–gels (Kram1) have all been reported. These processes are potentially important for commercial superconducting coatings on silicon (Kram1), on yttrium-stabilized zirconia (YSZ), on $SrTiO_3$ (Coop2, Gupta), and on MgO (Gupta, Rice1).

E. SINGLE CRYSTALS

The bulk properties of oxide superconductors are averages over components parallel and perpendicular to the Cu–O planes. In addition, for orthorhombic samples there is an averaging over properties that differ for the a and b directions in this plane. This in-plane anisotropy is especially pronounced for the YBa* 123 structure in which the Cu–O–Cu–O chains lie along the b axis. The best way to understand these materials is through experiments on perfect single crystals. Unfortunately, untwinned YBa* crystals are not available so the a,b anisotropy cannot be resolved. Tetragonal superconductors should not have this twinning problem. In this work twinned monocrystals will be referred to as single crystals.

A number of experiments have been carried out on monocrystals such as X-ray diffraction (e.g., Borde, Hazen, Lepag, Siegr, Onoda), magnetic studies (e.g., Crabt, Schn1, Worth), mechanical measurements (e.g., Cookz, Dinge), and micro-Raman spectroscopy (e.g., Hemle). In this section we will briefly describe how such crystals are made. The December 1987 issue of the *Journal of Crystal Growth* was devoted to superconductors.

Millimeter-size $(La_{1-x}Sr_x)_2CuO_4$ single crystals were grown in a molten copper oxide flux (Kawa1). Another basic technique employs other fluxes (Haned, Taka4, Zhou1), namely, PbF_2, B_2O_3, PbO, PbO_2, with the risk of possible Pb contamination. LaSr* crystals were also grown by the solid phase reaction using a hot press of pellets (Iwazu) and rapid quenching of a nonstoichiometric melt (Satoz).

Small single crystals of $YBa_2Cu_3O_{7-\delta}$ have been prepared from a sintered powder which was formed into a pellet and then heated, first in a reducing atmosphere and then in an oxidizing one at 925°C. Annealing a stoichiometric mixture also produced monocrystals (Liuzz). Millimeter-size crystals were grown by melting a stoichiometric mixture of $YBa_2Cu_3O_{7-\delta}$ plus excess CuO at 1150°C followed by holding at 900°C for 4 days (Damen, see also Fine1).

A gold crucible on a gold or alumina sheet was used to obtain free-standing $(1 \times 2 \times 0.1 \text{ mm})$ single crystals of YBa* (Kaise, Kais1, Holtz). A charge of 2 g was heated in air at 200°C/hr and held at 975°C for 1.5 hr, then it was cooled to 400°C at 25°C/hr. The molten charge creeps and forms single crystals and twins on the surfaces. The larger crystals formed in the space between the bottom of the crucible and the gold support sheet.

A detailed account has appeared of the preparation of a 123 compound single crystal by the flux method (Zhou1). The flux mole ratio $BaO_2:CuO$ was between 1:3 and 2:5, and the nutrient $Y_2O_3:BaO_2:CuO$ mole ratios were 0.5:2:3. A multistep temperature process was employed. Black single crystals of YBa* were found at the bottom and at the edge between the wall and the bottom of the crucibles. Platinum crucibles seemed to contaminate the samples so alumina crucibles were recommended. Crystals as large as $2 \times 2 \times 0.3 \text{ mm}^3$ were reported. A similar technique was used to produce single crystals of YBa* and DyBa* as large as 4 mm (Schn1).

F. ALIGNED GRAINS

Clearly high-quality single crystals are important for understanding the physics of superconductors. However, much useful information about anisotropies can be obtained by studying the properties of aligned grains, which are much easier to fabricate.

A superconducting sample can be initially a collection of randomly oriented grains, but various techniques can be used to partially orient these grains so that the c axis lies preferentially in a particular direction. For example uniaxial compression tends to orient compacted grains, with compressed 90-μm particles exhibiting more alignment than compressed 10-μm particles (Glowa). Epoxy-embedded grains have been aligned under the influence of an applied magnetic field and pressure (Arend).

X-ray and magnetic measurements have been reported on aligned crystalline grains of YBa* (Farr1). Optical studies have also been made on aligned grains. The critical current density for samples cut parallel to the compression axis of such grains was nearly isotropic with respect to the direction of an applied magnetic field, and it was a factor of 6 smaller than that for the samples cut perpendicular to this axis (Glowa).

G. REACTIVITY

The oxide superconductors are not inert materials, but rather they are sensitive to exposure to certain gases and to surface contact with particular materials. Great care must be exercised to avoid contamination from water vapor and carbon dioxide in the atmosphere. In addition these materials are catalytic to oxygenation reactions, and these factors result in the occurrence of various chemical and other interactions, especially at elevated temperatures. The granular and porous nature of the materials has an accelerating effect on such reactions.

Samples of YBaCuO may degrade in a matter of days when exposed to an ordinary ambient atmosphere; they react readily with liquid water, acids, and electrolytes, and moderately with basic solutions. The reaction with water (Barns, Kishi, Yanzz) produces nonsuperconducting cuprates. The effects of acetone and other organics (McAnd) have been determined, and stable carboxyl groups have been found in the YBaCuO lattice (Parmi).

Hydrogen enters the YBaCuO lattice at elevated temperatures and forms a solid solution. Low concentrations have very little effect and high concentrations degrade the superconducting properties (Berni, Reill, Yang3). The effects of exposure to oxygen at elevated temperature and oxidation have been discussed several places in this review (e.g., Blend, Engle, Tara3).

The foregoing evidence for the reactivity of the oxide superconductors makes it necessary to consider methods of passivation or protecting them from long-term degradation. An epoxy coating was found to provide some protection (Barns). Coating the surface with metals can be deleterious since metals such as Fe (Gaoz1, Hillz, Weave) and Ti (Meye1) react with the surface of LaSrCuO or YBaCuO. There is evidence for the passivation of the surface of LaSr* with gold (Meyer).

H. THERMOGRAVIMETRIC ANALYSIS

Thermogravimetric analysis (TGA) consists of monitoring the weight of a sample during a heating or cooling cycle. For example, one might determine the oxygen content of a superconducting material by measuring its weight change in an oxidizing (O_2 or air) or reducing (e.g., 4% H_2 in Ar) atmosphere. Typical procedures consist of heating or cooling at 20°C/min. The relative accuracy of the method is about 0.005 (Ongz1). Many workers (e.g., Beye3, Hauck, Huan1, John4, Leez7, Maruc, Ohish, Ongz1, Tara7, Zhuzz) are now using TGA or differential thermal analysis (DTA) routinely during their sample preparation procedures.

I. CHECKS ON QUALITY

After a sample has been prepared it is necessary to check its quality as a superconductor. Most investigators employ the four-probe resistivity check to determine whether it superconducts, and at what temperature it transforms to the superconducting state. A sharp, high T_c transition is an indicator of a high-quality sample. Another widely used quality control method is the determination of the magnetic susceptibility of the specimen. Good quality is indicated by a sharp, high T_c transition with both the flux exclusion and flux expulsion close to $-1/4\pi$. This is, in a sense, a more fundamental check on quality since the value of the susceptibility far below the transition temperature is a good indicator of the fraction of the sample that is superconducting (see Section III-D).

In addition to its superconducting properties, it is also of interest to know the chemical composition and the structure of the specimen. The nominal composition is deduced from the relative proportions of the various cations in the starting material. Chemical analysis and some more sophisticated techniques such as XPS, electrospectroscopic chemical analysis (ESCA), and an electron microprobe that is favorable for low-atomic-weight elements are applicable here. Most investigators only report the cation concentrations in the specimen. Oxygen content is much more difficult to determine, but is important to know. Rutherford back-scattering experiments (John1, Wuzz1, Wuzz4) can provide oxygen contents, and metallography characterizes grain sizes.

The structures of the oxide superconductors described in Chapter VI are easily checked by the X-ray powder pattern method. Many articles list the lattice constants a, b, c of samples and mention whether they are tetragonal ($a = b \neq c$) or orthorhombic ($a \approx b \neq c$). Narrow lines and the absence of spurious signals indicate a good, single-phase sample. Typical X-ray diffraction powder patterns for LaSr* (Skelt) and YBa* presented in Figs. V-3 and V-4, respectively, may be used to compare with patterns obtained from freshly prepared samples.

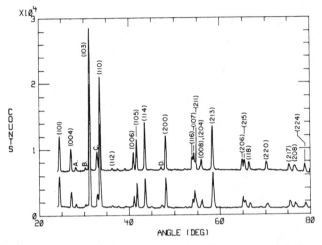

Fig. V-3. Room-temperature (upper curve) and 24-K (lower curve) X-ray diffraction powder patterns of $(La_{0.925}Ba_{0.075})_2CuO_4$ (Skelt).

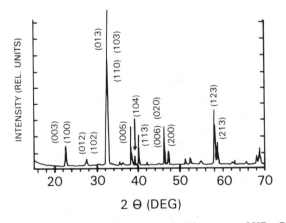

Fig. V-4. Room-temperature X-ray diffraction powder pattern of $YBa_2Cu_3O_7$. (Provided by C. Almasan, J. Estrada, and W. E. Sharp.)

J. RESISTIVITY MEASUREMENT

A measurement of the resistance $R(T)$ or resistivity $p(T)$ of a material versus the temperature is the principal technique employed to determine when a material becomes superconducting. The transition temperature manifests itself by a sharp drop in resistivity to zero. The simplest way to make this measurement is to apply a voltage across the sample and measure the current flow through it, but such a two-probe method (Baszy) is not very satisfactory, and is seldom used. Most resistivity determinations are made with the four-probe technique to be described below, although more sophisticated arrangements such as a six-probe method (Kirsc) can also be used. The fabrication of low-resistance contacts by silver glazing has been reported (Vand2). These researchers pointed out the importance of a low-contact resistance ($\rho < 10 \ \mu\Omega/mm^2$ at 77 K) for making transport J_C measurements.

The specimen resistance as a function of temperature is generally determined in a suitable cryostat by attaching leads or electrodes to it in the standard four-probe configuration. Two leads or probes carry a known constant current I into and out of the specimen, and the other two leads measure the potential drop between two equipotential surfaces resulting from the current flow. For superconducting specimens the leads are often arranged in a linear configuration, with the contacts for the input current on the ends, and those for the measurement voltage near the center.

VI

CRYSTALLOGRAPHIC STRUCTURES

A. INTRODUCTION

To properly understand the mechanisms that bring about the superconducting state in particular materials it is necessary to know the structures of the compounds that exhibit this phenomenon. Single-crystal structure studies have been carried out to determine the dimensions of the unit cell, the locations of the atoms in this cell, electronic charge distributions, and the possible presence of atomic irregularities. Neutron powder diffraction has also provided much of the detailed structure information found in this chapter (e.g., Antso, Beech, Cappo, Coxzz, Davi1, Dayzz, Greed, John4, Jorge, Jorg1, Paulz, Torar, Vakni, Yamag, Yanz2). More routine X-ray powder pattern measurements which can identify a known structure and provide the unit cell dimensions are useful for checking the quality of samples, as was explained in Section V-I.

The numerical values of quantities such as lattice parameters and bond lengths show some variation in the literature, and many of our quoted values will be typical ones. Much of the quantitative structural information is organized in the tables.

In the beginning of this chapter we will introduce the perovskite structure and indicate how it is related to the oxide superconductors. Then we will describe the 21 structure of LaSrCuO and the 123 structure of YBaCuO, we will show how each is generated from a perovskite prototype, and we will clarify its layering scheme. The chapter will end with descriptions of the structures of the newer high-transition-temperature bismuth and thallium compounds.

B. PEROVSKITES

Much has been written about the oxide superconductor compounds being perovskite types, so we will begin with a description of the perovskite structure. This will permit us to develop some of the notation to be used in describing the structures of the superconductors themselves.

1. Cubic Form

Above 200°C barium titanate crystallizes in the perovskite structure, which is cubic, so the three lattice parameters are all equal (i.e., $a = b = c$). The unit cell contains one formula unit $BaTiO_3$ and the atoms are located in the following special positions (Wyck2, p. 390):

$$
\begin{array}{lll}
Ba & (1a) & \frac{1}{2},\frac{1}{2},\frac{1}{2} \\
Ti & (1b) & 0,0,0 \\
O & (3c) & 0,0,\frac{1}{2};\ 0,\frac{1}{2},0;\ \frac{1}{2},0,0
\end{array}
\qquad (VI-1)
$$

where we have employed the crystallographic notation (1a) for an a-type lattice site which contains one atom, (3c) for a c-type lattice site which contains three atoms, and so on. Each atomic position is given by three coordinates, such as $0,0,\frac{1}{2}$ for the oxygen located at $x = 0, y = 0, z = 0.5a$. This arrangement corresponds to placing a titanium atom on each apex, a barium atom in the body center, and an oxygen atom on the center of each edge of the cube, as illustrated on Fig. VI-1. We see from the figure that the barium atoms are 12-fold coordinated and the titaniums have sixfold (octahedral) coordination. The lattice constant or length of the unit cell is $a = 4.0118$ Å at 201°C. The crystallographic space group is $Pm3m$, O_h^1.

An alternate way to represent this structure, which is commonly used in solid-state texts and in crystallography monographs (e.g., Wyck2), is to locate the

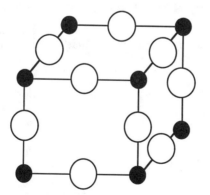

Fig. VI-1. Perovskite cubic unit cell showing titanium on the apices and oxygen in the edge-centered positions. Barium, which is in the body center, is not shown.

origin at the barium site; this places titanium in the center and the oxygens on the centers of the cube faces. The representation (Eq. VI-1) given above is more convenient for comparison with the structures of the oxide superconductors.

The compound $LaBaCu_2O_5$ was found to have a cubic perovskite subcell with the lattice parameter $a = 3.917$ Å (Sishe).

2. Tetragonal Form

At room temperature barium titanate is tetragonal with the unit cell dimensions $a = 3.9947$ Å and $c = 4.0336$ Å, which is close to cubic. For this lower symmetry the oxygens are assigned to two different sites, a single site along the side edges and a twofold one at the top and bottom. The atomic positions (Wyck2, p. 401)

$$\begin{array}{lll} \text{Ba} & \frac{1}{2}, \frac{1}{2}, 0.488 & \\ \text{Ti} & 0,0,0 & \\ \text{O(1)} & 0,0,0.511 & \text{(VI-2)} \\ \text{O(2)} & 0,\frac{1}{2},-0.026; \frac{1}{2},0,-0.026 & \end{array}$$

are shown in Fig. VI-2. The distortions from the ideal structure of Fig. VI-1 are exaggerated on this sketch. We will see later that a similar distortion occurs in the YBaCuO structure. The cubic and tetragonal atom arrangements (VI-1) and (VI-2) are compared in Table VI-1, and we see from this table that the deviation from cubic symmetry is actually quite small.

3. Orthorhombic Form

When barium titanate is cooled below 5°C it undergoes a transition with a further lowering of the symmetry to the orthorhombic space group Amm2, C_{2v}, and

TABLE VI-1. Comparison of Atom Positions of $BaTiO_3$ in Its Cubic, Tetragonal and Orthorhombic Forms[a]

		Cubic and Tetragonal		Cubic	Tetragonal	Orthorhombic
Group	Atom	x	y	z	z	z
TiO_2	Ti	0	0	1	1	1
	O	0	$\frac{1}{2}$	1	0.974	1
	O	$\frac{1}{2}$	0	1	0.974	1
BaO	O	0	0	$\frac{1}{2}$	0.511	$\frac{1}{2}$
	Ba	$\frac{1}{2}$	$\frac{1}{2}$	$\frac{1}{2}$	0.488	$\frac{1}{2}$
TiO_2	Ti	0	0	0	0	0
	O	0	$\frac{1}{2}$	0	−0.026	0
	O	$\frac{1}{2}$	0	0	−0.026	0

[a] The x and y coordinates are the same for both positions. The orthorhombic form z coordinates are also given (Wyck2, pp. 390, 401, 405).

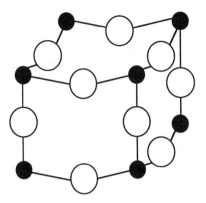

Fig. VI-2. Perovskite tetragonal unit cell showing the puckering of the Ti–O layers.

an enlargement of the unit cell to accommodate two formula units $(BaTiO_3)_2$. The enlarged cell is rotated by $45°$ relative to the higher-temperature ones, as shown on Fig. VI-3, and therefore its a and b lattice parameters are larger by the factor $\sqrt{2}$. The three lattice constants are $a = 5.669 = 4.009\sqrt{2}$ Å, $b = 5.682 = 4.018\sqrt{2}$ Å, and $c = 3.990$ Å. There are no longer any special sites, and the atomic positions are (Wyck2, p. 405):

$$
\begin{array}{lll}
\text{Ba} & (2a) & 0,\tfrac{1}{2},\tfrac{1}{2}; \tfrac{1}{2},0,\tfrac{1}{2} \\
\text{Ti} & (2b) & 0,u+\tfrac{1}{2},0; \tfrac{1}{2},u,0 \quad \text{with} \quad u = 0.510 \\
\text{O(1)} & (2a) & 0,u+\tfrac{1}{2},\tfrac{1}{2}; \tfrac{1}{2},u,\tfrac{1}{2} \quad \text{with} \quad u = 0.490 \\
\text{O(2)} & (4e) & u,v+\tfrac{1}{2},0; -u,v+\tfrac{1}{2},0; u+\tfrac{1}{2},v,0; -u+\tfrac{1}{2},v,0 \\
& & \quad \text{with} \quad u = 0.253,\ v = 0.237
\end{array}
\tag{VI-3}
$$

where $u = 0$ for Ba.

One should note that in Eq. (VI-3) Ba and O(1) are in the same (2a) type of site with different values of the parameter u. Figure VI-3 shows the coordinates of the atoms in the orthorhombic cell drawn using the approximation $\approx \tfrac{1}{2}$ for 0.490 and 0.510 and $\approx \tfrac{1}{4}$ for 0.253 and 0.237.

A comparison of Eqs. VI-1 to VI-3 indicates that the transformation from cubic to tetragonal involves only shifts in the z coordinates of atoms, while the orthorhombic phrase differs from the cubic one only through shifts in atom positions within x,y planes (see Table VI-1).

4. Atom Arrangements

The ionic radii of Ba^{2+} (1.34 Å) and O^{2-} (1.32 Å) are almost the same, and together they form a face-centered cubic (fcc) close-packed lattice with the smaller Ti^{4+} ions (0.68 Å) located in octahedral holes. The octahedral holes of a close-packed oxygen lattice have a radius of 0.545 Å, and if these holes were empty the lattice parameter would be $a = 3.73$, as shown on Fig. VI-4a. If each

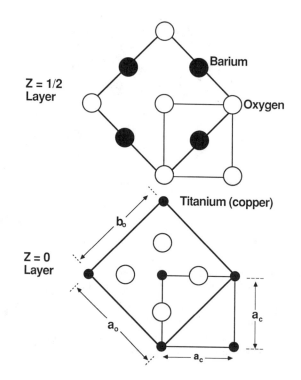

Fig. VI-3. Atom positions of perovskite when the monomolecular tetragonal unit cell is expanded to the bimolecular orthorhombic cell with new axes at 45° with respect to the old ones.

titanium were to move the surrounding oxygens apart to its ionic radius when occupying the hole, as shown on Fig. VI-4b, the lattice parameter a would be 4.00 Å. The observed cubic ($a = 4.012$ Å) and tetragonal ($a = 3.995$ Å, $c = 4.034$ Å) lattice parameters are close to these values, indicating a pushing apart of the oxygens. The tetragonal distortion illustrated on Fig. VI-2 and the orthorhombic distortion of Eq. (VI-3) constitute attempts to achieve this through an enlarged but distorted octahedral site. This same mechanism is operative in the oxide superconductors.

C. BARIUM–LEAD–BISMUTH OXIDE

In 1983 Mattheiss and Hamann referred to the 1975 "discovery by Sleight et al. of high temperature superconductivity" of the compound $BaPb_{1-x}Bi_xO_3$ in the composition range $0.05 \leq x \leq 0.3$ with T_c up to 13 K (Matt7, Sleig). Many consider this system, which disproportionates $2\ Bi^{4+} \rightarrow Bi^{3+} + Bi^{5+}$ in going from the metallic to the semiconducting state, as a predecessor to the LaSrCuO system.

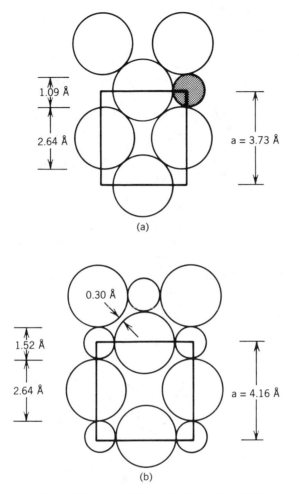

Fig. VI-4. The $z = 0$ plane of the perovskite unit cell showing (a) the size of the octahedral hole and (b) the pushing apart of the oxygens by the presence of a transition ion in the hole. For each case the oxygen and hole sizes are indicated on the left and the lattice parameter is given on the right.

Crystal structure determinations indicate that the metallic compound $BaPbO_3$ has the cubic perovskite structure with $a = 4.273$ Å (Wyck2, p. 391), and at room temperature semiconducting $BaBiO_3$ is monoclinic with $a = 6.136 = 4.339\sqrt{2}$ Å, $b = 6.181 = 4.371\sqrt{2}$ Å, $c = 8.670 = 4.335 \times 2$ Å, and $\beta = 90.17°$ (Coxz1, Coxz2). The latter is quite close to orthorhombic ($\beta = 90°$), and has what might be called a pseudocubic perovskite lattice parameter ($a = 4.35$ Å). These two compounds form a solid solution series $BaPb_{1-x}Bi_xO_3$ which at room temperature changes as a function of increasing x from orthorhombic to tetragonal to orthorhombic to monoclinic. Superconductivity appears in the tetragonal phase, and the metal-to-insulator transition occurs at the tetragonal-to-orthorhombic phase boundary $x \approx 0.35$ (Matt7, Slei1).

D. PEROVSKITE-TYPE SUPERCONDUCTING STRUCTURES

In their first report on high-temperature superconductors Bednorz and Müller referred to their samples as "metallic, oxygen deficient . . . perovskite like mixed valent copper compounds." Subsequent work has confirmed that the new superconductors do indeed have these characteristics. In this section we will comment on their perovskite-like aspects.

1. Atom Sizes

In the oxide superconductors Cu replaces the Ti^{4+} ions (0.68 Å) of perovskite, and in most cases retains the CuO_2 layering with two oxygens per copper in the layer. Other cationic replacements tend to be Bi, Ca, La, Sr, Tl, and Y for the larger Ba, forming "layers" containing only one oxygen or none per cation. We see from the following list of ionic radii

$$
\begin{array}{ll}
Cu^{2+} & 0.72 \text{ Å} \\
Bi^{5+} & 0.74 \text{ Å} \\
\\
Y^{3+} & 0.94 \text{ Å} \\
Tl^{3+} & 0.95 \text{ Å} \\
Bi^{3+} & 0.96 \text{ Å} \\
Ca^{2+} & 0.99 \text{ Å} \\
\\
Sr^{2+} & 1.12 \text{ Å} \\
La^{3+} & 1.14 \text{ Å} \\
\\
Ba^{2+} & 1.34 \text{ Å} \\
O^{2-} & 1.32 \text{ Å}
\end{array}
\qquad \text{(VI-4)}
$$

that there are four size groups, with all other cations significantly smaller then the Ba of perovskite. The common feature of CuO_2 layers that are planar or close to planar establishes a fairly uniform lattice size in the a,b plane. The parameters of the compounds LaSrCuO ($a = b = 3.77$ Å), YBaCuO ($a = 3.83$ Å, $b = 3.89$ Å), BiSrCaCuO ($a = b = 3.82$ Å), and TlBaCaCuO ($a = b = 3.86$ Å) are all between the ideal fcc oxygen lattice value of 3.73 Å and the perovskite one of 4.01 Å.

Table VI-2 gives the ionic radii of the positively charged ions of various elements of the periodic table. These radii are useful for estimating changes in lattice constant when ionic substitutions are made in existing structures. They also provide some insight into which types of substitutions will be most favorable.

TABLE VI-2. Ionic Radii in Angstroms of Selected Elements for Various Positive Charge States[a]

Z	Element	+1	+2	+3	+4	+5	+6
			Alkali				
3	Li	0.68					
11	Na	0.97					
19	K	1.33					
37	Rb	1.47					
55	Cs	1.67					
			Alkaline earths				
4	Be	0.44	0.35				
12	Mg	0.82	0.66				
20	Ca	1.18	0.99				
38	Sr		1.12				
56	Ba	1.53	1.34				
			Group III				
5	B	0.35		0.23			
13	Al			0.51			
31	Ga	0.81		0.62			
49	In			0.81			
81	Tl	1.47		0.95			
			Group IV				
6	C				0.16		
14	Si	0.65			0.42		
32	Ge		0.73		0.53		
50	Sn		0.93		0.71		
82	Pb		1.20		0.84		
			Group V				
15	P			0.44		0.35	
33	As			0.58		0.46	
51	Sb	0.89		0.76		0.62	
83	Bi	0.98		0.96		0.74	
			Chalcogenides				
16	S				0.37		0.30
34	Se	0.66			0.50		0.42
52	Te	0.82			0.70		0.56
			First transition series ($3d^n$)				
21	Sc			0.81			
22	Ti	0.96	0.94	0.76	0.68		
23	V		0.88	0.74	0.63	0.59	
24	Cr	0.81	0.89	0.63			0.52
25	Mn		0.80	0.66	0.60		

TABLE VI-2. (continued)

Z	Element	+1	+2	+3	+4	+5	+6
26	Fe		0.74	0.64			
27	Co		0.72	0.63			
28	Ni		0.69				
29	Cu	0.96	0.72				
30	Zn	0.88	0.74				

Second transition series (4d^n)

Z	Element	+1	+2	+3	+4	+5	+6
39	Y			0.94			
40	Zr	1.09			0.79		
41	Nb	1.00			0.74	0.69	
42	Mo	0.93			0.70		0.62
43	Tc						
44	Ru				0.67		
45	Rh			0.68			
46	Pd		0.80		0.65		
47	Ag	1.26	0.89				
48	Cd	1.14	0.97				

Third transition series (5d^n)

Z	Element	+1	+2	+3	+4	+5	+6
72	Hf				0.78		
73	Ta					0.68	
74	W				0.70		0.62
75	Re				0.72		
76	Os				0.88		0.69
77	Ir				0.68		
78	Pt		0.80		0.65		
79	Au	1.37		0.85			
80	Hg	1.27	1.10				

Rare earths (4f^n)

Z	Element	+1	+2	+3	+4	+5	+6
57	La	1.39		1.14			
58	Ce	1.27		1.07	0.94		
59	Pr			1.06	0.92		
60	Nd			1.04			
61	Pm			1.06			
62	Sm			1.00			
63	Eu			0.98			
64	Gd			0.62			
65	Tb			0.93	0.81		
66	Dy			0.92			
67	Ho			0.91			
68	Er			0.89			
69	Tm			0.87			
70	Yb			0.86			
71	Lu			0.85			

[a] Three anion radii are 1.32 for O^{2-}, 1.33 for F^-, and 1.84 for S^{2-} (*Handbook of Chemistry and Physics*).

80

2. Unit Cell Stacking

Three and four fundamental fcc unit cells stack vertically to form the superconducting unit cells of YBaCuO and LaSrCuO, respectively, with some oxygens
removed in the process. This causes the vertical height or c parameter of the unit
cell to be less than that expected for the stacking of perovskite cells:

$$
\begin{aligned}
\text{YBaCuO:} \quad & c \approx 11.7 \text{ Å}, \ 3c_{fcc} = 11.19 \text{ Å}, \ 3c_{per} = 12.03 \text{ Å} \\
\text{LaSrCuO:} \quad & c \approx 13.18 \text{ Å}, \ 4c_{fcc} = 14.92 \text{ Å}, \ 4c_{per} = 16.04 \text{ Å}
\end{aligned}
\quad \text{(VI-5)}
$$

Similar stackings occur in the BiSrCaCuO and TlBaCaCuO compounds.

E. LANTHANUM-COPPER OXIDE

The structure of LaSrCuO, $(La_{1-x}M_x)_2CuO_{4-\delta}$, called the 21 structure, where M
is usually Sr or Ba, is tetragonal in some cases and orthorhombic in others. We
will describe the tetragonal case first and then the orthorhombic distortion of it.
The structures will be described in terms of the prototype compound La_2CuO_4
corresponding to $x = \delta = 0$ in the above expression, keeping in mind that in the
superconducting compounds themselves some of the La atoms are replaced by a
divalent cation such as Sr or Ba. Since lanthanum has a charge of $+3$ and oxygen is -2, it follows that all of the copper is divalent ($+2$) when $x = 0$, and some
becomes trivalent for $x > 0$.

The compound La_2CuO_4 itself is considered to be nonsuperconducting, but
some investigators claim that it or portions of it do exhibit superconductivity,
perhaps of a filimentary type (Beill, Coop1, Dvora, Gran1, Pick1, Shahe, Skelt,
Skel1, Skel2).

1. Tetragonal Form

The tetragonal LaSrCuO superconductors crystallize in what is called the
K_2NiF_4 structure with space group $I4/mmm$, D_{4h}^{17} and two formula units per unit
cell (e.g., Burns, Coll1, Hirot, Mossz, Onoda; Wyck3, p. 68). The copper atoms
and one of the oxygen types O(1) are in special positions and the remaining atoms are all in general positions, with a single undetermined parameter associated with the z coordinate. The positions are

$$
\begin{array}{lll}
\text{La} & (4e) & 0,0,u; \ 0,0,-u; \ \tfrac{1}{2},\tfrac{1}{2},u+\tfrac{1}{2}; \ \tfrac{1}{2},\tfrac{1}{2},-u+\tfrac{1}{2} \\
\text{Cu} & (2a) & 0,0,0; \ \tfrac{1}{2},\tfrac{1}{2},\tfrac{1}{2} \\
\text{O(1)} & (4c) & 0,\tfrac{1}{2},0; \ \tfrac{1}{2},0,0; \ \tfrac{1}{2},0,\tfrac{1}{2}; \ 0,\tfrac{1}{2},\tfrac{1}{2} \\
\text{O(2)} & (4e) & 0,0,v; \ 0,0,-v; \ \tfrac{1}{2},\tfrac{1}{2},v+\tfrac{1}{2}; \ \tfrac{1}{2},\tfrac{1}{2},-v+\tfrac{1}{2}
\end{array}
\quad \text{(VI-6)}
$$

with $u = 0.362$ and $v = 0.182$. Typical lattice dimensions are $a = b = 3.77$ Å,
$c = 13.18$ Å. Table VI-3 gives more details on the atom positions and Fig. VI-
5a provides a sketch of this 21 structure. Table VI-4 lists the measured lattice

TABLE VI-3. Atom Positions of Regular and Alternate La_2CuO_4 Structure, Both of Which Correspond to Space Group $I4/mmm$, D_{4h}^{17a}

Complex	Ideal z	Regular Structure					Alternate Structure				
		Atom	Site	x	y	z	Atom	Site	x	y	z
CuO_2	1	O(1)	4c	$\frac{1}{2}$	0	1	O(1)	4c	$\frac{1}{2}$	0	1
		O(1)	4c	0	$\frac{1}{2}$	1	O(1)	4c	0	$\frac{1}{2}$	1
		Cu	2a	0	0	1	Cu	2a	0	0	1
OLa	$\frac{5}{6} = 0.833$	La	4e	$\frac{1}{2}$	$\frac{1}{2}$	0.862	La	4e	$\frac{1}{2}$	$\frac{1}{2}$	0.862
		O(2)	4e	0	0	0.818					
							O(2)	4d	0	$\frac{1}{2}$	$\frac{3}{4}$
							O(2)	4d	$\frac{1}{2}$	0	$\frac{3}{4}$
LaO	$\frac{2}{3} = 0.667$	O(2)	4e	$\frac{1}{2}$	$\frac{1}{2}$	0.682					
		La	4e	0	0	0.638	La	4e	0	0	0.638
O_2Cu	$\frac{1}{2}$	O(1)	4c	0	$\frac{1}{2}$	$\frac{1}{2}$	O(1)	4c	0	$\frac{1}{2}$	$\frac{1}{2}$
		O(1)	4c	$\frac{1}{2}$	0	$\frac{1}{2}$	O(1)	4c	$\frac{1}{2}$	0	$\frac{1}{2}$
		Cu	2a	$\frac{1}{2}$	$\frac{1}{2}$	$\frac{1}{2}$	Cu	2a	$\frac{1}{2}$	$\frac{1}{2}$	$\frac{1}{2}$
LaO	$\frac{1}{3} = 0.333$	La	4e	0	0	0.362	La	4e	0	0	0.362
		O(2)	4e	$\frac{1}{2}$	$\frac{1}{2}$	0.318					
							O(2)	4d	$\frac{1}{2}$	0	$\frac{1}{4}$
							O(2)	4d	0	$\frac{1}{2}$	$\frac{1}{4}$
OLa	$\frac{1}{6} = 0.167$	O(2)	4e	0	0	0.182					
		La	4e	$\frac{1}{2}$	$\frac{1}{2}$	0.138	La	4e	$\frac{1}{2}$	$\frac{1}{2}$	0.138
CuO_2	0	O(1)	4c	$\frac{1}{2}$	0	0	O(1)	4c	$\frac{1}{2}$	0	0
		O(1)	4c	0	$\frac{1}{2}$	0	O(1)	4c	0	$\frac{1}{2}$	0
		Cu	2a	0	0	0	Cu	2a	0	0	0

[a] Superconducting compounds crystallize in the regular structure (Oguch; see also Onoda). The ideal z values in column 2 are for the prototype perovskite.

constants for tetragonal LaSrCuO superconductors with various values of x, y, and δ in the formula $(La_{1-x}Sr_x)_{2-y}CuO_{4-\delta}$.

2. Alternate Tetragonal Form

In the previous section we discussed the tetragonal structure which is adopted by LaSrCuO superconductors. It has a variant (Hutir, Oguch) called the Nd_2CuO_4 structure in which the oxygens O(2) are in special sites (4d) instead of the general (4e) sites in the same space group, corresponding to

$$O(2) \quad (4d) \quad 0,\tfrac{1}{2},\tfrac{1}{4}; \tfrac{1}{2},0,\tfrac{1}{4}; \tfrac{1}{2},0,\tfrac{3}{4}; 0,\tfrac{1}{2},\tfrac{3}{4} \qquad \text{(VI-7)}$$

The remaining atoms are in the positions given by Eq. (VI-6) and listed in Table VI-3, and the unit cell is sketched on the right-hand side of Fig. VI-5. This structure tends to be unstable relative to its K_2NiF_4 counterpart, and is not known to superconduct.

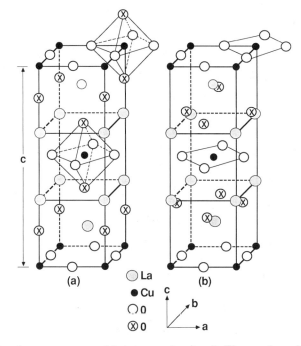

Fig. VI-5. Lanthanum copper oxide tetragonal unit cell. The regular cell (*a*) associated with the superconducting compounds is shown on the left and the alternative one (*b*) is on the right (Oguch; see also Ohba1). The oxygens denoted by ⊗ have different positions in the two cells.

3. Orthorhombic Form

The 21 orthorhombic LaSrCuO structure (Longo) is related to its tetragonal analogue given by Eq. (VI-6) in the same way that the orthorhombic perovskite structure (VI-3) is related to its tetragonal (VI-2) and cubic (VI-1) forms. This means that the orthorhombic basis directions are at 45° relative to the tetragonal ones, and the number of formula units in the cell is doubled. The situation is similar to that described by Fig. VI-3, with $a = 5.363$ Å $= 3.792\sqrt{2}$ Å, $b = 5.409$ Å $= 3.825\sqrt{2}$ Å, $c = 13.17$ Å. Writing the a and b lattice parameters times $\sqrt{2}$ compensates for the new choice of axes and shows that the orthorhombic values are close to the tetragonal $a = 3.81$ Å given earlier. There is also very little change in c. Table VI-5 lists the measured lattice constants for several orthorhombic compounds. The anisotropy factors ANIS

$$\text{ANIS} = \frac{100 \, |b - a|}{0.5 \, (b + a)} \tag{VI-8}$$

listed in column 6 give the percentage deviation from tetragonality.

TABLE VI-4. Selected Lattice Parameters for $(R_{1-x}M_x)_2CuO_{4-\delta}$ Type Superconductors with Tetragonal Structure[a]

R–M	x	Lattice Parameters[b]		Ref.
		$a = b$ (Å)	c (Å)	
Y–Ba	0.4	3.828	12.68	Allge
La–Ba	0.05	3.782	13.168	Skelt
	0.075	3.7817	13.2487	Yuzzz
	0.075	3.787	13.31	Fujit
	0.1	3.791	13.35	Fujit
La–Sr	0.05	3.7839	13.211	Tara1
	0.05	3.78	13.25	Hidak
	0.063	3.7784	13.216	Tara1
	0.075	3.7793	13.2	Decro
	0.075	3.7771	13.226	Tara1
	0.075	3.776	13.234	Shelt
	0.075	3.772	13.247	Brunz
	0.087	3.7739	13.232	Tara1
	0.1	3.7739	13.23	Tara1
	0.1	3.777	13.2309	Przys
	0.112	3.7708	13.242	Tara1
	0.125	3.7685	13.247	Tara1
	0.132	3.7666	13.255	Tara1
	0.15	3.7657	13.259	Tara1

[a] The table is sorted by cations and then by increasing x, the dopant parameter (prepared by M. M. Rigney).
[b] The a and b lattice parameters were converted from measured values of a_0, b_0 of Fig. VI-3 through the expression $a = a_0/\sqrt{2}$, $b = b_0/\sqrt{2}$.

Copper atoms and one of the oxygen types O(1) are in special positions; the remaining two atoms La and O(2) are in general positions with a single undetermined parameter associated with the z coordinate. The space group is $Fmmm$, D_{2h}^{23}, and the positions of the atoms are as follows:

$$
\begin{array}{lll}
\text{La} & \text{(8i)} & 0,0,u; \ 0,\tfrac{1}{2},\tfrac{1}{2}+u; \ \tfrac{1}{2},0,\tfrac{1}{2}+u; \ \tfrac{1}{2},\tfrac{1}{2},u; \\
& & 0,0,-u; \ 0,\tfrac{1}{2},\tfrac{1}{2}-u; \ \tfrac{1}{2},0,\tfrac{1}{2}-u; \ \tfrac{1}{2},\tfrac{1}{2},-u \\
\text{Cu} & \text{(4a)} & 0,0,0; \ 0,\tfrac{1}{2},\tfrac{1}{2}; \ \tfrac{1}{2},0,\tfrac{1}{2}; \ \tfrac{1}{2},\tfrac{1}{2},0 \\
\text{O(1)} & \text{(8e)} & \tfrac{1}{4},\tfrac{1}{4},0; \ \tfrac{1}{4},\tfrac{3}{4},\tfrac{1}{2}; \ \tfrac{3}{4},\tfrac{1}{4},\tfrac{1}{2}; \ \tfrac{3}{4},\tfrac{3}{4},0 \\
& & \tfrac{1}{4},\tfrac{1}{4},\tfrac{1}{2}; \ \tfrac{1}{4},\tfrac{3}{4},0; \ \tfrac{3}{4},\tfrac{1}{4},0; \ \tfrac{3}{4},\tfrac{3}{4},\tfrac{1}{2} \\
\text{O(2)} & \text{(8i)} & 0,0,v; \ \ldots \text{(same as La with } v \text{ replacing } u)
\end{array}
$$

(VI-9)

where the parameters $u = 0.362$ and $v = 0.182$ have the same values as in the tetragonal case presented above. Since u and v are the same and the lattice constants are so close to the tetragonal values, the sketch of the tetragonal unit cell in Fig. VI-5a applies here also. Another work (Hirot, see also Onoda) assigned

TABLE VI-5. Selected Lattice Parameters for $(R_{1-x}M_x)_2CuO_{4-\delta}$ Type Superconductors with the Orthorhombic Structure[a]

R-M	x	a (Å)	b (Å)	c (Å)	ANIS	Ref.
La–Ba	0.02	3.786	3.811	13.17	0.66	Fujit
	0.075	3.786*	3.808*	13.257	0.58	Shelt
	0.075	3.798*	3.803*	13.234	0.13	Onoda
La–Ba	0.1	3.786*	3.824*	13.264	1.00	Hirot
La–Ca	0.075	3.772*	3.808*	13.168	0.95	Shelt

[a] ANIS is the anisotropy factor $100|b - a|/0.5(b + a)$ (prepared by M. M. Rigney).
[b] The a and b lattice parameters were converted from the measured values of a_0, b_0 of Fig. VI-3 through the expressions $a = a_0/\sqrt{2}$, $b = b_0/\sqrt{2}$.

$(La_{0.9}Ba_{0.1})_2O_4$ to the space group $Pccm$, D_{2h}^3 with $a = 5.354 = 3.786\sqrt{2}$ Å, $b = 5.408 = 3.824\sqrt{2}$ Å, and $c = 13.264$ Å.

4. Phase Transition

The compounds $(La_{1-x}M_x)_2CuO_4$ with M = Sr and Ba are orthorhombic at low temperatures and low M contents, and tetragonal otherwise, and superconductivity has been found on both sides of this transition (Baris, Bedn3, Birge, Dayzz, Dvora, Fujit, Gree1, Kangz, Koyam, Mihal, Paulz; see also Heldz). The prototype compound La_2CuO_4 itself also exhibits the tetragonal-to-orthorhombic transition. The phase diagram of Fig. VI-6 shows the tetragonal, orthorhom-

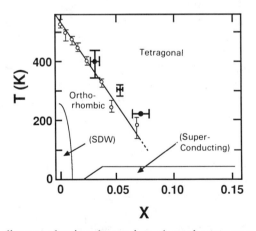

Fig. VI-6. Phase diagram showing data points along the tetragonal-to-orthorhombic transition line for $(La_{1-x}Ba_x)_2CuO_{4-\delta}$ (○, Fujit) and $(La_{1-x}Sr_x)_2CuO_4$ (●, Moret). The spin-density wave (SDW) and superconducting ● regions are indicated. These two compounds have about the same superconducting region.

bic, superconducting, and spin-density wave (SDW) regions for the barium compound (Fuji1), and data points for the strontium compound (Moret, More8). An alternate phase diagram has been proposed (Ahar1). Alkaline metal contents much larger than those shown on the figure (e.g., $x \approx 0.5$) can be non-superconducting. The SDW region occurs below the minimum concentration for the onset of superconductivity. Another work (Geise) showed that LaSr(0.04) undergoes a structural phase transition between 180 and 300 K.

5. Generation of LaSrCuO Structures

The LaSrCuO tetragonal structures may be visualized as being derived from four LaCuO$_3$ perovskite unit cells of the type illustrated in Fig. VI-1 stacked one above the other along the z or c axis. To generate La$_2$CuO$_4$ in the K$_2$NiF$_4$ structure the layers of CuO$_2$ atoms on the $z = \frac{1}{4}$ and $z = \frac{3}{4}$ levels of this four-cell stacking are removed, La and O are interchanged on two other layers, and the middle layer Cu atom is shifted from the edge to the center point $(\frac{1}{2}, \frac{1}{2}, \frac{1}{2})$ of the unit cell. Then the cell is compressed vertically from 14.9 to 13.2 Å (Table VI-4) to take up the space formerly occupied by the removed CuO$_2$ layers. Finally, the lanthanums along the c axis and the oxygens along the side edges are shifted vertically to accommodate the new atom arrangement.

To generate La$_2$CuO$_4$ with the Nd$_2$CuO$_4$ arrangement from this same four-cell stacking all of the oxygens on the vertical edges are removed, and two lanthanums are moved to edge sites. Copper is handled the same way as before, so in both cases the generated structure lacks two CuO$_2$ layers.

6. Layering Scheme of LaSrCuO

When we described the LaSrCuO structures we left out what is perhaps their most important characteristic, namely, their layered aspect. Lanthanum copper oxide may be looked upon as consisting of Cu–O layers of square-planar coordinated copper ions with lanthanum and O(2)-type oxygen ions populating the spaces between the layers. These Cu–O layers are stacked equally spaced, perpendicular to the c axis, as shown in Fig. VI-7, and their oxygens are aligned along the c axis, as indicated by the vertical dotted line on the left side of the figure. The copper ions, on the other hand, are not aligned vertically, but rather alternate between (000) and $(\frac{1}{2} \frac{1}{2} \frac{1}{2})$ sites in adjacent layers, as illustrated in Figs. VI-5 and VI-7.

The copper is actually octahedrally coordinated with oxygen, but the Cu–O distance of 1.9 Å in the CuO$_2$ planes is much less than the vertical distance of 2.4 Å between copper and the oxygens above and below, as shown in Fig. VI-8. When the structure is distorted orthorhombically the Cu–O spacings in both the planes and the c direction remain quite close to their tetragonal counterparts.

The copper ions and the O(1)-type oxygens in the planes are both in special sites in the tetragonal and orthorhombic forms, in accordance with Eqs. (VI-6) and (VI-9), and as a result the plane is perfectly flat in both cases. When the

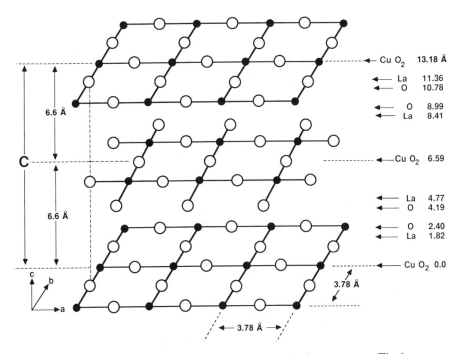

Fig. VI-7. Layering scheme of the LaSrCuO superconducting structure. The layers are perpendicular to the c axis.

structure is tetragonal the square-planar arrangement is also perfect, and of course the planes are perfectly parallel to each other. These characteristics of the planes could influence the superconducting properties.

The copper–oxygen planes are bound together by Cu–O and La–O bonds, as indicated on Fig. VI-5, and Fig. VI-8 shows the spacial arrangement of the CuO_6 octahedra. This figure also makes clear how the copper ions alternate along the c axis. The superconducting properties are probably less influenced by the way the planes are bound together than by the internal characteristics of the planes themselves.

F. YTTRIUM–BARIUM–COPPER OXIDE

The YBaCuO compounds such as $Y_{1-x}Ba_{2-y}Cu_3O_{7-\delta}$, like their LaSrCuO counterparts, come in tetragonal and orthorhombic varieties, and both will be described in turn. Then we will show how to generate the structures from their perovskite prototypes, we will explain the layering scheme, and finally related defect structures will be discussed.

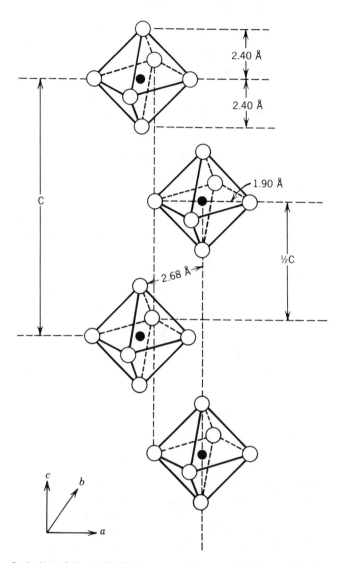

Fig. VI-8. Ordering of the LaSrCuO copper atoms and their associated octahedra of oxygen nearest neighbors along the c axis. The LaSrCuO structure consists of alternately displaced octahedra with axes parallel to c.

In much of the early work the formula $YBa_2Cu_3O_{9-\delta}$ was used for YBaCuO because there are nine oxygens in the prototype perovskite structure. When the crystallographers showed that the 14-atom unit cell of YBaCuO contains 8 oxygen sites, the formula $Y_1Ba_2Cu_3O_{8-\delta}$ began being widely used, and finally when structure refinements demonstrated that one of the oxygen sites is systematically vacant, the more appropriate expression $Y_1Ba_2Cu_3O_{7-\delta}$ was introduced, and it is the one that we will use throughout this work.

The orthorhombic phase is ordinarily superconducting. There are, however, some reported exceptions: (1) doping with gallium in copper chain sites can induce the orthorhombic-to-tetragonal transformation with T_c remaining as high as 81 K (Xiao3); (2) replacing one oxygen by sulfur in $EuBa_2Cu_3O_{7-\delta}$ induces this phase transformation with a small change in T_c from 92 to 85 K (Feln2); and (3) replacing one oxygen with two fluorines to form $YBa_2Cu_3O_6F_2$ could produce a tetragonal structure with all eight oxygen sites occupied and an enhanced T_c (Feln1, Ovshi).

1. Tetragonal Form

The tetragonal $YBa_2Cu_3O_{7-\delta}$ structure, which is stable above about 650°C with $\delta > 0.5$, was assigned to space group $P4/mmm$, D_{4h}^1 (Jorge, see Eagle, Lepag). Earlier assignments were $P4m2$, D_{2d}^5 or possibly $P4mm$, C_{4v}^1 (Hazen). There is one formula unit per unit cell. Yttrium and one copper atom are in special positions, and the remaining atoms are all in general positions with a single undetermined parameter associated with the z coordinate of each:

$$
\begin{array}{llll}
\text{Y} & \text{(1d)} & \frac{1}{2},\frac{1}{2},\frac{1}{2} & \\
\text{Ba} & \text{(2h)} & \frac{1}{2},\frac{1}{2},u; \frac{1}{2},\frac{1}{2},-u & u = 0.1914 \\
\text{Cu(t)} & \text{(1a)} & 0,0,0 & \\
\text{Cu(m)} & \text{(2g)} & 0,0,v; 0,0,-v & v = 0.3590 \quad \text{(VI-10)} \\
\text{O(t)} & \text{(2f)} & 0,\frac{1}{2},0; \frac{1}{2},0,0 & \\
\text{O(m,m')} & \text{(4i)} & 0,\frac{1}{2},w; \frac{1}{2},0,w & \\
& & 0,\frac{1}{2},-w; \frac{1}{2},0,-w & w = 0.3792 \\
\text{O(b)} & \text{(2g)} & 0,0,x; 0,0,-x & x = 0.1508 \\
\end{array}
$$

The u, v, w, and x parameters (from Jorge) are used in column 10 of Table VI-6. The unit cell dimensions are $a = 3.9018$ Å, $c = 11.9403$ Å (Jorge), and those reported by other investigators are listed in Table VI-7. Oxygen site O(t) may be only partly occupied. The atom positions are given in Table VI-6 and a sketch of this structure is presented on the right side of Fig. VI-9. The oxygen sites in the basal plane at $z = 0$ are about half occupied in a random or disordered manner. The lack of planarity of the CuO_2 layers immediately below and above the yttrium atom is reminiscent of tetragonal perovskite, which is sketched in Fig. VI-2.

The semiconducting compound $YBa_2Cu_3O_6$ was also assigned to $P4/mmm$, D_{4h}^1 (Borde, Renau, Swinn, Torar), as shown in the table. This is a tetragonal version of orthorhombic $YBa_2Cu_3O_7$ formed by removing the chain oxygens so that none remain on the basal plane. A claim has been made (Relle) that the "ideal" perovskite z values shown in column 6 of Table VI-6 provide an adequate fit to X-ray powder diffraction data from $YBa_2Cu_3O_7$.

The compounds $La_3Ba_3Cu_6O_{14+\delta}$ (Davil, see also Golb1), $(La_{0.4}Ba_{0.6})_6$ $Cu_6O_{14-\delta}$ (Iwaz3), and $LaBa_2Cu_{3-x}O_{7-\delta}$ (Nakai) have been reported as isostructural with tetragonal $YBa_2Cu_3O_8$.

TABLE VI-6. Atom Positions of $YBa_2Cu_3O_{7-\delta}$ in its Orthorhombic (Superconducting) and Tetragonal Forms[a]

Layer	Atom	Site	x	y	Ideal z	z	Average[b]	Jorge	Jorge	Hazen	Borde[c]
CuO₂	O(t)	1e	$\frac{1}{2}$	0	1	$1+w$	1	1	1	1.03	—
	Cu(t)	1a	0	0	1	1	1	1	1	1.00	1
	O(t′)	1b	0	$\frac{1}{2}$	1	$1-w$	1	1	1	0.97	—
BaO	O(b)	2q	0	0	$\frac{5}{6} = 0.8333$	$1-z$	0.8432	0.8458	0.8492	0.85	0.8460
	Ba	2t	$\frac{1}{2}$	$\frac{1}{2}$	$\frac{5}{6} = 0.8333$	$1-u$	0.8146	0.8105	0.8086	0.81	0.8079
CuO₂	Cu(m)	2q	0	0	$\frac{2}{3} = 0.6667$	$1-v$	0.6445	0.6426	0.6410	0.64	0.6395
	O(m′)	2r	0	$\frac{1}{2}$	$\frac{2}{3} = 0.6667$	$1-x$	0.6219	0.6196	0.6208	0.62	0.6206
	O(m)	2s	$\frac{1}{2}$	0	$\frac{2}{3} = 0.6667$	$1-y$	0.6210	0.6233	0.6208	0.61	0.6206
Y	O(y)	—	0	0	$\frac{1}{2}$	$\frac{1}{2}$	—	—	—	—	—
	Y	1h	$\frac{1}{2}$	$\frac{1}{2}$	$\frac{1}{2}$	$\frac{1}{2}$	$\frac{1}{2}$	$\frac{1}{2}$	$\frac{1}{2}$	$\frac{1}{2}$	$\frac{1}{2}$
CuO₂	O(m)	2s	$\frac{1}{2}$	0	$\frac{1}{3} = 0.333$	y	0.3790	0.3767	0.3792	0.39	0.3794
	O(m′)	2r	0	$\frac{1}{2}$	$\frac{1}{3} = 0.333$	x	0.3781	0.3804	0.3792	0.38	0.3794
	Cu(m)	2q	0	0	$\frac{1}{3} = 0.333$	v	0.3555	0.3574	0.3590	0.36	0.3605
BaO	Ba	2t	$\frac{1}{2}$	$\frac{1}{2}$	$\frac{1}{6} = 0.1667$	u	0.1854	0.1895	0.1914	0.19	0.1921
	O(b)	2q	0	0	$\frac{1}{6} = 0.1667$	z	0.1568	0.1542	0.1508	0.15	0.1540
CuO₂	O(t)	1b	0	$\frac{1}{2}$	0	w	0	0	0	0.03	—
	Cu(t)	1a	0	0	0	0	0	0	0	0	0
	O(t′)	1e	$\frac{1}{2}$	0	0	$-w$	0	0	0	−0.03	—
δ							0	$0 < \delta < 0.5$	$0.5 < \delta < 1$	+0.5	+1.0
a (Å)							3.827	3.8591	3.9018	3.859	3.8715
b (Å)							3.882	3.9195	3.9018	3.859	3.8715
c (Å)							11.682	11.8431	11.9403	11.71	11.738

[a] Column 8 gives the average z values of several investigators for $\delta = 0$. Values are also given for $\delta = 0.5$, and 1.0. The ideal z is for the prototype perovskite $YBa_2Cu_3O_9$ shown on the left side of Fig. VI-11.

[b] The average assumes z of O(m) is greater than z of O(m′). Various authors differ on this point (e.g., Jorge).

[c] $Y_{0.9}Ba_{2.1}Cu_3O_6$.

TABLE VI-7. Selected Lattice Parameters for $RBa_2Cu_3O_{7-\delta}$ Type Copper Oxides with Tetragonal Structure[a]

R–M	δ	$a = b$ (Å)	c (Å)	Ref.
Y–Ba	0.5	3.859	11.71	Hazen
	0.5	3.859	11.71	Hemle
	—	3.87[b]	11.67	Ihar2
	0.0	3.87[b]	11.67	Hirab
Dy–Ba	0.72	3.8656	11.783	Tara3
Er–Ba	0.84	3.854	11.796	Tara3
Eu–Ba	0.41	3.86	11.73	Boolc
Gd–Ba	0.48	3.877	11.81	Tara3
Ho–Ba	0.87	3.8601	11.791	Tara3
Tm–Ba	0.93	3.8491	11.788	Tara3

[a] The table is sorted by cations (prepared by M. M. Rigney).
[b] The a and b lattice parameters were converted from the measured values a_0, b_0 of Fig. VI-3 through the expressions $a = a_0/\sqrt{2}$, $b = b_0/\sqrt{2}$.

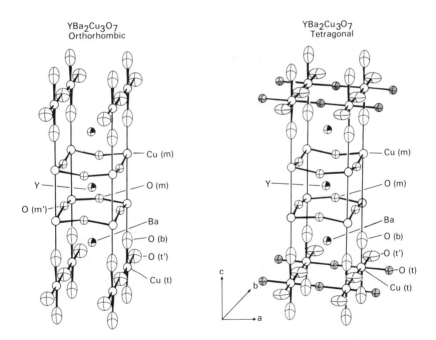

Fig. VI-9. Sketches of the orthorhombic (left) and tetragonal (right) yttrium–barium–copper oxide unit cells. (Adapted from Jorg1.) Oyxgens are randomly dispersed over the basal plane sites in the tetragonal structure. Thermal vibration ellipsoids are shown for the atoms.

2. Orthorhombic Form

Orthorhombic $YBa_2Cu_3O_8$ with the 123 structure was assigned to the space group *Pmmm*, D_{2h}^1 (Antso, Beech, Benoz, Cales, Cappo, Coxzz, Greed, Jorge, Siegr, Yosh3, Yanz2, Youzz) with one formula unit per unit cell and the representative lattice parameters $a = 3.827$ Å, $b = 3.882$ Å, and $c = 11.682$ Å. Yttrium, one copper, and two oxygens are in special positions, and the remaining atoms are all in general positions with a single undetermined parameter associated with the z coordinate:

$$
\begin{array}{llll}
Y & (1h) & \frac{1}{2},\frac{1}{2},\frac{1}{2} & \\
Ba & (2t) & \frac{1}{2},\frac{1}{2},u; \frac{1}{2},\frac{1}{2},-u & u = 0.1854 \\
Cu(t) & (1a) & 0,0,0 & \\
Cu(m) & (2q) & 0,0,v; 0,0,-v & v = 0.3555 \\
O(t) & (1b) & \frac{1}{2},0,0 & \\
O(t') & (1e) & 0,\frac{1}{2},0 & \\
O(m') & (2r) & 0,\frac{1}{2},x; 0,\frac{1}{2},-x & x = 0.3790 \\
O(m) & (2s) & \frac{1}{2},0,y; \frac{1}{2},0,-y & y = 0.3781 \\
O(b) & (2q) & 0,0,z; 0,0,-z & z = 0.1568
\end{array}
\qquad \text{(VI-11)}
$$

The u, v, x, y, and z parameters correspond to the average atom positions given in column 8 of Table VI-6. Lattice dimensions a, b, and c and atom positions for several structure determinations are given in Table VI-6. Table VI-8 lists lattice parameters for a number of orthorhombic YBaCuO compounds. Variable temperature crystallographic data are also available (Antso, Cappo, Hewa1, Jorge, Jorg1, Momin, Renau). Sketches of this structure are presented on the left side of Fig. VI-9 (see also Stei1). We see from this figure that the O(t) oxygen site is empty, which corresponds to the presence of –Cu–O(t')–Cu(t')–O– chains along the b direction. The vacancy of the O(t) site causes the unit cell to compress slightly along a to render $a < b$. The compound $TmBa_2Cu_3O_{7-\delta}$ (Andr1) and other rare earth analogues (Lepa1) are isostructural with $YBa_2Cu_3O_{7-\delta}$.

Table VI-9 gives the bond distances and bond angles of this structure (Beech, Benoz, Borde, Cales, Coxzz, Greed, Hazen, Lepag, Siegr, Yanz2) and their temperature dependence has also been reported (Antso, Cappo).

A transmission electron microscope examination of $YBa_2Cu_3O_{7-\delta}$ in the superconducting state indicated that it is orthorhombic with the space group *Pm2m*, C_{2v}^1 and the lattice constants $a = 3.80$, $b = 3.86$, and $c = 11.55$. The a and b axes alternate across an antiphase boundary which runs parallel to the [110] direction.

A yttrium-rich phase of YBaCuO was found to have the structure *Pnma*, D_{2h}^{16} with $a = 13.5$ Å, $b = 6.3$ Å, and $c = 7.6$ Å (Eagle). $GdBa_2Cu_3O_{7-\delta}$ has $a = 3.909$, $b = 3.849$, and $c = 11.682$ Å with the following possible space groups: *Pmmm*, D_{2h}^1; *Pmm2*, C_{2v}^1; *P222*, D_2^1 (Xuzz1). $YBa_2Cu_3O_{7-\delta}$ has also been assigned

TABLE VI-8. Selected Lattice Parameters for $RM_2Cu_3O_{7-\delta}$ Type Superconductors with Orthorhombic Structure[a]

R–M	δ	a (Å)	b (Å)	c (Å)	ANIS	Ref.
Y–Ba	0.62	3.85	3.86	11.78	0.26	Kuboz
	0.57	3.85	3.87	11.77	0.52	Kuboz
	0.47	3.84	3.88	11.75	1.04	Kuboz
	0.28	3.8237	3.8874	11.657	1.65	Tara3
	0.19	3.8231	3.8864	11.6807	1.64	Benoz
	0.15	3.8282	3.8897	11.6944	1.59	Bonn1
	0.1	3.8591	3.9195	11.8431	1.55	Jorge
	0.1	3.83	3.89	11.7	1.55	Kuboz
	0	3.8124	3.8807	11.6303	1.75	Cappo
	0	3.825	3.886	11.660	1.58	Relle
	0	3.856	3.870	11.666	0.36	Siegr
	0	3.825	3.883	11.68	1.50	Ginle
	0	3.825	3.883	11.68	1.50	Ventu
	0	3.816	3.892	11.682	1.97	Greed
	0	3.84	3.88	11.63	1.04	Ding1
	0	3.82	3.88	11.67	1.56	Crabt
	0	3.817	3.882	11.671	1.69	Coxzz
	0	3.8271	3.8771	11.7086	1.30	Larb1
	0	3.83	3.89	11.71	1.55	Worth
Dy–Ba	0	3.828	3.886	11.66	1.50	Mapl1
	0	3.941	3.894	11.673	1.37	Yamad
	0.18	3.828	3.889	11.668	1.58	Tara3
Er–Ba	0	3.844	3.885	11.532	1.06	Kuzzz
	0	3.845	3.884	11.53	1.01	Lynnz
	0	3.813	3.874	11.62	1.59	Mapl1
	0	3.832	3.88	11.639	1.24	Yamad
	0.18	3.815	3.884	11.659	1.79	Tara3
Eu–Ba	−0.1	3.8449	3.9007	11.704	1.40	Tara3
	0	3.8152	3.8822	11.6502	1.74	Golbe
	0	3.843	3.897	11.7	1.40	Mapl1
	0	3.851	3.901	11.746	1.29	Yamad
Gd–Ba	−0.08	3.840	3.899	11.703	1.52	Tara3
	0	3.836	3.894	11.62	1.50	Mapl1
	0	3.845	3.898	11.732	1.37	Yamad
Ho–Ba	0	3.845	3.886	11.547	1.06	Kuzzz
	0	3.8253	3.8856	11.6578	1.56	Leez2
	0	3.821	3.886	11.66	1.69	Mapl1
	0	3.841	3.883	11.676	1.09	Yamad
	0.29	3.822	3.888	11.670	1.71	Tara3
Lu–Ba	0	3.835	3.886	11.531	1.32	Kuzzz
	0	3.791	3.859	11.57	1.78	Mapl1
Nd–Ba	−0.16	3.8546	3.9142	11.736	1.53	Tara3
	0	3.867	3.906	11.71	1.00	Mapl1
	0	3.873	3.902	11.761	0.75	Yamad

TABLE VI-8. (continued)

R–M	δ	a (Å)	b (Å)	c (Å)	ANIS	Ref.
				Lattice Parameters		
Sm–Ba	−0.11	3.855	3.899	11.721	1.13	Tara3
	0	3.843	3.906	11.72	1.63	Mapl1
	0	3.867	3.909	11.75	1.08	Yamad
Tm–Ba	0	3.836	3.885	11.529	1.27	Kuzzz
	0	3.845	3.881	11.618	0.93	Yamad
	0	3.802	3.878	11.63	1.98	Mapl1
	0.35	3.810	3.882	11.656	1.87	Tara3
Yb–Ba	0.29	3.7989	3.8727	11.650	1.92	Tara3
	0	3.834	3.884	11.531	1.30	Kuzzz
	0	3.798	3.87	11.61	1.88	Mapl1
	0	3.832	3.83	11.61	0.05	Yamad

[a] The table is sorted by cations and then by decreasing oxygen deficiency parameter, δ. ANSI is the anisotropy factor $100|b − a|/0.5(b + a)$ (prepared by M. M. Rigney).

TABLE VI-9. Selected Bond Distances and Angles in $YBa_2Cu_3O_7$, Where n is Number of Equivalent Bonds[a]

Bond	Distance (Å)	n	Mean (Å)
Cu(t)–O(t′)	1.941	2	1.886
–O(b)	1.831	2	
Cu(m)–O(b)	2.285	1	
–O(m)	1.931	2 }	1.943
–O(m′)	1.955	2 }	
Ba–O(t′)	2.891	2	
–O(m)	2.976	2	
–O(m′)	2.963	2	2.864
–O(b)	2.747	4	
Y–O(m)	2.404	4	
–O(m′)	2.383	4	2.394

Configuration	Angle (deg)
Cu(t)–O(t′)–Cu(t)	180
Cu(m)–O(m)–Cu(m)	163.6
Cu(m)–O(m′)–Cu(m)	164.0

[a] The bond distances are averages of those reported by various investigators (Beech, Benoz, Borde, Cappo, Coxzz, Greed, Lepag, Siegr); the angles are from Coxzz.

to the orthorhombic space group $Pmm2$, D_{2v}^1 with $a = 3.820$ Å, $b = 11.688$ Å, and $c = 3.893$ Å (Beye1), and to $Pm2m$, C_{2v}^1.

3. Temperature Factors

The structure refinements provided temperature factors $B = 8\pi^2 \langle u^2 \rangle$ (Beech, Benoz, Cappo, Coxzz, Greed, Lepag, Siegr) which are a measure of the mean square displacement $\langle u^2 \rangle$ in Å^2 of an atom about its equilibrium position due to thermal vibrations, and these are listed in Table VI-10. We see from the table that there is a great deal of scatter in the temperature factors reported by various investigators. This is in sharp contrast to the close agreement among these same investigators on the atom positions. The vibrations themselves are anisotropic (Antso, Cappo, Youzz), and the values listed in the table may be looked upon as averages over thermally excited normal modes. The extent of the atomic vibrations in the x, y, and z directions is indicated in Fig. VI-9 by ellipsoids (Jorge). We see from the figure that the light oxygen atoms O(t') and O(b) bonded to chain coppers Cu(t) on the basal plane undergo larger amplitude vibrations than the oxygens O(m) and O(m') on the CuO$_2$ planes.

4. Phase Transition

The compound YBaCuO is tetragonal at high temperatures and undergoes a second-order (Frei1) order–disorder transition at about 700°C to the low-temperature orthorhombic phase (Bakke, Beyer, Eatou, Iwaz2, Jorge, Jorg1, Jorg2, Jorg3, Sagee, Schul, Torar, Vant2). Quenching can produce the tetragonal phase sketched on the right side of Fig. VI-9 at room temperature. The oxygens are disordered on the basal ($z = 0$) plane sites in the high-temperature phase and ordered to form chains at low temperature, as indicated on the two figures. This occurs because the two oxygen sites O(t) and O(t'), which are equivalent and randomly occupied in the tetragonal phase, become inequivalent in the orthorhombic phase, where all of the basal plane oxygens reside on O(t'). A superlattice associated with this ordering has been observed (Vant2).

Figure VI-10 shows the fractional site occupancy of the oxygens in the basal plane as a function of the heating temperature of the sample in an oxygen atmosphere (Jorge). The central curve in the orthorhombic region gives the mean of the fractional occupancies of the a and b sites. This curve also gives the value of δ in the formula YBa$_2$Cu$_3$O$_{7-\delta}$. One should note that the low-temperature compound ($T \approx 25°C$) of Fig. VI-10 corresponds to the formula YBa$_2$Cu$_3$O$_{6.9}$. Site occupancies were also obtained for heating in different partial pressures of oxygen. An anomaly found in the orthorhombic distortion of YBa* at the superconducting transition (Hornz) was interpreted as evidence for anisotropic pairing.

The orthorhombic 123 structure is the superconducting phase of YBaCuO. The tetragonal phase can be obtained at room temperature by quenching from above the phase transition, and it is found to be semiconducting. Ordinarily it does not form a superconductor (Chen2, Kwok2), but some exceptions are mentioned in Section VI-F.

TABLE VI-10. Temperature Factors $B = 8\pi^2 \langle u^2 \rangle$ for Mean Square Displacements $\langle u^2 \rangle$ of Atoms of YBa$_2$Cu$_3$O$_7$ About Their Equilibrium Positions[a]

Groups		Beech	Benoz	Cappo	Coxzz	Greed	Jorge	Lepag	Siegr
CuO$_2$	O(t)	—	—	—	1.6	—	—	—	2.32
	Cu(t)	0.55	0.50	0.38	0.2	0.69	1.4	0.9	1.10
	O(t')	1.73	1.35	2.4	1.6	0.59	—	4.2	2.00
BaO	O(b)	0.78	0.67	0.93	0.5	1.32	—	1.6	1.30
	Ba	0.65	0.54	0.59	0.4	0.78	1.7	0.51	0.84
CuO$_2$	Cu(m)	0.49	0.29	0.51	0.3	0.81	1.50	0.5	0.67
	O(m')	0.55	0.37	0.31	0.4	0.38	1.3	0.2	1.20
	O(m)	0.57	0.56	0.11	0.5	0.36	1.6	0.2	0.40
Y	Y	0.56	0.46	0.58	0.2	0.60	1.4	0.7	0.53
CuO$_2$	O(m)	0.57	0.56	0.11	0.5	0.36	1.6	0.2	0.40
	O(m')	0.55	0.37	0.31	0.4	0.38	1.3	0.2	1.20
	Cu(m)	0.49	0.29	0.51	0.3	0.81	1.50	0.5	0.67
BaO	Ba	0.65	0.54	0.59	0.4	0.78	1.7	0.51	0.84
	O(b)	0.78	0.67	0.93	0.5	1.32	—	1.6	1.30
CuO$_2$	O(t)	—	—	—	1.6	—	—	—	2.32
	Cu(t)	0.55	0.50	0.38	0.2	0.69	1.4	0.9	1.10
	O(t')	1.73	1.35	2.4	1.6	0.59	—	4.2	2.00

[a]The results of several investigations are shown; some (Antso, Borde, Cappo, Jorge) give anisotropic values.

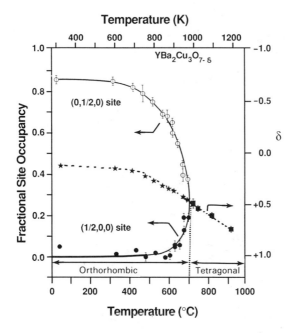

Fig. VI-10. Dependences of the fractional occupancies of the $(0, \frac{1}{2}, 0)$ (top) and $(\frac{1}{2},0,0)$ (bottom) sites (scale on left) and of the oxygen content parameter δ (center curve, scale on right) on the quench temperature. This latter curve is the average of the occupancies of the two sites. (Adapted from Jorge.)

5. Oxygen-Site Nomenclature

Various workers use different numbering schemes for the oxygen sites, and several of them are compared in Table VI-11. We have adopted the more mnemonic convention, given in column 3 of the table and illustrated in Fig. VI-9, of using the letters t and t′ for top (and equivalent bottom) copper oxide layer, and m and m′ for the median copper oxide layers, with O(b) denoting the oxygen on the barium level.

6. Generation of YBaCuO Structure

The YBaCuO tetragonal structure may be visualized as being derived from three prototype fcc oxygen unit cells of the type illustrated on Fig. VI-1, stacked one above the other along the z or c axis, as shown in the center of Fig. VI-11. Column 6 of Table VI-6 gives the z parameter values for this ideal case. To generate the YBaCuO tetragonal unit cell the barium centered in the middle cube is replaced by yttrium, and the oxygens on the edges of this middle cube are removed, as indicated on the left side of Fig. VI-11. To take up the space of the removed oxygens and that arising from the smaller size of yttrium, the center

TABLE VI-11. Notations Used for Atoms in YBa$_2$Cu$_3$O$_{7-\delta}$

Group	Plane	Present Work[a]	Site (Pmmm)	Beech, Cappo, Coxzz, Siegr, Yauzz, Youzz	Benoz, Hazen, Jorge	Antso	Borde[b]	Greed	Lepag
CuO	Top plane	O(t)	1e	[O(5)]	—	—	—	—	—
		Cu(t)	1a	Cu(1)	Cu(1)	Cu(1)	Cu(1)	Cu(1)	Cu(1)
		O(t′)	1b	O(4)	O(1)	O(1)	O(1)	O(1)	O(1)
BaO	Upper barium plane	O(b)	2q	O(1)	O(4)	O(2)	O(1)	O(2)	O(3)
		Ba	2t	Ba	Ba	Ba	Ba	Ba	Ba
CuO$_2$	Upper median plane	Cu(m)	2q	Cu(2)	Cu(2)	Cu(2)	Cu(2)	Cu(2)	Cu(2)
		O(m′)	2r	O(3)	O(3)	O(3)	O(2)	O(4)	O(2)
		O(m)	2s	O(2)	O(2)	O(4)	O(2)	O(3)	O(2)
Y	Yttrium center plane	Y	1h	Y	Y	Y	Y	Y	Y
CuO$_2$	Lower median plane	O(m)	2s	O(2)	O(2)	O(4)	O(2)	O(3)	O(2)
		O(m′)	2r	O(3)	O(3)	O(3)	O(2)	O(4)	O(2)
		Cu(m)	2q	Cu(2)	Cu(2)	Cu(2)	Cu(2)	Cu(2)	Cu(2)
BaO	Lower barium plane	Ba	2t	Ba	Ba	Ba	Ba	Ba	Ba
		O(b)	2q	O(1)	O(4)	O(2)	O(1)	O(2)	O(3)
CuO	Bottom plane	O(t)	1b	[O(5)]	—	—	—	—	—
		Cu(t)	1a	Cu(1)	Cu(1)	Cu(1)	Cu(1)	Cu(1)	Cu(1)
		O(t′)	1e	O(4)	O(1)	O(1)	—	O(1)	O(1)

[a] Oxygen sites O(t) and O(m) are along the shorter a axis ($x = \frac{1}{2}$, $y = 0$).
[b] Y$_{0.9}$Ba$_{2.1}$Cu$_3$O$_6$.

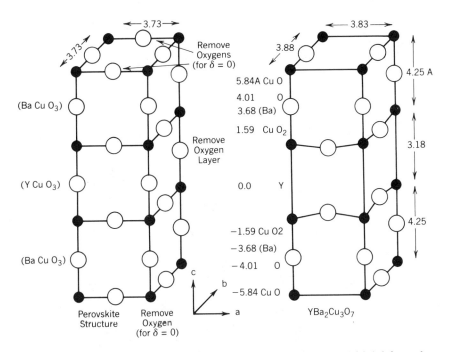

Fig. VI-11. Generation of the yttrium–barium–copper oxide unit cell (right) from three stacked $BaCuO_3$ perovskite unit cells (left) by the removal of the oxygens on the yttrium level where Y replaces Ba.

cube is compressed along the c direction. Finally, the vertical edge oxygens are moved along c toward the apical Cu(t) ions. This copper ion Cu(t) is sixfold coordinated for $YBa_2Cu_3O_8$ and square-planar coordinated for $YBa_2Cu_3O_7$ with Cu–O distances of 1.94 Å in the basal xy plane and 1.83 Å vertically along c, as shown in Fig. VI-12. The two other coppers, Cu(m) and Cu(m$'$), exhibit fivefold pyramidal coordination, as indicated in Fig. VI-14, with Cu–O spacings of 1.94 Å in the basal plane and 2.29 Å vertically. One should note that the final $\delta = 0$ compound only differs in composition from the prototype perovskite on the left side of Fig. VI-11 by the deficiency of two oxygens.

7. Layering Scheme of YBaCuO

In Section VI-E-6 we discussed the CuO_2 layering scheme of the LaSrCuO superconductors. The layering scheme of the YBaCuO case, which is illustrated in Fig. VI-13, is somewhat more complicated than the LaSrCuO case. There is a threefold layering sequence with the two median planes adjacent to the yttrium much closer together (3.18 Å) than they are to the basal plane (4.25 Å), as indicated in the figure. The basal plane copper ions Cu(t) are coupled to the median layer coppers Cu(m) through oxygens, and such coupling does not exist between the two median planes. The basal plane coppers and oxygens are in

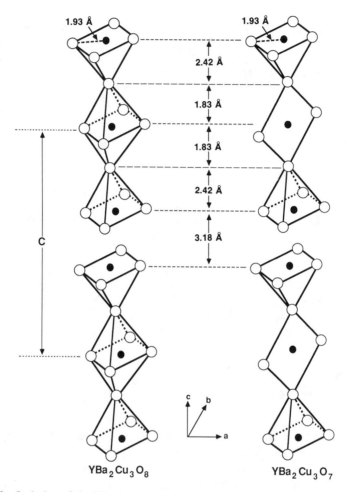

Fig. VI-12. Ordering of the YBaCuO copper atoms and their associated oxygen nearest neighbors along the *c* axis. The arrangement is a stacking of pyramid-octahedron-inverted-pyramid groups in the tetragonal structure (left) and of pyramid–square-planar–inverted-pyramid groups in the orthorhombic structure (right).

special sites so the plane is perfectly flat, as shown. In contrast to this the median plane coppers and oxygens are both in general sites with slightly different z parameters, so the median planes have a puckered appearance with a thickness of about 0.23 Å, as indicated in Figs. VI-9 and VI-13.

The case illustrated in Fig. VI-13 corresponds to $YBa_2Cu_3O_{7-\delta}$ with $\delta = 0$, so the oxygen sites along the *a* direction of the basal plane are all empty and those along the *b* direction are all occupied. This produces Cu–O–Cu–O chains along the *b* direction, as shown in the figure. The missing oxygens cause the coppers to move slightly closer together along *a*, thereby inducing the orthorhombic distor-

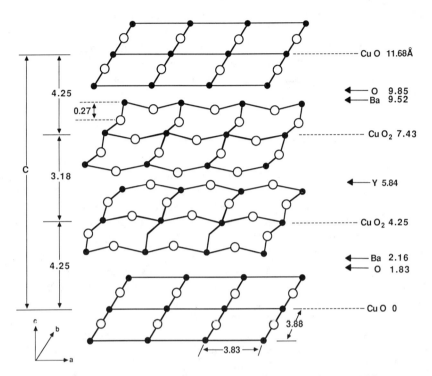

Fig. VI-13. Layering scheme of the YBaCuO superconducting (orthorhombic) structure. The layers are perpendicular to the c axis. The extent of puckering of the median CuO_2 layers is indicated.

tion with $a < b$. When the oxygen content increases ($\delta < 0$), oxygens begin to occupy the vacant sites along a.

8. YBaCuO Defect Structures

Since yttrium has a charge of $+3$, barium is $+2$, and oxygen is -2, it follows that for the hypothetical compound $YBa_2Cu_3O_8$ which has $\delta = -1$, all of the copper is trivalent ($+3$). This compound cannot be prepared because of the strong tendency toward oxygen deficiency. We have seen that the compound $YBa_2Cu_3O_7$ with $\delta = 0$ has all of its oxygen loss in the basal plane where linear O-Cu-O coordination replaces square planar CuO_4. The oxygen vacancies are ordered and hence the copper ions form $(Cu-O)_n$ chains in this plane. This $\delta = 0$ case corresponds to an average copper charge of 2.33, which suggests a mixture of $+2$ and $+3$ copper valence states. Samples with $\delta = 0.5$ (e.g., Hazen) have an average copper charge of 2.0. This subject of copper valence has been discussed at greater length in Section III-J.

An extra layer of yttrium atoms (Ourma) or an extra or double CuO plane (Zand1) constitute common planar defects in the YBaCuO structure. The presence of such a planar defect may be visualized as enlarging a unit cell on the right-hand side of Fig. VI-11 from, in the yttrium case, a threefold Ba, Y, Ba sequence to a fourfold Ba, Y, Ba, Y sequence, with this sequence continuing indefinitely in the horizontal direction. Other unit cells are left unchanged. Square planar CuO_2 has been found in the structures of $(Y_{1-x}Ba_x)_3Cu_2O_6$ and $(Y_{1-x}Ba_x)_3Cu_2O_7$ (Kitaz).

G. OTHER LaSrCuO AND YBaCuO STRUCTURES

The system $La(Ba_{1-x}La_x)_2Cu_3O_{7+\delta}$ has the region of solubility $0.125 < x < 0.25$. These compounds are disordered isomorphs of the orthorhombic $YBa_2Cu_3O_7$ structure (Segre). The highest $T_c = 60$ K occurs for $x = 0.065$ and $x = 0$, the latter stoichiometric compound $LaBa_2Cu_3O_{7-\delta}$ being outside the range of solubility. The compounds $La_2SrCu_2O_{6.2}$, $La_4BaCu_5O_{13}$, and $La_5SrCu_6O_{15}$ are not superconducting (Torr1).

Studies of related superconducting compounds such as $(La_{1.6}Ba_{0.4})CuO_{4-\delta}$, $(La_{0.8}Ba_{0.2})CuO_{4-\delta}$, and $(Y_{0.8}Ba_{0.2})CuO_{4-\delta}$ (Kirs1, Kirs2), and $(Y_{0.4}Ba_{0.6})CuO_3$ (Luozz) have been reported. See also Bedn4.

The green semiconducting phase Y_2BaCuO_5 which is often found admixed with superconducting YBa* is orthorhombic and was assigned to the space group $Pbnm$, D_{2h}^{16} or $Pna2_1$, C_{2v}^9 (Hazen, Kitan, Mansf, Mich2, Raozz, Rossz) with $a = 7.1$ Å, $b = 12.2$ Å, and $c = 5.6$ Å.

The compound La_2CuO_4 was identified as the first member ($n = 1$) of the homologous series of composition $R_{n+1}M_nO_{3n+1}$ with R = La and M = Cu in the present case (Davie). Several members of the series were prepared with n in the range from 1 to 6, and their crystallography consisted of slabs of $(LaCuO_3)_n$ groups containing CuO_6 octahedra with a perovskite-type structure separated by layers of LaO with the La and O atoms in an NaCl-type structure, as shown on Fig. VI-14. The perovskite $LaCuO_3$, which plays the role of the limiting structure of $La_{n+1}Cu_nO_{3n+1}$ as $n \to \infty$, is shown for comparison.

Other oxide types have also been mentioned in the literature, such as the possible lower symmetry space groups of the R_2MO_4 structure arising from rigid octahedral tiltings at phase transitions (Hatch).

H. BISMUTH-STRONTIUM-CALCIUM-COPPER OXIDE

Early in 1988 two new superconducting systems were discovered which have transition temperatures considerably above those attainable with the YBaCuO compounds, namely, the bismuth- (Chuz2, Maeda, Zand2) and the thallium- (Gaoz2, Hazel, Sheng, Shen1) based materials. In this section we will discuss the structure of BiSrCaCuO, and in the next we will treat TlBaCaCuO.

The 2212 compound $Bi_2(Sr,Ca)_3Cu_2O_{8+\delta}$ crystallizes in the same tetragonal space group $I4/mmm$, D_{4h}^{17} as La_2CuO_4 with two formula units per unit cell and

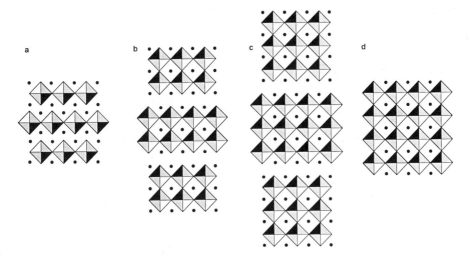

Fig. VI-14. Idealized representations of the structures of the series of compounds $La_{n+1}Cu_nO_{3n+1}$ with (a) $n = 1$, (b) $n = 2$, (c) $n = 3$, and (d) $LaCuO_3$ ($n = \infty$). The squares are CuO_6 octahedra and the solid circles denote La atoms. In (a)–(c) the tetragonal cell is projected down [010] and the c axis is vertical (Davie).

the lattice parameters $a = 3.817$ Å, $c = 30.6$ Å (Tara9). The parameters for the atoms are:

$$
\begin{array}{llll}
\text{Ca} & \text{(2a)} & & \\
\text{Sr} & \text{(4e)} & u = 0.1097 & \\
\text{Bi} & \text{(4e)} & u = 0.3022 & 87\% \text{ occupancy} \\
\text{Bi} & \text{(4e)} & u = 0.2681 & 13\% \text{ occupancy} \\
\text{Cu} & \text{(4e)} & u = 0.4456 & \\
\text{O(1)} & \text{(8g)} & u = 0.446 & \\
\text{O(2)} & \text{(4e)} & u = 0.375 & \\
\text{O(3)} & \text{(4e)} & u = 0.205 & \\
\text{O(4)} & \text{(4d)} & & 6.5\% \text{ occupancy}
\end{array}
\qquad \text{(VI-12)}
$$

where the atom positions for sites (2a) and (4e) are given above in Eq. VI-6, those for site (4d) are given by Eq. VI-7, and the remaining site (8g) has the following atom positions:

$$
\text{(8g)} \quad 0,\tfrac{1}{2},u; \ 0,\tfrac{1}{2},-u; \ \tfrac{1}{2},0,u; \ \tfrac{1}{2},0,-u; \\
\tfrac{1}{2},0,u+\tfrac{1}{2}; \ \tfrac{1}{2},0,-u+\tfrac{1}{2}; \ 0,\tfrac{1}{2},u+\tfrac{1}{2}; \ 0,\tfrac{1}{2},-u+\tfrac{1}{2} \qquad \text{(VI-13)}
$$

Table VI-12 gives more details on the atom positions. Superlattice structures have been reported along a and b (Iqba6).

TABLE VI-12. Atom Positions of $Bi_2Sr_2CaCu_2O_{8+\delta}$ Structure with Two Formula Units per Unit Cell[a]

Complex	Vertical Position	Atom	Site	x	y	z
Ca	30.6	Ca	2a	0	0	1.0
O_2Cu	29.0	O(1)	8g	0	$\frac{1}{2}$	0.9460
	29.0	O(1)	8g	$\frac{1}{2}$	0	0.9460
	28.9	Cu	4e	$\frac{1}{2}$	$\frac{1}{2}$	0.9456
SrO	27.2	Sr	4e	0	0	0.8903
	26.8	O(2)	4e	$\frac{1}{2}$	$\frac{1}{2}$	0.8750
OBi	24.5	Bi	4e	$\frac{1}{2}$	$\frac{1}{2}$	0.8022
	24.3	O(3)	4e	0	0	0.7950
BiO_2	23.5	Bi'	4e	$\frac{1}{2}$	$\frac{1}{2}$	0.7681
	23.0	O(4)	2d	0	$\frac{1}{2}$	$\frac{3}{4}$
	23.0	O(4)	2d	$\frac{1}{2}$	0	$\frac{3}{4}$
	22.4	Bi'	4e	0	0	0.7319
BiO	21.6	O(3)	4e	$\frac{1}{2}$	$\frac{1}{2}$	0.7050
	21.4	Bi	4e	0	0	0.6978
OSr	19.1	O(2)	4e	0	0	0.6250
	18.7	Sr	4e	$\frac{1}{2}$	$\frac{1}{2}$	0.6097
CuO_2	17.0	Cu	4e	0	0	0.554
	17.0	O(1)	8g	0	$\frac{1}{2}$	0.554
	17.0	O(1)	8g	$\frac{1}{2}$	0	0.554
Ca	15.3	Ca	2a	$\frac{1}{2}$	$\frac{1}{2}$	$\frac{1}{2}$
CuO_2	13.6	O(1)	8g	$\frac{1}{2}$	0	0.4460
	13.6	O(1)	8g	0	$\frac{1}{2}$	0.4460
	13.6	Cu	4e	0	0	0.4456
OSr	11.9	Sr	4e	$\frac{1}{2}$	$\frac{1}{2}$	0.3903
	11.5	O(2)	4e	0	0	0.3750
BiO	9.25	Bi	4e	0	0	0.3022
	9.03	O(3)	4e	$\frac{1}{2}$	$\frac{1}{2}$	0.2950
BiO_2	8.20	Bi'	4e	0	0	0.2681
	7.65	O(4)	2d	$\frac{1}{2}$	0	$\frac{1}{4}$
	7.65	O(4)	2d	0	$\frac{1}{2}$	$\frac{1}{4}$
	7.10	Bi'	4e	$\frac{1}{2}$	$\frac{1}{2}$	0.2319
OBi	6.27	O(3)	4e	0	0	0.2050
	6.05	Bi	4e	$\frac{1}{2}$	$\frac{1}{2}$	0.1978
SrO	3.83	O(2)	4e	$\frac{1}{2}$	$\frac{1}{2}$	0.1250
	3.36	Sr	4e	0	0	0.1097
O_2Cu	1.66	Cu	4e	$\frac{1}{2}$	$\frac{1}{2}$	0.0544
	1.65	O(1)	8g	$\frac{1}{2}$	0	0.0540
	1.65	O(1)	8g	0	$\frac{1}{2}$	0.0540
Ca	0	Ca	2a	0	0	0

[a] The space group is $I4/mmm$. D_{4h}^{17}. The unit cell dimensions are $a = 3.817$ Å, $c = 30.6$ Å (Tara9).

TABLE VI-13. Atom Positions of $Tl_2Ba_2CaCu_2O_8$ Structure Belonging to Space Group $I4/mmm$, D_{4h}^{17} with Two Formula Units per Unit Cell[a]

Complex	Vertical Position	Atom	Site	x	y	z
Ca	29.32	Ca	2a	0	0	1.0
O_2Cu	27.76	O(1)	8g	0	$\frac{1}{2}$	0.9469
	27.76	O(1)	8g	$\frac{1}{2}$	0	0.9469
	27.74	Cu	4e	$\frac{1}{2}$	$\frac{1}{2}$	0.946
BaO	25.75	Ba	4e	0	0	0.8782
	25.04	O(2)	4e	$\frac{1}{2}$	$\frac{1}{2}$	0.8539
OTl	23.06	Tl	4e	$\frac{1}{2}$	$\frac{1}{2}$	0.7864
	22.91	O(3)	16n	0.104	0	0.7815
TlO	21.07	O(3)	16n	0.604	$\frac{1}{2}$	0.7185
	20.92	Tl	4e	0	0	0.7136
OBa	18.94	O(2)	4e	0	0	0.6461
	18.23	Ba	4e	$\frac{1}{2}$	$\frac{1}{2}$	0.6218
CuO_2	16.24	Cu	4e	0	0	0.5540
	16.22	O(1)	8g	0	$\frac{1}{2}$	0.5531
	16.22	O(1)	8g	$\frac{1}{2}$	0	0.5531
Ca	14.66	Ca	2a	$\frac{1}{2}$	$\frac{1}{2}$	$\frac{1}{2}$
CuO_2	13.10	O(1)	8g	$\frac{1}{2}$	0	0.4469
	13.10	O(1)	8g	0	$\frac{1}{2}$	0.4469
	13.08	Cu	4e	0	0	0.4460
OBa	11.09	Ba	4e	$\frac{1}{2}$	$\frac{1}{2}$	0.3782
	10.38	O(2)	4e	0	0	0.3539
TlO	8.40	Tl	4e	0	0	0.2864
	8.25	O(3)	16n	0.604	$\frac{1}{2}$	0.2815
OTl	6.41	O(3)	16n	0.104	0	0.2185
	6.26	Tl	4e	$\frac{1}{2}$	$\frac{1}{2}$	0.2136
BaO	4.28	O(2)	4e	$\frac{1}{2}$	$\frac{1}{2}$	0.1461
	3.57	Ba	4e	0	0	0.1218
O_2Cu	1.58	Cu	4e	$\frac{1}{2}$	$\frac{1}{2}$	0.0540
	1.56	O(1)	8g	$\frac{1}{2}$	0	0.0531
	1.56	O(1)	8g	0	$\frac{1}{2}$	0.0531
Ca	0	Ca	2a	0	0	0

[a] The unit cell dimensions are $a = 3.8550$ Å, $c = 29.318$ Å (Subra).

I. THALLIUM-BARIUM-CALCIUM-COPPER OXIDE

The compound $Tl_2Ba_2CaCu_2O_8$ crystallizes in the same tetragonal space group $I4/mmm$, D_{4h}^{17} as the bismuth compound described above, with two formula units per unit cell and the lattice parameters $a = 3.8550$ Å, $c = 29.318$ Å (Subra). The atoms are at the following sites:

$$
\begin{array}{lll}
\text{Ca} & \text{(2a)} & \\
\text{Tl} & \text{(4e)} & u = 0.21359 \\
\text{Ba} & \text{(4e)} & u = 0.12179 \\
\text{Cu} & \text{(4e)} & u = 0.0540 \\
\text{O(1)} & \text{(8g)} & u = 0.0531 \\
\text{O(2)} & \text{(4e)} & u = 0.1461 \\
\text{O(3)} & \text{(16n)} & v = 0.604, \qquad u = 0.2815
\end{array}
\qquad \text{(VI-14)}
$$

where the atom positions for sites (2a), (4e), and (8g) are the same as in the previous section. The remaining $\frac{1}{4}$ occupied site (16n) has two parameters v and u, and the following possible atom positions:

$$
\begin{aligned}
\text{(16n)} \quad & 0,v,u;\ 0,v,-u;\ 0,-v,u;\ 0,-v,-u;\ \tfrac{1}{2},v+\tfrac{1}{2},u+\tfrac{1}{2}; \\
& \tfrac{1}{2},v+\tfrac{1}{2},-u+\tfrac{1}{2};\ 0,-v+\tfrac{1}{2},u+\tfrac{1}{2};\ 0,-v+\tfrac{1}{2},-u+\tfrac{1}{2} \\
& v,0,u;\ v,0,-u;\ -v,0,u;\ -v,0,-u;\ v+\tfrac{1}{2},\tfrac{1}{2},u+\tfrac{1}{2}; \\
& v+\tfrac{1}{2},\tfrac{1}{2},-u+\tfrac{1}{2};\ -v+\tfrac{1}{2},\tfrac{1}{2},u+\tfrac{1}{2};\ -v+\tfrac{1}{2},\tfrac{1}{2},-u+\tfrac{1}{2}
\end{aligned}
\qquad \text{(VI-15)}
$$

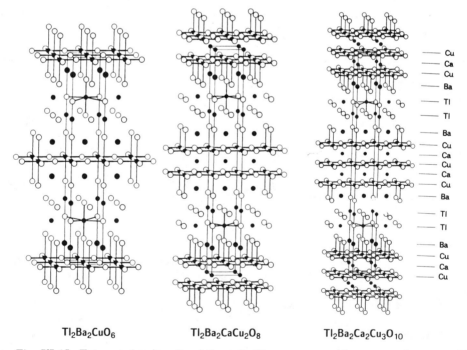

$Tl_2Ba_2CuO_6$ $Tl_2Ba_2CaCu_2O_8$ $Tl_2Ba_2Ca_2Cu_3O_{10}$

Fig. VI-15. Tetragonal unit cells of three thallium-copper oxide superconductors, $Tl_2Ba_2Ca_nCu_{n+1}O_{6+2n}$ with, from left to right, $n = 0, 1, 2$. Metal atoms are shaded and Cu-O bonds are shown (Tora2).

Table VI-13 gives more details on the atom positions and Fig. VI-15 presents a sketch of the structure.

J. SITE SYMMETRIES

Some experiments such as electron spin resonance of transition ions, optical spectroscopy, and infrared spectroscopy provide data that depend upon the symmetry of particular lattice sites. Table VI-14 lists for reference purposes the site symmetries at all of the lattice sites in the various structures mentioned above. These symmetries are given in both the International and Schoenflies notations for point groups.

TABLE VI-14. Point Symmetries at Lattice Sites of Various Compounds (with 100% stoichiometry)[a]

Space Group	Site	Atom	Point Symmetry	
		Cubic Perovskite BaTiO₃		
$Pm3m$, O_h^1	(1a)	Ba	$m3m$	O_h
	(1b)	Ti	$m3m$	O_h
	(3c)	O	$4/mmm$	D_{4h}
		Orthorhombic Perovskite BaTiO₃		
$Amm2$, C_{2v}^{14}	(2a)	Ba,O(1)	mm	C_{2v}
	(2b)	Ti	mm	C_{2v}
	(4e)	O(2)	m	C_s
		Tetragonal La₂CuO₄		
$I4/mmm$, D_{4h}^{17}	(2a)	Cu	$4/mmm$	D_{4h}
	(4c)	O(1)	mmm	D_{2h}
	(4d)	O(2) (alt.str.)	$\overline{4}m2$	D_{2d}
	(4e)	La,O(2)	$4mm$	C_{4v}
		Orthorhombic La₂CuO₄ (Longo)		
$Fmmm$, D_{2h}^{23}	(4a)	Cu	mmm	D_{2h}
	(8e)	O(1)	$2/m$	C_{2h}
	(8i)	La,O(2)	mm	C_{2v}
		Tetragonal YBa₂Cu₃O₈ (Borde, Jorge)		
$P4/mmm$, D_{4h}^1	(1a)	Cu(t)	$4/mmm$	D_{4h}
	(1d)	Y	$4/mmm$	D_{4h}
	(2f)	O(t)	mmm	D_{2h}
	(2g)	Cu(m), O(b)	$4mm$	C_{4v}
	(2h)	Ba	$4mm$	C_{4v}
	(4i)	O(m) = O(m')	mm	C_{2v}

TABLE VI-14. (continued)

Space Group	Site	Atom	Point Symmetry	
		Tetragonal YBa$_2$Cu$_3$O$_8$ (Hazen)		
$P\bar{4}m2$, D_{2d}^5	(1a)	Cu(t)	$\bar{4}2m$	D_{2d}
	(1c)	Y	$\bar{4}2m$	D_{2d}
	(2e)	Cu(m),O(b)	mm	C_{2v}
	(2f)	Ba	mm	C_{2v}
	(2g)	O(t),O(m),O(m′)	mm	C_{2v}
		Orthorhombic YBa$_2$Cu$_3$O$_8$[b]		
$Pmmm$, D_{2h}^1	(1a)	Cu(t)	mmm	D_{2h}
	(1b)	O(t′)	mmm	D_{2h}
	(1e)	O(t)	mmm	D_{2h}
	(1h)	Y	mmm	D_{2h}
	(2q)	Cu(m),O(b)	mm	C_{2v}
	(2r)	O(m′)	mm	C_{2v}
	(2s)	O(m)	mm	C_{2v}
	(2t)	Ba	mm	C_{2v}
		Orthorhombic Y$_2$BaCuO$_5$ (Hazen)		
$Pbnm$, D_{2h}^{16}	(4c)	Cu,Ba,Y(1),Y(2),O(3)	m	C_s
"Green Phase"	(8d)	O(1), O(2)	1	C_1
		Tetragonal Bi$_2$Sr$_2$CaCu$_2$O$_8$ Tl$_2$Ba$_2$CaCu$_2$O$_8$ (Tara9, Subra)		
$I4/mmm$, D_{4h}^{17}	(2a)	Ca	$4/mmm$	D_{4h}
	(4d)	O(4)	$\bar{4}m2$	D_{2d}
	(4e)	Ba,Bi,Cu,O,Sr,Tl	$4mm$	C_{4v}
	(8g)	O	—	—
	(16n)	O	—	C_s

[a] In particular, sites such as Cu(t) in YBa$_2$Cu$_3$O$_7$ with nearest-neighbor oxygens missing have point symmetries lower than those given.
[b] See Table VI-6.

K. STRUCTURAL ORIGIN OF SUPERCONDUCTIVITY

Various types of evidence presented throughout this review support the contention that the copper oxide planes play a crucial role in the origin of the superconductivity of the LaSrCuO and YBaCuO compounds. It has also been proposed that the CuO chains are required for the superconductivity of YBaCuO (Bard1, Tora1, Vand1), perhaps through coupling to the CuO$_2$ planes (Engl2, Murp2). Opinions of this type were widely held prior to the discovery of the bismuth and thallium compounds described in the sections above. These new materials are tetragonal with all of the oxygen sites occupied on the CuO$_2$ planes, and hence no chains are present. The commonalities of these various superconductor types have been discussed (Pool5).

VII

OTHER STRUCTURAL PROPERTIES

A. INTRODUCTION

The previous chapter described the crystallographic structures of the LaSrCuO, YBaCuO, BiSrCaCuO, and TlBaCaCuO superconducting materials. Technical details such as atom positions, lattice spacings, and nearest-neighbor environments were stressed. Some additional structural aspects of these same materials will be treated in the present chapter, such as atom replacements, defects, anisotropies, and lattice instabilities.

We mentioned in Sections II-C and II-G that the older transition metal superconductors were quite sensitive to stoichiometry. The superconducting properties of the newer oxide materials are not especially dependent on small changes in their cation ratios such as Y:Ba or La:Sr, or even M:Cu where M is a divalent or trivalent cation. Some are, however, quite sensitive to the extent of oxygen deficiency and also probably to the distribution of copper valence states. We will begin the chapter with a discussion of this topic, after which anisotropies and instabilities will be treated. Then we will examine the questions of atom substitutions, the isotope effect and high pressures. The chapter will end with a section on elastic and mechanical properties.

B. OXYGEN DEFICIENCY

The newer oxide superconductors are oxygen deficient in the sense that some or many of the crystallographic lattice sites allotted to oxygen are vacant, and the oxygen atoms associated with these sites can be readily removed and reintercalated (Tara7). The oxygen deficiency is a critical factor in determining the superconducting properties. This deficiency is associated with the bonding configurations and valence states of the copper ions that are present. Small losses of oxygen lower the average copper charge while large removals of oxygen can appreciably change the nearby copper coordination which, at stoichiometry, consists of four, five, or six nearest neighbor Cu–O bonds. Withdrawing oxygen decreases this number of adjacent bonds. We will discuss these effects for the LaSrCuO and YBaCuO cases. Oxygen deficiency determinations and characterizations have been carried out using various instrumental techniques such as differential thermal and thermal gravimetric analyses (e.g., Beye3, Ongz1, Tara2, Tara7), solid-state electrochemical cells (Ohana), neutron diffraction (Beech, Greed), and photoemission spectroscopy (Stoff, Schro).

1. Oxygen Loss

The compound $YBa_2Cu_3O_{7-\delta}$ reversibly exchanges oxygen with the surrounding atmosphere in the temperature range 600–1100 K. This process is an activated one with activation energies quoted between 0.7 and 1.65 eV (Burge, Halle, Strob, Tuzzz). When oxygen is lost during sample preparation the lattice parameters a, b, and c of YBaCuO tend to change in a regular manner (Cava3, Kuboz) even near the tetragonal to orthorhombic transformation temperature which occurs between 620°C for 2% O_2 and 700°C for 100% O_2 in the oven atmosphere (Jorg1, Nakam).

Samples of $YBa_2Cu_3O_{7-\delta}$ were prepared (Tara3; see also Cava2) with δ in the range from 0 to 0.9, and they had the properties given in Table VII-1. Figure VII-1 shows the tetragonal-to-orthorhombic transition at $\delta = 0.6$ with $a = b$ for $\delta > 0.6$, and the lattice parameter c increases continuously with decreasing oxygen content. The magnetization data of Fig. VII-2 indicate that the superconducting transition is highest and sharpest for $\delta = 0$. The magnetization has the same low temperature limiting value of $\approx -5 \times 10^{-4}$ emu for all oxygen contents from $\delta = 0$ to $\delta = 0.59$, as indicated on Fig. VII-2, while the data in Table VII-1 show that the Meissner effect decreases continuously from 35% for $\delta = 0$ to 11% for $\delta = 0.59$. The susceptibility data above T_c fit the Curie Weiss law (vide Section VIII-D-1)

$$\chi = N\mu^2/3k(T + \theta) \qquad \text{(VII-1)}$$

with the magnetic moments μ and Weiss constants θ listed in Table VII-1. The average moment μ per copper ion, ≈ 0.30 for the superconducting samples, may be compared with the spin-only values $2(S(S + 1))^{1/2}$ of 2.83 for Cu^{3+}, 1.73 for Cu^{2+}, and 0 for Cu^+. This moment increases to 0.37 and 0.62 for the $\delta = 0.66$ and $\delta = 0.9$ nonsuperconducting samples, respectively. In addition, the room temperature resistivity increases for $\delta > 0$ (Cava2, Cava3).

A procedure was developed for converting an oxygen-deficient n-type tetragonal phase to a p-type orthorhombic phase, accompanyed by a decrease in resistivity of almost a factor of a million. This could be of technological importance

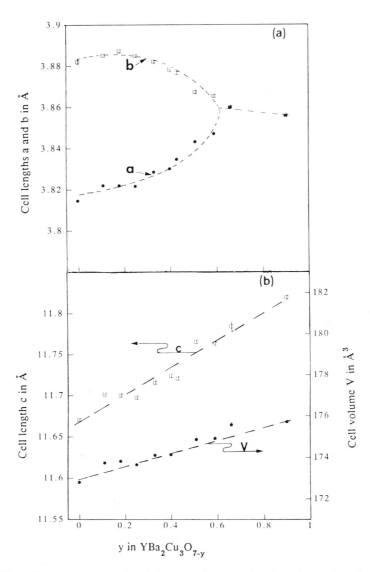

Fig. VII-1. The lattice parameters a, b, c and the unit cell volume V as a function of the oxygen content for the $YBa_2Cu_3O_{7-y}$ series. The dashed lines are a guide for the eye. Note that the structural transition occurs near $\delta = y = 0.6$ (Tara8).

112

TABLE VII-1. Crystal Data, Meissner Effect, Effective Magnetic Moment μ, Curie Weiss Temperature θ, and rms Deviation of Data from the Curie–Weiss Law for $YBa_2Cu_3O_{7-\delta}$ (Tara3)

δ	a (Å)	b (Å)	c (Å)	ANIS (%)	V (Å)^3	Meissner (%)	μ per Mole of Cu	θ (K)	rms (%)	T_c (K)
0	3.8146	3.8819	11.670	1.75	172.80	35	0.29	−21	0.15	92
0.11	3.8221	3.8851	11.701	1.63	173.75	28	—	—	—	64
0.18	3.8220	3.8872	11.700	1.69	173.83	23	—	—	—	—
0.25	3.8217	3.8848	11.697	1.64	173.66	30	0.30	−30	0.57	57
0.33	3.8286	3.8820	11.715	1.39	174.12	27	—	—	—	52
0.435	3.8347	3.8765	11.720	1.08	174.14	27	0.30	−25	0.50	50
0.51	3.8433	3.8573	11.765	0.36	174.87	21	0.30	−24	0.47	37
0.59	3.8471	3.8654	11.763	0.47	174.93	11	0.31	−20	0.19	22
0.66	3.8600	—	11.786	0	175.58	0	0.37	−20	0.8	0
0.9	3.8559	—	11.819	0	175.73	0	0.62	−11	0.39	0

^{a}Anisotropy factor ANIS $= 100|b - a|/\frac{1}{2}(b + a)$ from Eq. VI-8.

112

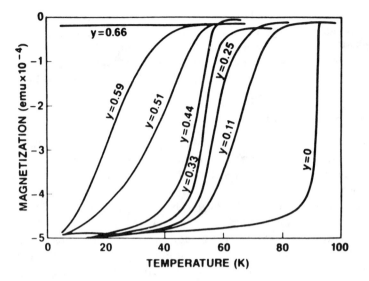

Fig. VII-2. Temperature-dependent magnetization for the $YBa_2Cu_3O_{7-y}$ series. The curves shown are for warming the samples in a 1 mT field after they have been cooled in zero field. T_c is defined at the midpoint (50% of the total diamagnetic shift) (Tara3).

because it permits oxygen to be restored at temperatures compatible with device processing (Bagle).

2. Ordered Oxygen Vacancies in YBaCuO

In Section VI-F-2 we showed that the oxygen deficiency in $YBa_2Cu_3O_{7-\delta}$ for $\delta = 0$ is concentrated in the CuO_2 basal plane layer between the barium ions where half of the oxygens tend to be missing in an ordered manner (Camps, Jorge, Ronay, Sarik, Schu2, Segre, Viege). The copper ions form linear chains of $(-Cu-O-)_n$ along the b axis, as indicated in Fig. VI-13, and they are square planar (CuO_4) coordinated in the $b-c$ plane. The long-range order associated with these chains extends over dimensions of about 500 Å (Ourma). When δ is not equal to zero the chains cannot be perfect and oxygen defects can cause lattice deformation (Vante, Vant1). In the high-temperature tetragonal phase the oxygen vacancies are disordered. If there is no oxygen loss, reversible order–disorder transitions may be produced by thermal cycling (Vante). The oxygens in the remaining two planes near yttrium are not effected and retain their CuO_5 pyramidal coordination. The transition is second order with a small hysteresis due to twin boundary kinetics (Frei1). Thus we see that the large oxygen deficiency has a very pronounced effect on both the copper charge state and on its nearest-neighbor coordination. It should be recalled (Section II-C, Table II-2) that ordered vacancies had also been observed in the older transition ion superconductors.

We just described the long-range order associated with the oxygen vacancies that form chains in the $\delta = 0$ compound (Ourm2). A short-range or partial ordering of these vacancies has been deduced from the diffuse scattering seen in electron diffraction photographs with $0 < \delta < 1$ (Chail). Long-range order reappears in the limit of $\delta = 1$, where all of the basal plane oxygen sites are empty.

3. Copper Charge States

The average copper charge is calculated by assuming that all of the La and Y ions are trivalent, all of the alkaline earth ions such as Ba and Sr are divalent, and each oxygen has the charge $2-$. This is reasonable on chemical grounds, as explained in Section III-J.

The prototype compound La_2CuO_4 has all of its copper in the common valence state Cu^{2+}, and our assumption given above is surely valid for it. When alkaline earth ions such as Sr replace some of the lanthanum to form $(La_{1-x}Sr_x)_2CuO_4$ the conventional charge assumption requires some of the copper to change to the trivalent state. When in addition there is an oxygen deficiency denoted by δ corresponding to the structural formula $(La_{1-x}Sr_x)_2CuO_{4-\delta}$, the copper is driven to a lower average valence state. Thus substituting strontium and removing oxygen have opposite effects on the copper charge, and all of the copper remains divalent for the special case $x = \delta$. In the more general case when positive values are assigned to the parameter y of $(La_{1-x}Sr_x)_{2-y}CuO_{4-\delta}$, the La,Sr content decreases and this leads to an increase in the average copper charge. Typical parameter values are $x = 0.075$ or 0.1, $y \approx 0$ and $\delta \approx 0$, so the copper coordination is not much disturbed, just its valence. The sample with $x = 0.075$ is the best and has become the standard with the special symbol LaSr*.

Stoichiometric $YBa_2Cu_3O_8$ has, in theory, all of its copper in the Cu^{3+} state. It has not been synthesized because attempts to do so result in oxygen-deficient compounds such as $YBa_2Cu_3O_{7-\delta}$ with oxygen contents closer to 7 ($|\delta| \ll 1$). By our argument the compound $YBa_2Cu_3O_7$ has $\frac{2}{3}$ of its copper in the state Cu^{2+} and $\frac{1}{3}$ as Cu^{3+}. Experimentally fabricated samples of $Y_{1-x}Ba_{2-y}Cu_3O_{7-\delta}$ can have a range of x, y, and δ values with various Cu^{3+}/Cu^{2+} ratios. Synchrotron radiation photoemission measurements indicate that reducing the oxygen content by vacuum annealing results in the reduction of copper to Cu^+.

C. DEFECTS

The extent and significance of defects in the oxide superconductors has attracted a great deal of attention, and in this section we will summarize and comment on the results that have been reported.

1. Microscopy Studies

Electron microscopy (EM) and high-resolution electron microscopy (HREM) studies have been reported on the superconducting oxides (Chau2, Chau3,

Damen, Douse, Hervi, Huxfo, Naray, Vante, Vant1, Yinhu, Yuech, Zhouz). Overstoichiometry as well as substoichiometry and cationic disorder have been detected (Domen). This technique is particularly good for studying grain boundaries and growth habits. One such study reported that most grain boundaries involve a facet with the c axis normal to the boundary (Nakah). Field-ion microscopy was used to examine $YBa_2Cu_3O_{7-\delta}$, and an imaging atom probe analysis of field-ion tips made from the ceramic oxides revealed a composition that varied from sample to sample (Kello).

2. Defect Types

The oxide materials exibit various types of defects extending from individual missing atoms on the 0.1-nm scale, to faults extending over tens or hundreds of nanometers, to grain imperfections in the micrometer range. Another classification is into bulk and surface defects. Examples of the former are stacking faults (Naray) and the interpolation of extra sheets or planes of atoms (Ourma, Zandl; cf. Section VI-F-8). The material can accommodate such faults in response to stoichiometric constraints. Surface defects have been detected by thermally stimulated luminescence measurements in, for example, Ho- and Gd-substituted materials (Cooke). Structural perfection can be important for high critical currents.

3. Twinning

The a and b dimensions of the $YBa_2Cu_3O_{7-\delta}$ orthorhombic unit cell are so close together, $(b - a)/\frac{1}{2}(a + b) \approx 0.01$, that a and b occasionally interchange directions during sample preparation, crystal growth, or cooling down through the tetragonal to orthorhombic transition. This phenomenon is called twinning, and in YBaCuO it has been observed by electron microscopy (e.g., Brokm, Pande, Ravel, Sarik, Zandb). The twin plane is the 110 plane and their spacing has been reported to be about 80 nm (Hewat). Samples can differ considerably in their twinning density. Many YBaCuO samples exhibit related structural irregularities such as oriented domains, moirè patterns, and doubling or quadrupling of the unit cell (Defon, Hervi).

The compound YBaCuO is known to exhibit various anisotropic properties and parameters such as coherence length, penetration depth, resistivity, critical current, and critical fields which differ in their values in the ab plane and along the c direction, as is shown in Section VII-D. This means that YBaCuO acts like an axially symmetric crystal. If these same quantities were to be measured using true nontwinned single crystals or films oriented in the ab plane, most of them would surely exhibit values that differ along the a and b directions due to the presence of Cu–O chains along b but not along a. Twinning has the effect of providing a type of averaging of the experimental values corresponding to the a and b directions, thereby inhibiting the ability to detect the true orthorhombic anisotropy of the samples.

It has been suggested that nontwinned crystals or films could provide super-conductors with superior properties. Another viewpoint is that increasing the density of twinned regions to an optimum value can enhance the superconduct-ing properties. For example twin boundaries can act as vortex pinning sites, and such pinning serves to increase the critical current density. The BiSrCaCuO and TlBaCaCuO superconductors are tetragonal, and hence do not have this twin-ning propensity.

4. Atomic Scale Defects

Even inside superconducting grains there can be atomic scale defects (Ravea, Tsue1) and internal Josephson junctions which might be responsible for the "glassy state" in these materials (Deuts).

5. Radiation Damage

Irradiation with neutrons (Atobe, Umeza, Will1) and ion implantation (He, O, As; Clark, Stof1, White) adversely affect superconducting properties by reduc-ing and broadening T_c. The formation of a thin amorphous layer at grain bound-aries accounts for the degradation of T_c (Clar1, Clar2). Neutron irradiation was found to enhance the magnitude and reduce the anisotropy of the critical mag-netization of YBa*. Stimulated thermal luminescence studies of X-irradiated $Ho_{1.5}Ba_{1.5}Cu_2O_x$ and $GdBa_2Cu_3O_{7-\delta}$ detected surface emissions attributed to oxygen instability (Cooke).

D. ANISOTROPY

The oxide superconductors are very anisotropic in many of their properties, and this is most easily demonstrated by studying single crystals and epitaxial films. BiSrCaCuO is the most anisotropic and TlBaCaCuO is the least anisotropic of the copper oxide superconductors. Structure determinations, magnetometry, Raman spectroscopy (e.g., Hemle) and transport properties of thin films (Su-zuk) has employed single crystals. However, most past scientific work has been carried out with polycrystalline or amorphous specimens, so the measurements involve averages over the directionally dependent properties.

It was noted above that $YBa_2Cu_3O_{7-\delta}$ single crystals are extensively twinned, so it has not been possible to study the in-plane (a,b) anisotropies of this class of material.

1. LaSrCuO

The magnetization of a monocrystal of $(La_{0.925}Sr_{0.075})_2CuO_4$ measured in the temperature range from 5 to 42 K with the applied magnetic field parallel and perpendicular to the Cu–O or *ab* planes (Iwazu) followed the same curve when the sample was cooled down through the transition temperature. In contrast to this, parallel and perpendicular field measurements made during heating after zero-field cooling differed from each other, as shown in Fig. VII-3. The results indicated an anisotropy in the critical magnetic fields, and this anisotropy was used to estimate the anisotropy in the penetration depth

$$\Lambda_{\parallel} \approx 4000 \text{ Å}, \qquad \Lambda_{\perp} \approx 600 \text{ Å} \qquad \text{(VII-2)}$$

in support of the two-dimensional character of the LaSrCuO. In the normal state the resistivity of LaSrCuO perpendicular to the *ab* planes it is 20 times larger than within or parallel to the planes (Hidak)

$$\rho_{\parallel} = 1 \times 10^{-3} \, \Omega \text{ cm}, \qquad \rho_{\perp} = 1.9 \times 10^{-2} \, \Omega \text{ cm} \qquad \text{(VII-3)}$$

Below 30 K both resistivities dropped abruptly and became zero at 3.8 K. This result means that charge mobility is much larger in the Cu–O planes than it is perpendicular to these planes. Hidaka et al. also found $H_{c2\parallel}/H_{c2\perp} > 5$.

The apparent discrepancy between the far IR and tunneling experiments on superconducting LaSrCuO has been explained in terms of the anisotropy of the

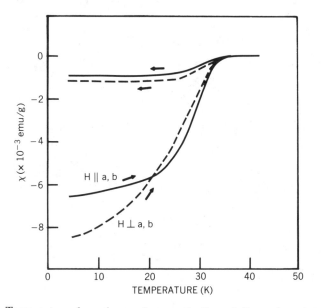

Fig. VII-3. Temperature dependence of magnetization of $(La_{0.925}Sr_{0.075})_2CuO_4$ single crystal for warm-up after zero-field cooling (lower curves) and field-cooling process (upper curves). The results are shown for the magnetic field parallel to (——) and perpendicular to (---) the *a,b* planes. (Redrawn from Iwazu.)

energy gap (Maeka). Far IR spectra have been interpreted in terms of a model of randomly oriented grains highly conducting in two directions and nonconducting in the third (Herrz).

2. YBaCuO

Anisotropies have been studied in $YBa_2Cu_3O_{7-\delta}$ oriented polycrystalline pellets (Otter) and single crystals. The resistivity in the c direction increased by 30–50% (Hage1, Tozer), slightly above T_c, before dropping rapidly to zero while the in-plane resistivity continually decreased until the transition point, as indicated in Fig. VII-4. The zero-field cooled diamagnetism was essentially 100%, but the Meissner flux expulsion was rather small, varying between $\approx 4\%$ in the parallel and $\approx 17\%$ along the perpendicular orientations. Other investigators have reported much larger flux expulsions (cf. Section VIII-B). Magnetization hysteresis loops are shown in Fig. VII-5 for the two sample orientations (Ding1, Worth), and the temperature variation of the upper critical fields H_{C2} for the two orientations (Ding1, Taji1) are presented in Fig. VII-6. We see from the figure that for single-crystal $EuBa_2Cu_3O_7$ the critical field slope $-dH_{C2}/dT = 3.8$ T/K for H parallel to the ab plane is almost 10 times the value $-dH_{C2}/dT = 0.41$ for H directed in the c direction, perpendicular to this plane. Induced critical current densities at various temperatures as a function of the applied magnetic field parallel and perpendicular to the Cu–O planes are sketched on Fig. VII-7. Comparison tables have been published of measured parameters (critical currents, critical field-temperature derivatives, resistivities), and derived parameters (coherence length, critical fields at 0 K, penetration depths) parallel and perpendicular to the Cu–O planes (Hikit, Salam, Worth, Wortl). Such anisotropic parameters are listed in several of the tables.

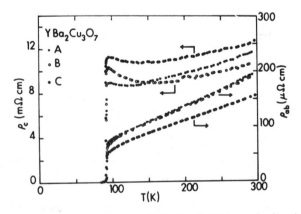

Fig. VII-4. Resistivity of current flow parallel to the CuO planes (ρ_{ab}) and perpendicular to them, that is, along the c axis (ρ_c). Data are shown for three samples A, B, and C. Note that $\rho_{ab} < \rho_c$ (Hage1).

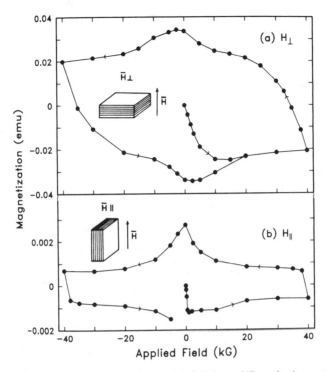

Fig. VII-5. Magnetization hysteresis loops at 4.5 K for a YBa$_*$ single crystal with the CuO planes oriented (*a*) perpendicular and (*b*) parallel to the applied field (Ding1).

The upper critical field anisotropy is $H_{C2\parallel}/H_{C2\perp} \approx 10$ for rare-earth substituted RBa$_2$Cu$_3$O$_{7-\delta}$ where R = Y, Nd, Eu, Gd, Dy, Ho, Er, and Tm (e.g., Orlan). Epitaxial films of YBa$_2$Cu$_3$O$_{7-\delta}$ had anisotropies in H_{C2} between 2 and 3 (Chau1). YBa$_2$Cu$_3$O$_{7-\delta}$ exhibited strong anisotropic flux pinning (Section VIII-B-2) with critical magnetization current densities of 1.4×10^6 A/cm^2 (Crabt).

E. INSTABILITY

The layered oxide materials have three related instabilities: the normal to superconducting, the tetragonal to orthorhombic, and the metal to insulator. Each region of instability is associated with a phase transformation from one stable state to another. It has been known for a long time that high transition temperatures in electron–phonon-mediated superconductivity often occur in materials that are structurally unstable or close to the lattice stability threshold (Matt1, Ginzb, Hase1, Machi). In particular, the onset of superconductivity in compounds with oxygen octahedra often occurs near metal-to-insulator phase boundaries (Chuz1, Chuz3, Phil2). A number of authors have pointed out that the LaSrCuO (e.g., Flemi, Hase1, Jorg3), LaBaCuO (e.g., Chuz1, Fujit), and

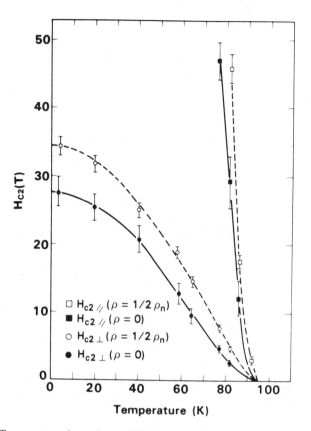

Fig. VII-6. Temperature dependence of H_{c2} for a $EuBa_2Cu_3O_{7-\delta}$ crystal oriented with the CuO planes parallel (right) and perpendicular (left) to the applied field (Taji1).

YBaCuO (e.g., Raozz) compounds are in the instability category, as is also the prototype La_2CuO_4. Jahn Teller instabilities or lattice distortions have also been discussed (Aokiz, Egami).

1. LaSrCuO

In the La–Sr–Ba system it was suggested (Flemi) that the role of the Sr or Ba is to spoil the Fermi surface nesting and the Peierls instability (Fento, Hutir) while leaving the states near the Fermi level with large deformation potentials which can provide strong electron phonon coupling. Superconductivity may coexist with Fermi surface instability and possibly a CDW in the La–Sr,Ba–Cu oxide system (Uchi1).

The prototype compound La_2CuO_4 is susceptible to a Peierls or frozen-in charge-density wave distortion that opens a semiconductor gap over the entire Fermi surface and spoils the potential superconductivity (Stavo; Hase1, Machi).

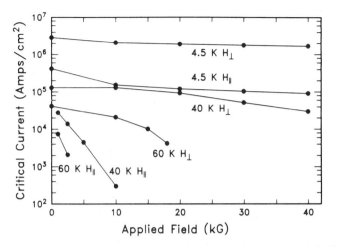

Fig. VII-7. Induced critical-current densities at various temperatures calculated from magnetization hysteresis loops as a function of magnetic field applied parallel and perpendicular to the CuO planes (Ding1).

The Sr and Ba substitutions prevent this spoiling by suppressing the CDW and preserving the strong electron–phonon coupling that is needed for a high T_c (Matt6, Yuzzz). La_2CuO_4 has a van Hove singularity just at the Fermi energy and a nearly square Fermi surface with a half-filled Cu(3d)–O(2p) σ band that may be important in a lattice instability and in producing a strong electron–phonon coupling (Oguch). The electronic structure near the Fermi energy is very sensitive to the positions of oxygen atoms situated outside the Cu–O planar network. The van Hove saddle point singularity has a strong influence on superconducting properties (Xuzzz).

A study of the optical reflectivity plasma spectra of $(La_{1-x}M_x)_2CuO_4$ where M = Ba, Ca, and Sr showed that the electronic state is close to a just half-filled two-dimensional band (Tajim). In such a state there can be a strong electron–phonon interaction which favors a CDW or a strong electron–electron interaction which favors a SDW. The latter instability could induce superconductivity. The weakening and eventual disappearance of the 680-cm^{-1} vibrational line for increasing x was associated with a CDW suppression (Stavo). A cusp found at 250 K on the resistivity versus temperature curve of La_2CuO_4 was interpreted as indicative of an electronic instability and most probably the formation of a SDW (Uchi2).

2. YBaCuO

It has been pointed out that the room-temperature resistivity of $YBa_2Cu_3O_{7-\delta}$ is on the borderline of the metal to nonmetal transition, and this may be conducive to high-temperature superconductivity (Raozz). On the other hand, the vibrational spectroscopic evidence for lattice instabilities is not conclusive (cf. Section IX-E-1).

3. Ferroelectricity

The high-temperature superconductor compounds are related to perovskite ferroelectric types (Cross), and the discoverers of the first such materials were aided in their quest by their experience with ferroelectric phase transitions in oxides (Mull1). The associated large dielectric constant may provide a channel for the enhanced electron pairing interaction (Chuz1, Zhon1).

F. SUBSTITUTIONS

Two of the high-temperature superconductor types LaSrCuO and YBaCuO each contain a group IIIB ion (Y, La), an alkaline earth (Ca, Sr, Ba), copper, and oxygen. An important question that arises concerns which of the constituent atoms are essential and which can be replaced by related or perhaps not so related atoms. We will examine atomic substitutions involving both $(R_{1-x}M_x)_2CuO_{4-\delta}$ with copper constituting $\frac{1}{3}$ of the cations and $RM_2Cu_3O_{7-\delta}$ with copper accounting for $\frac{1}{2}$ of the total cation content. Partial atomic substitutions and total replacements at all atom positions will be discussed. A concluding section will comment on the significance of the results.

Some articles present data on a series of rare-earth ions replacing La in LaSrCuO or Y in YBaCuO and others report only particular rare-earth-substituted compounds. Several papers involve members of the first transition series and other atoms replacing Cu, and some substitutions for O will also be presented. The magnetic behavior of the rare earths substituted in YBaCuO will be discussed in Section VIII-D-3.

1. Rare Earths in LaSrCuO

Several authors substituted Nd (namely, Crab1, Kwok2) and various rare-earth elements R for lanthanum in $(La_{1-x-y}R_ySr_x)_2CuO_{4-\delta}$ (namely, Dikoz, Fueki, Grove, Haseg, Hoso1, Kish1, McKin, Ogita, Phata, Tara5), and this causes a lowering of the transition temperature, as the normalized resistivity plots of Fig. VII-8 demonstrate. Room-temperature resistivities for this series of compounds varied between 1600 and 2540 $\mu\Omega$ cm, and the ratio p_{300K}/p_{50K} varied between 2.9 and 4.6. In some cases double substitutions such as Ba_xSr_{y-x} for lanthanum were employed (e.g., Hosoy). Figure VII-9 shows how T_c varies across the rare-earth series.

2. Rare Earths in YBaCuO

Many investigators have studied rare earth substitutions in the YBaCuO system (namely, Chuz2, Fiskz, Horzz, Kurih, Kuzzz, Lepa1, Maple, Marcu, McKin,

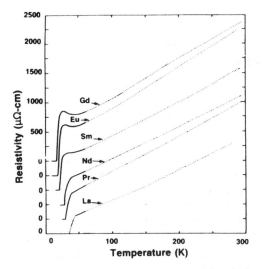

Fig. VII-8. Resistivity versus temperature for rare-earth (R) substitutions in the compound $(La_{0.8}Sr_{0.1}R_{0.1})_2 CuO_{4-\delta}$ (Tara5).

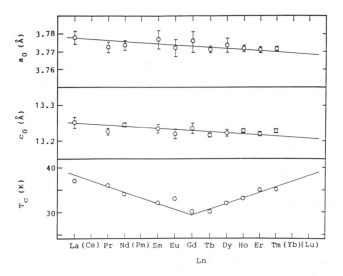

Fig. VII-9. Variations of tetragonal lattice parameters a_0 and c_0 and ac susceptibility onset T_c for the various rare earth ion (R) substitutions $(La_{0.85}Sr_{0.1}R_{0.05})_2CuO_{4-\delta}$ (Kish1).

Murph, Namzz, Parki, Tana1, Tara4, Torra, Xiao1, Yamad, Yangz, Yang1, Yang2), and the subject has been reviewed (Tara2). Experimental data on particular rare earths replacing Y in YBaCuO have been reported in individual articles, and some examples are dysprosium (Qiru1), erbium (Ayyub, Golbe, Hayri, Qiru3, Zuozz), europium (Boolc), gadolinium (Escu2, Podda, Qiru2, Thomp, Zuozz), holmium (Hayri, Kago1, Leez2, Podda, Thom2, Zuozz), lanthanum (Cheva, Leez3, Segre), lutetium (Raych), neodymium (Escu2), praseodymium (Dalic, Soder), scandium (Shizz, Zhao1), samarium (Csach, Garci, Hajko), thulium (Andr1, Neume), and ytterbium (Grove, Qiru3). This list of individual elements does not include articles that involve several rare-earth substitutions.

Typically all of the yttrium was replaced by a rare-earth ion, and this total replacement produced lattice constant changes (Eagle, Horzz, Tara4, Yang2), as shown on Fig. VII-10. The results summarized in Table VII-2 demonstrate that most of the rare earths produce transition temperatures of 85-95 K, and several of them (Eu, Dy, Ho, Tm, and Yb) achieved narrow transitions with T_c above 90 K. Four rare-earth elements (Ce, Pr, Pm, and Tb) did not form superconductors. Also included in the table are values of the resistivity at 300 K and at the temperature of the onset of superconductivity together with lattice constants and susceptibility data. Figure VII-11 presents a typical resistivity versus temperature plot for various substitutions (Maple, McKin, Murph, Tara2–Tara5, Yangz).

3. Alkaline Earths

Many measurements have been made on barium-substituted and some on calcium-substituted $La_2CuO_{4-\delta}$ (Hosoy, Terak), and data on these compounds are found in several places throughout this review. The Sr^{2+} ion, which has a size (1.12 Å) close to that of La^{3+} (1.14 Å), produces higher transition temperatures and three times the diamagnetic susceptibility in $(La_{1-x}M_x)_2CuO_{4-\delta}$ than the larger Ba^{2+} (1.34 Å) ion (Bedn2).

When strontium is substituted for barium in the system $Y(Ba_{1-x}Sr_x)_2Cu_3O_{7-\delta}$, the transition temperature drops from 90 K to about 78 K in the range from $x = 0$ to 0.75 (Qirui, Wuzz2). Meissner flux exclusion is about 45% from $x = 0$ to $x = 0.5$ and drops off thereafter (Vealz). Therefore, the $x = 0$ composition is the best.

4. Paramagnetic Substitutions for Copper

The series of compounds $YBa_2(Cu_{0.9}M_{0.1})_3O_{7-\delta}$ where M is a member of the first transition series of elements was fabricated with the oxygen deficiency parameter δ undetermined (Xiaoz). All of the transition elements reduced T_c, but to a different extent, as shown on the resistivity plots of Fig. VII-12. The susceptibility χ above T_c is described by a temperature-independent part χ_0 and a paramagnetic part, as is explained in Section VIII-D-1. The depression of T_c correlates with the magnetic moment of the substituted transition ion, the larger the mo-

TABLE VII-2. Effect of Substituting Various Atoms in R and M Sites of $RM_2Cu_3O_{7-\delta}$ [a]

R	M	T_c (K)	ΔT (K)	a (Å)	b (Å)	c (Å)	ρ_{onset} μΩ·cm	ρ_{300}/ρ_{onset}	$-\chi_{DS}$	χ_{ME}/χ_{DS} (%)	Ref.
Y	Ba	93.3	2.1	3.86	3.87	11.67	260	2.3	10.9	52	Avg*
La	Ba	59.2	17.9	3.94	3.95	11.97	—	—	—	—	Avg*
Ce	Ba	—	—	—	—	—	—	—	—	—	—
Pr	Ba	—	—	3.886	3.912	11.710	—	—	—	—	Avg*
Nd	Ba	78.3	28	3.882	3.96	11.746	—	—	3.1	48	Avg*
Pm	Ba	—	—	—	—	—	—	—	—	—	—
Sm	Ba	88.6	11.9	3.83	3.85	11.74	—	—	19.1	35	Avg*
Eu	Ba	91.1	8.3	3.87	3.88	11.75	740	2.5	14.2	29	Avg*
Gd	Ba	90.9	7.4	3.845	3.898	11.732	—	—	8.5	18	—
Tb	Ba	—	—	—	—	—	—	—	—	—	—
Dy	Ba	91.8	2.3	3.87	3.89	11.60	—	—	5.5	27	Avg*
Ho	Ba	91.1	4.3	3.84	3.87	11.62	—	—	6.1	33	Avg*
Er	Ba	90.7	4.9	3.838	3.862	11.586	—	—	12.8	14	Avg*
Tm	Ba	90.5	1.7	3.840	3.883	11.574	—	—	13.9	28	Avg*
Yb	Ba	89.3	1.6	3.833	3.882	11.571	—	—	10.7	29	Avg*
Lu	Ba	72.6	13	3.83	3.88	11.63	—	—	5.7	3.5	Avg*
$Y_{0.75}Sc_{0.25}$	Ba	91	5	3.84	3.86	11.74	1860	2.15	—	—	Murph
$Y_{0.5}Sc_{0.5}$	Ba	90	4	—	—	—	—	—	—	—	Engle
$Y_{0.5}La_{0.5}$	Ba	87	10	3.86	3.88	11.69	4200	1.91	—	—	Murph
$Y_{0.25}Eu_{0.75}$	Ba	95	2	3.85	3.87	11.75	800	2.6	—	—	Murph
$Y_{0.1}Eu_{0.9}$	Ba	94.5	2	3.85	3.87	11.77	320	3.0	—	—	Murph
$Pr_{0.1}Eu_{0.9}$	Ba	82	5	3.86	3.88	11.76	1860	1.93	—	—	Murph
Eu	Ba	95	2.5	3.85	3.88	11.77	740	2.5	—	—	Murph
$Eu_{0.75}Sc_{0.25}$	Ba	93	5	3.87	3.89	11.77	3000	1.73	—	—	Murph
La	Ba	60	29	3.93	—	—	39000	1.04	—	—	Murph
Y	$Ba_{0.75}Sr_{0.25}$	87	5	3.82	3.83	11.67	1260	2.60	—	—	Murph
La	$Ba_{0.5}Ca_{0.5}$	79	3	—	—	—	2100	2.5	—	—	Murph

[a] The rows give from left to right the R and M atoms; the transition temperature T_c and its width ΔT; the three lattice parameters a,b,c; resistivity data for the onset, ρ_{onset}, the ratio with room-temperature value ρ_{300}; and the diamagnetic shielding (DS) and Meissner (ME) susceptibility. The rows with an asterisk (*) contain transition temperatures T_c, ΔT and lattice parameters a,b,c averaged from Engle, Horzz, Kurih, Kuzzz, Maple, McKin, Tara3, Yamad, Yangz, Kish2, and Smrck.

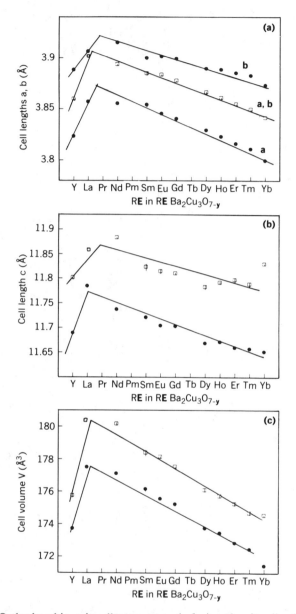

Fig. VII-10. Orthorhombic unit-cell parameters (a,b,c) and unit-cell volume (V) as a function of the rare earth R in the $RBa_2Cu_3O_{7-\delta}$ series for the as-prepared samples ($\bullet$) and vacuum-annealed samples ($\square$) are shown. The lines drawn are a guide for the eye (Tara4).

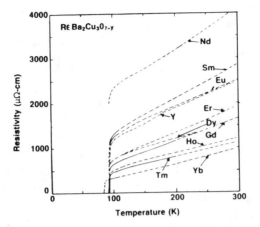

Fig. VII-11. Resistivity as a function of temperature for $YBa_2Cu_3O_{7-\delta}$ and compounds for which a rare earth has been substituted for the Y (Tara2).

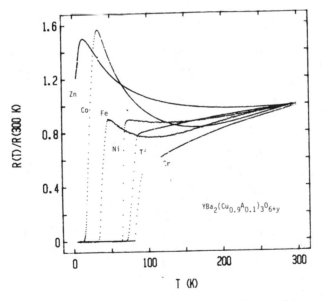

Fig. VII-12. Temperature dependence of the normalized resistance of YP_{γ} $(Cu_{0.9}A_{0.1})_3O_{7-\delta}$, where A = Ti, Cr, Fe, Co, Ni, and Zn (From Xiaoz).

ment, the lower the T_c value, as shown by the data in Fig. VII-13. Comparing with Fig. II-2, we see that there are two maxima for T_c, but they do not occur for the same number of valence electrons. The Cu maximum at $N_e = 11$ on Fig. VII-13 is beyond the region of the curves shown on Fig. II-2. Others (Felne, Maen2, Osero) have published similar results for YBaCuO and also for LaSr-CuO (Haseg).

An extensive examination was made of the effect of substituting paramagnetic Ni^{2+} and diamagnetic Zn^{2+} ions (M) for copper ions in $(La_{0.925}Sr_{0.075})_2$-$Cu_{1-x}M_xO_4$ and $YBa_2(Cu_{1-x}M_x)_3O_{7-\delta}$ (Csach, Tara3, Tara6, Tara8, Thiel). Changes in lattice constants, transition temperature, susceptibility, and upper critical fields were determined. Variations in unit-cell dimensions may result from the release of the Cu^{2+} Jahn Teller distortion. The substitution of Ni in the YBaCuO system degrades the superconductivity to a much lesser extent than in the LaSrCuO case (Tara3, Tara8). This lessened effect relative to LaSrCuO may result because all of the Ni in the YBaCuO compound is not on the particular Cu–O layers where the Cooper pairs reside. The Cu–O layers in the LaSrCuO compound, on the other hand, are all equivalent and contribute equally to the superconductivity.

5. Nonmagnetic Substitutions for Copper

Zinc and Ga have definite valence states of $+2$ and $+3$ and ionic radii 0.74 and 0.62, respectively (cf. Table VI-2), which reflect those of Cu^{2+} and Cu^{3+}. There-

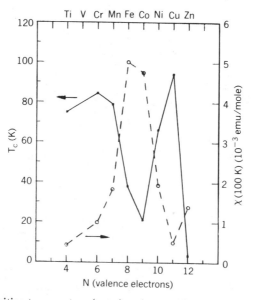

Fig. VII-13. Transition temperature (——) and magnetic susceptibility at 100 K (---) of $YBa_2(Cu_{0.9}M_{0.1})_3O_{7-\delta}$, where M is a 3d transition element, as a function of the number of valence electrons (from Xiaoz).

fore Zn should replace Cu^{2+} on the $Cu-O_2$ planes, and 15% replacement is enough to completely destroy the superconductivity. In like manner Ga should substitute for Cu^{3+} on the Cu-O chain layer, and substitutions up to 20% reduce T_c by only 30%. In addition gallium contents above 4% induce the tetragonal-to-orthorhombic transformation (Xiao3).

The substitution of Ag, which is ordinarily monovalent, for Cu in $(La_{0.9}Sr_{0.1})_2Cu_{1-x}Ag_xO_{4-\delta}$ depressed T_c by 50% at the $x = 0.15$ level (Malik), much less than the effect of Ni or Zn.

Partial replacement of Cu by monovalent Ag in $YBa_2Cu_3O_{7-\delta}$ depressed T_c (Tomyz), and in one report total replacement lowered the onset T_c to 50 K and increased the transition width to 30 K (Panz1). This replacement increased the Cu^{3+}/Cu^{2+} ratio, presumably for charge balance, although an increase in oxygen vacancies may also occur. Ag replacement was found to enhance the critical current density by a factor of 15 (Kungz).

6. Substitutions for Oxygen

Several studies have been made with fluorine replacing part of the oxygen (Ovshi, Tonou). The structure of $YBa_2Cu_3O_6F_2$ may be tetragonal with fluorines occupying all of the $O(l)$ and $O(t')$ sites on the $z = 0$ basal plane and oxygen occupying all of its other sites (cf. Table VI-6). This arrangement would eliminate oxygen site vacancies and chains, with all of the copper ions in square-planar nearest-neighbor coordination. Implanted fluorine and its diffusion were also studied (Kiste, Tesme, Xianr).

Zero resistance has been claimed at $T_c = 155$ K for $YBa_2Cu_3O_6F_2$ (Ovshi) and at 148.5 K with fluorine implantation (Xianr).

The replacement of part of the oxygen by sulphur can form the compound $YBa_2Cu_3O_6S$, which has a sharper transition without changing T_c. The compound displays the full Meissner effect (Felne). Total oxygen replacement has not (yet) been achieved, but partial replacement by either F or S is not necessarily destructive and can enhance the superconducting properties (Feln1). Sulphur can also transform the crystal to the tetragonal form.

7. Other Substitutions

Aluminum substitution in $Y_{1-x}Al_xBa_2Cu_3O_{7-\delta}$ for $0 < x < 0.85$ (Franc) and in BiSrCaCuO (Chuz5) lowers T_c. Another report showed that small quantities of aluminum ($x = 0.05, 0.08, y \approx 0.35$) in the system $(Y_{1-x-y}Ba_xAl_y)_2CuO_4$ do not effect the onset T_c (Escu1).

8. Influence of Substitutions on Superconductivity

We have seen that substitutions of magnetic ions for La or Y have very little effect on the superconductivity while, in general, substitutions for Cu have a destructive effect. This supports the belief that the superconducting quasi-particles or Cooper pairs are associated with the Cu–O layers, and that the La, Ba, and Y layers are not directly involved in the superconductivity mechanism.

In ordinary or low-temperature superconductors the presence of magnetic ions destroys the superconductivity quite strongly (Matt1), while in the LaSrCuO and YBaCuO cases the destruction is selective, depending on how close the substitutions are to the Cooper-pair layers (Haseg, Hoso1).

It is doubtless significant that a high T_c superconducting compound can be obtained with all of the Cu replaced by Ag and with much of it replaced by Ga. It is also significant that F and S substitutions for O are not necessarily deleterious.

G. ISOTOPE EFFECT

The traditional test for phonon-mediated superconductivity is the isotope effect (Section III-G-2), whereby for an element T_c is related to the atomic mass M through the expression

$$T_c M^\alpha = \text{const} \qquad\qquad \text{(VII-4)}$$

with the isotope effect coefficient $\alpha = \frac{1}{2}$ for electron–phonon coupling in elemental superconductors. For compounds with many atoms it is difficult to predict how T_c depends on the mass of a particular isotope in a complex structure, but isotope effects are observed in compounds. Several investigators report that ^{18}O substitution on the two sites of LaSrCuO produce the positive isotope effect $0.1 < \alpha < 0.35$, as shown in Fig. VII-14 (Batl1, Batl2, Cohe1, Falte). In contrast, the results for YBaCuO have been partly negative. The replacement of ^{16}O by ^{18}O in $YBa_2Cu_3O_7$ and $EuBa_2CuO_{7-\delta}$ gave $\alpha \approx 0$ (Batlo, Batl2, Bourn, Cohe1, Morri). The absence of an isotope effect with enriched ^{63}Cu, ^{65}Cu, ^{135}Ba, and ^{138}Ba was also reported (Bour4). Nevertheless, there were reports (Katay, Leary) of a decrease in T_c of 0.2-0.5 K by the 70–90% replacement of ^{16}O by ^{18}O, as indicated in Fig. VII-15. The same magnitude of shift was found for the four oxide superconductors $BaBi_{0.25}Pb_{0.75}$ with $T_c = 11$ K, $(La_{0.925}Ca_{.075})_2CuO_4$ with $T_c = 20$ K, $(La_{0.925}Sr_{.075})_2CuO_4$ with $T_c = 37$ K, and $YBa_2Cu_3O_7$ with $T_c = 92$ K, with a small dependence on T_c (Stacy). In addition, Raman phonon frequencies do exhibit the expected 4% mass dependent shifts (Batlo).

A positive result of α close to $\frac{1}{2}$ is strong support for a phonon-mediated BCS mechanism, as was noted in Section IV-B-2. The isotope effect coefficient increases slowly with coupling strength (Mars2). Some transition ion superconductors such as Os and Nb_3Sn have unusually small isotopic mass dependencies, and for others such as Ru and Zr, T_c is independent of M (Table II-1), so negative results do not preclude the operation of a phonon mechanism. Also, other mechanisms such as excitons, resonant valence bonds, and the librational model of e-e pairing (Hardy) are predicted to exhibit little or no isotopic shift.

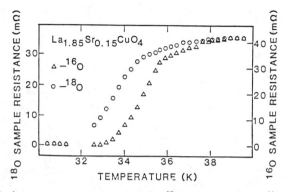

Fig. VII-14. Resistance versus temperature for ^{18}O-substituted and ^{16}O portions of the same LaSr$_*$ pellet (Cohe1).

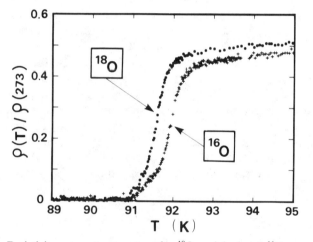

Fig. VII-15. Resistivity versus temperature for ^{18}O-enriched and ^{16}O samples of YBa$_*$ (Katay).

H. EFFECT OF PRESSURE

In this section we will be concerned with changes in the transition temperature, its width, the pressure coefficient dT_c/dP and the change in lattice parameters brought about by the application of pressure to the samples. One of the early milestones of the new superconductivity era was the attainment of $T_c = 52$ K in LaSrCuO under high pressure (Chuzz).

The pressure dependence of the high T_c superconductors was examined within the framework of the standard BCS electron–phonon theory (Gries) and it was concluded that the data are consistent with a two-dimensional BCS model

(Labbe), resonant valence bond models (Cyrot, Fukuy), and a bipolaron model (Alexa).

1. LaSrCuO

Increasing the pressure on a LaSrCuO-type superconductor has the effect of raising its transition temperature and sometimes broadening the range over which the transition takes place (Allg1, Chan3, Erski, Mihal, More7). Figure VII-16 shows the pressure dependence of the change in transition temperature $\Delta T_c = T_C(P) - T_C(0)$ from the zero-pressure value $T_C(0)$ for $La_{2-x}Sr_xCuO_4$ with three values of x (Schi1). Figure VII-17 shows how the pressure coefficient dT_C/dP depends on x. Other workers determined dT_c/dP for $(La_{1-x}M_x)_2CuO_4$ and found: 0.17 K/kbar for Ba, $x = 0.075$ (Kuris); 0.29 K/kbar for Sr, $x = 0.1$ (Polit); 0.13, 0.20, and 0.13 K/kbar, respectively for Ca, Sr, Ba with $x = 0.075$ (Shelt); 0.12 K/kbar for Sr (Przys); and 1 and 0.2 K/kbar for Ba and Sr, respectively (Chuz1, Satoz). The pressure coefficient dT_c/dP exhibits a saturation effect whereby it levels off to a fairly constant value at high pressures, above 10 or 20 kbar (Chuzz, Shelt). The rate at which pressure raises T_c far exceeds that of previously known superconductors (Chuzz, Chuz1, Chuz2).

At room temperature hydrostatic pressure linearly decreases the resistivity of LaSrCuO by about 0.15%/kbar up to 20 kbar (Kuris). The lattice parameters a and c of $(La_{1-x}Sr_x)_2CuO_{4-\delta}$ with $x = 0.1$ and 0.3, respectively, decreased linearly with pressure by 2% between atmospheric pressure and 70 kbar (Takah, Terad). Meanwhile the c/a ratio remained fairly constant over this pressure range.

Pressure was claimed to enhance the Cu^{3+}/Cu^{2+} ratio in these compounds (Wuzzz). It also decreased the transition temperature of the tetragonal-to-orthorhombic transformation, and the orthorhombic phase was suppressed above 15 kbar (More7).

2. YBaCuO

In YBaCuO-type materials the sharpness of the transition decreases at high pressure, but the transition temperature itself is not very sensitive to the pressure, as shown in Fig. VII-18. The temperature derivative dT_c/dP was found to be positive (0.07 K/kbar, Schir; 0.043 K/kbar, Dries), negative (-0.25 K/kbar, Murat), or very little affected (Horz1). The normal-state resistivity decreases with pressure by 1.27%/kbar (Murat). This is in contrast to the LaSrCuO system, which has a large positive temperature derivative.

3. BiSrCaCuO

Increasing the pressure increases the resistivity of BiSrCaCuO in the normal state, as indicated in Fig. VII-19, but it has very little effect on T_c, as shown in Fig. VII-20 (Chuz5). This is true both for the pronounced transformation to the superconducting state near 78 K (T_{c2} on Fig. VI-20), and for the less prominent

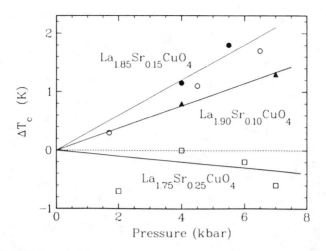

Fig. VII-16. Shift in transition temperature ΔT_c versus pressure for $x = 0.1$ (▲), $x = 0.15$ (● ○), and $x = 0.25$ (□) in $La_{2-x}Sr_xCuO_4$. The curves are straight lines through the data from several runs on each sample, but only points from a single run are shown, except for the $La_{1.85}Sr_{0.15}CuO_4$, where an indication of the scatter involved is shown by data from two runs. The dashed horizontal line is the zero reference (Schi1).

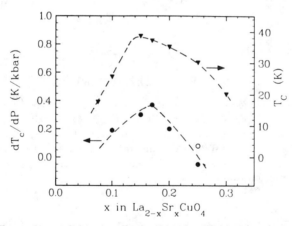

Fig. VII-17. Dependence of the superconducting transition temperature T_c at zero pressure (right legend) and the pressure coefficient dT_c/dP (left legend) on the strontium content x in $La_{2-x}Sr_xCuO_4$ (Schi1).

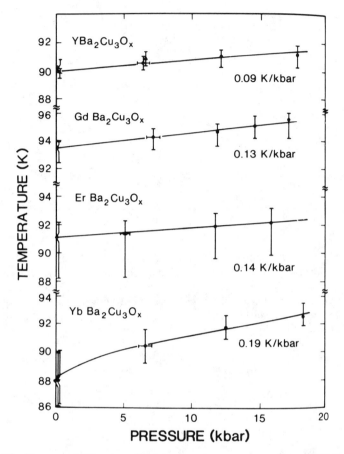

Fig. VII-18. Superconducting transition temperature versus pressure at 90 K for YBa_2-Cu_3O_x, $GdBa_2Cu_3O_x$, $ErBa_2Cu_3O_x$, and $YbBa_2Cu_3O_x$: (●) resistive midpoints determined with pressure increasing; (▲) decreasing pressure. The vertical error bars correspond to 90–10% resistive transition widths and the horizontal error bars to the estimated uncertainty in pressure. Note the linear increase in T_c for all samples except $YbBa_2$-$Cu_3O_{7-\delta}$ at the lowest pressures (Borge).

indication of a possible second superconducting transition near 116 K (T_{c1} on the figure).

I. ELASTIC AND MECHANICAL PROPERTIES

The copper oxide superconductor era began with the study of materials whose physical and chemical properties had not yet been characterized. This section will survey some of the elastic and ultrasonic measurements that have been made on these materials. Most of the data to be described were obtained on normal

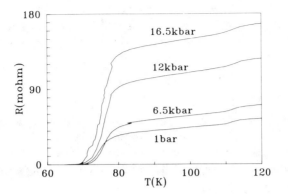

Fig. VII-19. Pressure dependence of the resistive transition (R versus T) of BiSrCaCuO (Chuz5).

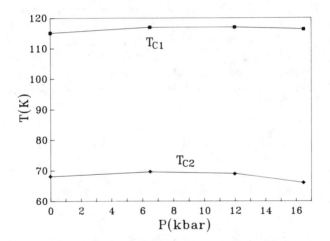

Fig. VII-20. Dependence on the pressure of the transition temperature T_{c2} from the main resistive drop near 69 K on Fig. VII-19, and of the transition temperature T_{c1} determined from the small drop in R near 116 K on Fig. VII-19 (Chuz5).

specimens far above the transition temperature. A short review is available which discusses these properties and compares changes in sound velocity and specific heat jumps (Laegr).

1. Lattice Size and Mass Density

One of the most important mechanical properties of these oxide materials is the change in lattice size and Cu–O neighbor distance under various conditions such as atomic substitution (Tara4, Tara5), oxygen content, temperature (Bisho, Boyce, Skelt), and pressure (Terad), as was described earlier in this chapter.

Typical densities of YBaCuO material are 5.56 and 4.30 g/cm³, and comparing these to the X-ray density 6.383 g/cm³, the calculated porosities $100(\rho - \rho_{Xray})/\rho_{Xray}$ are 12.9 and 32.6%, respectively (Blend, Mathi). The unit cell size is 174 Å³.

2. Thermal Expansion

The compound $YBa_2Cu_3O_7$ was found to expand continuously when heated from 4 to 1100 K (Bayo1, Salom). A change in slope occurred at 700°C owing to the onset of oxygen desorption. The average Grüneisen parameter was 3.1 for YBa* (Salom), and it was 1.7 near 30 K and 1.2 at room temperature for LaSr* (Collo).

3. Elastic Moduli

Young's modulus of LaSr* varies with the temperature in the manner shown in Fig. VII-21 (Bour2), and the bulk, Young, and shear moduli together with Poisson's ratio of YBa* have the temperature dependence shown in Fig. VII-22 (Ledbe–Ledb3). All moduli exhibit normal increases from room temperature down to T_c except Young's modulus for LaSr*, which anomalously decreases from 200 to 100 K. The application of a magnetic field in the superconducting region increases Young's modulus (Bour2, Datt3). Some adiabatic elastic prop-

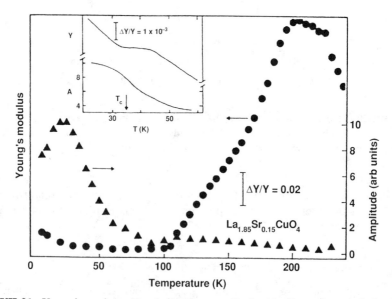

Fig. VII-21. Young's modulus Y and vibration amplitude A in $(La_{0.925}Sr_{0.075})_2CuO_4$. The latter is inversely proportional to the internal friction. The inset shows the superconducting transition region in detail. T_c is indicated by an arrow on the inset (Bour2).

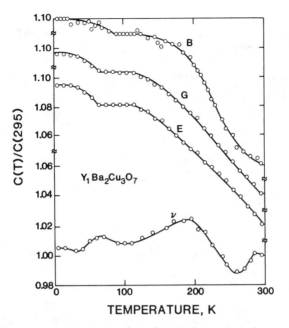

Fig. VII-22. Temperature dependence of relative elastic constants between 295 and 4 K. B denotes bulk modulus, E the Young modulus, G the shear modulus, and ν the Poisson ratio (Ledb1).

erties are given in Table VII-3 (Blend). The torsional modulus exhibits an unusually large change or softening at T_c (Mathi). The temperature dependence of the elastic modulus of superconducting $BaPb_{1-x}Bi_xO_3$ has also been studied (Hikat).

In a recent work on reentrant softening, Datta and coworkers argue that LaSr∗ and YBa∗ are neither dissimilar nor inconsistent with thermodynamic requirements. In both materials, as the temperature decreases the lattice softens

TABLE VII-3. Adiabatic Elastic Properties of YBa$_2$Cu$_3$O$_7$ (Blend)

Property	Units	Porosity = 12.9%		Porosity = 32.6%	
		Measured	Corrected for Porosity	Measured	Corrected for Porosity
Longitudinal wave speed	km/s	4.87	5.25	3.63	4.32
Shear wave speed	km/s	2.76	2.95	2.06	2.35
Shear modulus	GPa	42.4	55.6	18.3	35.2
Poisson ratio	—	0.264	0.268	0.263	0.290
Young's modulus	GPa	107	141	46.1	90.8
Bulk modulus	GPa	75.5	102	32.4	72.2

just above T_c, and this softening is offset below T_c by increased stiffness associated with the developing superconducting phase. This model agrees with the results from other measurements, and predicts a higher elastic stiffness in the normal state (Datt3, Viole).

The (100) and (001) planes of single crystal $YBa_2Cu_3O_{7-\delta}$ are preferred fracture planes. They have a toughness of 1.1 MPa $m^{1/2}$ and a hardness of 8.7 GPa (Cookz, Dinge). These values are between those of the perovskite $BaTiO_3$ (0.59 MPa/$m^{1/2}$, 5.9 GPa) and the spinel $MgAl_2O_4$ (1.2 MPa/$m^{1/2}$, 13.1 GPa).

There have been observations in YBa* of internal friction peaks and modulus minima (Yenin), and of pseudoelasticity involving ferroelastic loops at room temperature, 240, 220, and 170 K (Huimi).

4. Ultrasonics

Ultrasonic attenuation studies in YBa* show peaks at 250 and 160 K (Horil, Schen, Xuzzz, Yushe), which is indicative of first-order structural changes. These temperatures are close to those of the ferroelastic loops mentioned in the previous section. Another peak was observed just below $T_c = 91$ K. It was claimed that the change in attenuation between the normal and superconducting states is larger than that for other conventional metals (Schen). An acoustic study of LaSrCuO revealed a displacive phase transition, and when this phase boundary approached the superconducting one, T_c seemed to increase (Yosh2). Ultrasonic attenuation data of LaBa* suggest the presence of a possible glassy state (Fossh).

The dependence of the velocity of sound $v_S \approx 5 \times 10^5$ cm/sec on the temperature has been determined for LaSr* from 4 to 250 K (Bisho, Bish1, Horie). The ultrasound could only be propagated at low frequencies, 10–30 MHz. At higher frequencies the wavelength of the sound waves became comparable with the size of the sintered particles and Rayleigh scattering dominated. There was a softening or decrease in sound velocity as T_c was approached from above and a hardening or increase in v_S below T_c, the fractional changes being about 800 ppm.

Low-temperature (≈ 1 K) damping of sound waves in YBa* has been reported. The relative change in the sound velocity was observed to decrease linearly with the logarithm of the temperature in the range from 0.008 to 0.5 K. This was attributed to atomic oxygen tunneling; the tunneling system density of states was reported to be 4.2×10^{33}/erg cm^3 (Goldi).

Ultrasonic studies were reported on LaSrCuO (Horie, Ledb2, Yushe), YBaCuO (Almon, Almo1, Bish2, Esqui, Migl1, Xuzz3, Yenin), and HoBa* (Ledb4). Photoacoustic measurements were made on $(La_{0.9}Sr_{0.1})_2CuO_4$ (Sawan).

VIII

MAGNETIC PROPERTIES

A. INTRODUCTION

Superconductivity can be defined as the state of perfect diamagnetism, and consequently researchers have always been interested in the magnetic properties of superconductors. The first part of this chapter will discuss magnetization, and then we will treat critical fields which set limits on the magnetization. Next we will discuss how the specimens can be categorized in terms of traditional magnetic behavior such as diamagnetism and paramagnetism. Finally, we will summarize the data obtained from the various types of magnetic resonance and end with some comments on the interrelationship between magnetism and superconductivity.

B. SUSCEPTIBILITY AND MAGNETIZATION

The most important magnetic parameter measured is the magnetization M, which provides the susceptibility χ through the expression $M = 4\pi\chi H$. The magnetization becomes strongly negative below T_c, as shown in Fig. VIII-1 for LaSr*. In like manner the susceptibility becomes strongly negative below T_c, as shown in Fig. VIII-2 for YBa* (Datt1) and Fig. VIII-3 for $Tl_2Ba_2CaCu_2O_{8-\delta}$. For a perfect diamagnet $M = -H$, which means that $\chi = -1/4\pi$ (per unit volume in the cgs system). Oxide superconductors are not perfect and the measured value of χ depends upon how the experiment is carried out. The FC and ZFC notation on the figures will be explained below.

In this volume we use the cgs system. Some workers prefer the MKS or SI system whereby $M = \chi H$ and $\chi = -1$ for perfect diamagnetism.

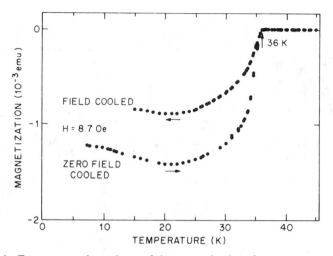

Fig. VIII-1. Temperature dependence of the magnetization of an oxygen-annealed sample of $(La_{0.9}Sr_{0.1})_2CuO_4$. The magnetization of the zero-field cooled sample is 60–70% of the ideal diamagnetic value (Cavaz.)

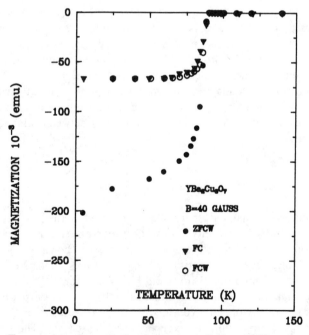

Fig. VIII-2. Temperature dependence of the susceptibility of YBa* showing the ZFC diamagnetic shielding and the FC Meissner effect in 4 mT. In the FC case both cool down (FC) and subsequent warm up (FCW) data are shown (Datt1).

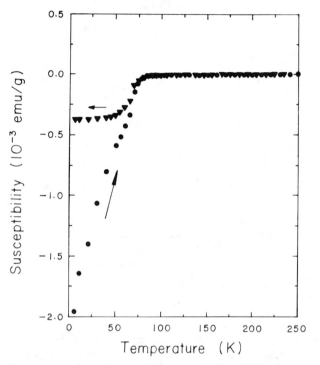

Fig. VIII-3. Temperature dependence of the susceptibility of TlBaCaCuO in a field of 2 mT. Field-cooled (▼) and ZFC (●) data points are shown (Sheng).

1. Measurement Techniques

Many susceptibility and magnetism measurements are carried out with a super-conducting quantum interference device, commonly called a SQUID magne-tometer, which is a dc-measuring instrument. The data illustrated in Fig. VIII-3 were obtained with a SQUID magnetometer. More classical techniques such as the vibrating sample magnetometer or occasionally the Guoy or Faraday bal-ances are also used.

Some workers measure the ac susceptibility (e.g., Dharz, Mali1; cf. Section VIII-B-5) with a low-frequency mutual-inductance bridge operating at, for ex-ample, 200 Hz. In one experiment the mutual-inductance signal provided a much broader transition than a dc magnetization measurement on the same sample (Moode), as shown in Figs. VIII-4 and VIII-5, respectively. There are also reports of ac susceptibility versus temperature curves being sharper than their dc counterparts (Tara1).

2. Type II Superconductivity

We explained in Section III-G that the oxide superconductors are Type II with the lower and upper critical fields H_{C1} and H_{C2}, respectively, defined in Fig.

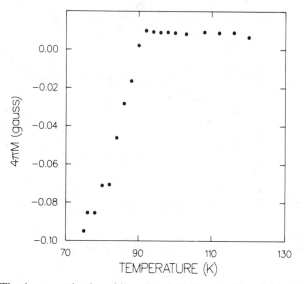

Fig. VIII-4. The dc magnetization of $(La_{0.9}Ba_{0.1})_2CuO_4$ cooled in a field of 1 mT illustrating the Meissner effect (Moode).

III-7. When the applied magnetic field is less than the lower critical field H_{C1}, an ideal Type II superconductor acts like a perfect diamagnet, and no magnetic flux penetrates it. Most susceptibility measurements are carried out in applied fields much less than this, sometimes as low as 0.5 μT (Crone) or 5 μT (Krusi), so the samples exhibit Type I behavior by excluding and expelling magnetic flux. When the applied field exceeds the lower critical field, the magnetic flux pene-

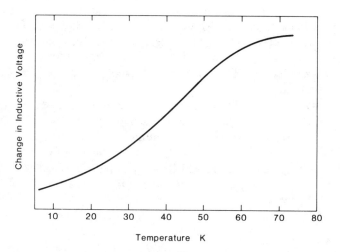

Fig. VIII-5. Temperature dependence of mutual-inductance signal of $(La_{0.9}Ba_{0.1})_2CuO_4$ (Moode).

trates, and the imaginary term χ'' of the ac susceptibility becomes measurable, giving rise to hysteresis loss.

3. Field Cooling and Zero-Field Cooling

The susceptibility at a particular magnetic field strength H and temperature T in the superconducting region depends upon the path in the H,T plane that the specimen follows to reach the particular field and temperature. Two types of measurements are customarily made to quantify this path dependence.

(a) The first measurement, called zero-field cooling (ZFC), consists of cooling the specimen below T_c in the absence of a magnetic field, and then turning on a small field, typically several millitesla, and measuring the susceptibility. Under this condition the superconductor exhibits what is called diamagnetic shielding or flux exclusion, which means that the magnetic field fails to penetrate the material. The measured volume susceptibility has a strong negative value, ideally $1/4\pi$ in magnitude for complete flux exclusion. In practice, the flux exclusion is generally not complete. For a precise comparison of the measured susceptibility with the ideal value, care should be taken to correct for the geometry via the demagnetization factor. After the initial ZFC determination at a low temperature, the susceptibility is often measured as the magnetic field is gradually increased, and data obtained with zero-field cooled samples followed by warming (ZFCW) are shown in Fig. VIII-2.

(b) The second measurement, called field cooling (FC), consists of turning on the magnetic field while the sample is still in the normal state, cooling it down below T_c in the presence of the applied field, and measuring the susceptibility at various temperatures during the cooling down process. Under this condition the superconductor exhibits what is referred to as the Meissner effect or flux expulsion. Sometimes after reaching a low temperature the susceptibility is then measured while the sample is gradually warmed. This is known as field-cooled warming (FCW), and the FCW data may follow the initial FC curve, as shown in Fig. VIII-2.

For the same magnetic field and the same temperature $T < T_c$, an oxide superconductor expels less flux when it is field cooled than it excludes when it is zero-field cooled, and the difference is called trapped flux or sometimes pinned flux. Thus the magnitude of the susceptibility is less for the Meissner effect than it is for diamagnetic shielding, as shown in Fig. VIII-2 where the ZFC data points are far below the corresponding FC and FCW data.

4. Diamagnetic Shielding and Meissner Effect

Meissner-effect or flux-expulsion experiments are ordinarily carried out along with measurements of diamagnetic shielding or flux exclusion on the same sample. In this section we will report on some representative data.

In typical susceptibility experiments on rare-earth-substituted $YBa_2Cu_3O_{7-\delta}$ (Yangz) the samples were cooled in zero field to ≈ 6 K then the field $H \approx 1$ mT was applied to measure χ. The authors estimated $\chi \approx 3 \times 10^{-3}$ emu/g for a perfect diamagnet, assuming a spherical shape with a demagnetization factor of $\frac{1}{3}$ and a porosity of $\frac{1}{2}$.

A sample of $YBa_2Cu_3O_7$ with a 5:1 ratio between flux exclusion and flux expulsion was compressed at 20–30 kbar to a claimed 100% of theoretical density, and this brought the measured flux expulsion to within about 11% of the theoretical value (Ventu). Remnent field effects could also have contributed to this result. Both 100% flux shielding and 95% flux expulsion were found in $YBa_2Cu_3O_7$ at 4.2 K (Larb1). Many other investigators have observed the flux-trapping effect in LaSrCuO (e.g., Tokum, Yosh1) and YBaCuO (e.g., Minam) systems. Similar flux-trapping effects are observed in both the BiSrCaCuO and TlBaCaCuO compounds.

Table VIII-1 lists the quantity $-4\pi \chi_{DS}$ extrapolated to $T = 0$ K for various superconductors. This has the value 1 for an ideal superconductor with $\chi_{DS} = -1/4\pi$, and it is less than 1 for most materials. The table expresses these quantities in percentages.

5. Real and Imaginary Susceptibilities

The magnetic susceptibility χ has a real or dispersion part χ' and an imaginary or absorption part χ''

$$\chi = \chi' + i\chi'' \qquad \text{(VIII-1)}$$

and most reported data involve dc measurements of χ'. In this section we will discuss the frequency, magnetic field, and porosity dependence of the ac susceptibility.

The frequency dependences of χ' and χ'' versus temperature for $YBa_2Cu_3O_{7-\delta}$ are shown in Fig. VIII-6 (Raozz). We see from the figure that increasing the frequency broadens and shifts the peak of the χ'' curve to lower temperatures, and it raises the χ' curves corresponding to less diamagnetic shielding.

The real and imaginary parts of the ac susceptibility of $YBa_2Cu_3O_{7-\delta}$ at 10, 100, and 1000 Hz in the superconducting region were reported to be virtually independent of the frequency but strongly dependent on the amplitude H_0 of the field, which varied from 1.4 μT to 0.7 mT (Goldf). Figure VIII-7 shows how the diamagnetic shielding becomes complete at low magnetic-field amplitudes, but less so at higher amplitudes. The imaginary part of the susceptibility exhibits a cusp where the real part undergoes a sharp diamagnetic change. This cusp and the position of the diamagnetic drop both shift to lower temperatures and broaden at higher fields, as shown in the figure. Similar field dependencies occur in YBaCuO (Raoz1) and LaSrCuO (Odazz), respectively. The increase in the area under the χ'' curve with increasing field indicates greater energy dissipation arising from flux fluctuations, and suggests that the material is not homogeneous.

TABLE VIII-1. Typical Examples of Diamagnetic Shielding Susceptibility χ_{DS} and Meissner Susceptibility χ_{ME} Normalized Relative to Perfect Diamagnetism Susceptibility $(-1/4\pi)$ and Ratio $\chi_{ME}/\chi_{DS}{}^a$

Material	$4\pi\chi_{DS}$ (%)	$4\pi\chi_{ME}$ (%)	χ_{ME}/χ_{DS} (%)	Comments	Ref.
LaSr$_*$	24	13	54	$H = 4$ mT	Allge
			35	Bulk	Renke
			75	Same, powdered	Renke
LaSr(0.1)	60–70	~40	~63		Cava2
	65–100				Espar
	80	30	37.5		Schl1
	80	40	50		Green
YBa$_*$	85–100	60–80	70–80		Freit
	70	12	17		Crone
	~100	20	20		Grant
	~100	0.1–71			Krusi
	100	95	95		Larb1
$Y_{0.5}Ba_{0.5}CuO_3$	30	3.5	12		Sunzz
$Y_{0.6}Ba_{0.4}CuO_4$	24–19	20–14	73	2–14 mT	Allge
$(Y_{0.8}Ba_{0.2})_2CuO_4$		6			Sunzz

^{a}All columns are given as a percentage. $(La_{1-x}M_x)_2CuO_4 = LaM(x)$, $LaSr(0.075) = LaSr_*$, $YBa_2Cu_3O_{7-\delta} = YBa_*$.

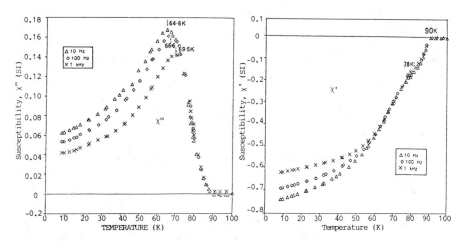

Fig. VIII-6. Temperature dependence of the real (χ') and imaginary (χ'') contributions to the ac susceptibility measured at the frequencies 10, 100, and 1000 Hz for YBaCuO in an applied field of 1 mT rms (Raoz1).

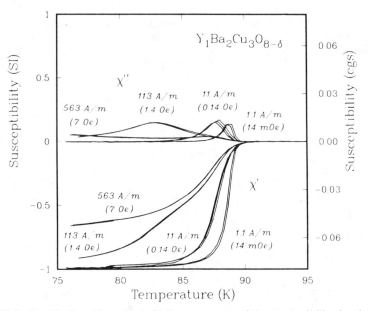

Fig. VIII-7. Real (χ') and imaginary (χ'') components of the susceptibility for the magnetic field amplitudes, 1.4 μT, 14 μT, 0.14 mT, 0.7 mT (1 mT = 10 Oe) at the frequencies, 10, 100, and 1000 Hz. Each curve is composed of straight-line segments connecting about 30 data points. The susceptibility is a strong function of the field amplitude, but irtually independent of frequency. Where the curves for different frequencies at constant field amplitude do not exactly overlap, the far right curve corresponds to 1000 Hz (Goldf).

Figure VIII-8 shows the dependence of $\chi'(T)$ and $\chi''(T)$ measured at 0.1 mT and 10 Hz on the preparation conditions and porosity of a sintered YBaCuO alloy (Raoz1, see also Chen3). The denser material had much better susceptibility characteristics. These results point out the important roles played by flux pinning on the surface, flux trapping due to inhomogeneities, and the porous nature of the materials. The data for the Dy analogue have also been reported (Ningz).

6. Magnetization

The previous two sections discussed the susceptibility of the superconducting state. Closely related to this is the magnetization M or magnetic moment per unit volume, which is given by

$$M = 4\pi\chi H \qquad\qquad \text{(VIII-2)}$$

The volume (mass) susceptibility is the magnetic moment per unit field per unit volume (per unit mass). In many materials χ is independent of the applied field

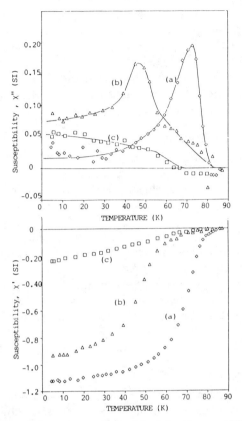

Fig. VIII-8. Temperature dependence of χ' and χ'' for the same YBaCuO used in Fig. VIII-6 sintered into dense (a), less dense (b), and porous material (c), but having the same T_c measured at 0.1 mT (rms) and 10 Hz (Raoz1).

H, and M is proportional to it. At high magnetic fields, however, the magnetization deviates from linearity and begins to saturate, that is, to approach a constant value as H increases. In addition, the magnetization can exhibit hysteresis, as explained below. Substitutions of several rare earths (R) in $RBa_2Cu_3O_{7-\delta}$-type compounds produced normalized FC magnetization $M(T)/M(T = 0)$ versus temperature T/T_c curves that were similar to those of superconducting lead (Namzz).

7. Magnetization Hysteresis

Many authors have observed hysteresis in the magnetization of superconductors, which means that the magnetization depends on the previous history of how magnetic fields were applied, and several examples will be described (e.g., Degro, Larb1, Marcu, Ousse, Pryz1, Sebek, Toku1, Xiao2).

Figure VIII-9 shows several hysteresis loops at 4.5 K at low field in the milli-tesla (10 G) range. The larger the magnetic field excursion, the more the loop is elongated horizontally (i.e., the diamagnetic coercivity is increased).

Figure VIII-10 shows the magnetic-field variation of the magnetization up to 110 mT (1100 G) at 4.8 K (Takag, see also Hikam). Applying the magnetic field to the ZFC sample gradually increases the magnetization until the lower critical field H_{C1} is reached at ≈ 40 mT (400 Oe), and beyond this point flux penetrates and the magnetization begins to decrease gradually. The field-cooling curve starting at a field of $\approx 3\ H_{C1}$ lies considerably below the initial ZFC one. This large hysteresis is indicative of flux pinning.

Figure VIII-11 presents several hysteresis loops at high field, in the tesla (10 kG) range (Sunzz). A high-field loop tends to be linear at high temperatures, and as the temperature is lowered it increases in area and is rotated in the M–H plane, as shown.

The magnetization hysteresis loops of LaSrCuO prompted the suggestion that the sample consists of superconducting grains coupled weakly by a network of Josephson contacts (Renke). At low fields macroscopic shielding is established by the Josephson currents. When the shielding currents exceed the critical Josephson values, flux can penetrate into the regions between the grains. At higher fields each grain is screened individually, and hence the shielded volume is reduced.

8. Flux Quantization

An rf SQUID magnetometer was used to observe flux quanta in $(Y_{0.6}Ba_{0.4})_2$-CuO_4, and Fig. VIII-12 gives a plot of the detector output showing integral numbers of flux quanta jumping in and out of the ring (Gough). Their results agreed within 3% with the known value of the flux quantum $\Phi_0 = hc/2e$. The factor of 2 in the denominator confirms that the carriers in the superconducting state are pairs of electrons, presumably Cooper pairs.

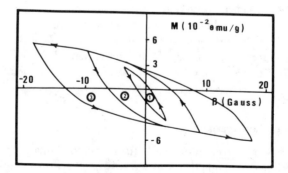

Fig. VIII-9. Very-low-field hysteresis loops at 4.5 K for a $(La_{0.9}Sr_{0.1})_2CuO_4$ sample cycled up to 20 G (2 mT) (Marcu).

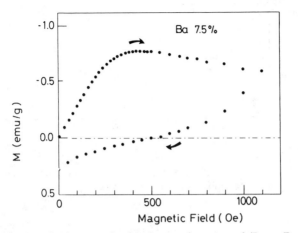

Fig. VIII-10. Medium-field magnetization hysteresis curves of $(La_{0.925}Ba_{0.075})_2CuO_4$ at 4.8 K for fields up to 1100 G (0.11 T) (Takag).

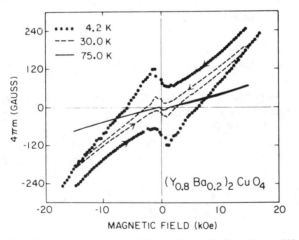

Fig. VIII-11. High-field hysteresis loops of $(La_{0.8}Ba_{0.2})_2CuO_4$ at three different temperatures over the range ± 20 kG (± 2 T) (Sunzz).

9. Comparison with Other Measurements

Magnetic susceptibility data ordinarily correlate well with resistivity measurements on the same sample, as was explained in Section III-D. In a typical narrow transition the resistivity drops sharply to zero at a temperature slightly below the onset of the susceptibility or magnetization transition, as shown in Fig. III-1. When the transition is broad the χ versus T and ρ versus T curves will overlap considerably. Some articles with susceptibility and resistivity curves for the same samples are, for LaSrCuO (e.g., Cavaz, Chuzz, Odazz, Razav, Reeve, Sampa,

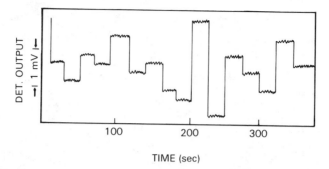

Fig. VIII-12. Output of a rf-SQUID magnetometer showing small integral numbers of flux quanta jumping in and out of the ring (Gough).

Yoshi), and for YBaCuO (e.g., Grant, Maple, Takab, Taka1, Yangz). Figure III-5 shows a specific heat measurement compared with resistivity and suscepti-bility data (Juno1, Kita1).

The reduced mass ratio $m_*/m = 4$ of the effective mass m_* to the free-elec-tron mass m was determined from susceptibility data (Green). This ratio may also be estimated from magnetization measurements.

C. CRITICAL FIELDS

The lower and upper critical fields, H_{C1} and H_{C2}, respectively, defined in Fig. III-8, are related to the thermodynamic critical field H_C through the Ginz-burg–Landau parameter κ, which is the ratio Λ/ξ of the penetration depth to the coherence length (Section III-G-5,7), through the expressions (Forga, Grant, Salam, Takag, Tinkh)

$$H_{C1} = \frac{H_C \ln \kappa}{\sqrt{2}\,\kappa}$$

(VIII-3)

$$H_{C2} = \sqrt{2}\,\kappa\, H_C$$

(VIII-4)

Multiplying these two together gives for H_C

$$H_C = \left[\frac{H_{C1} H_{C2}}{\ln \kappa}\right]^{1/2}$$

(VIII-5)

When anisotropy is taken into account, expressions similar to those given above can be obtained for the parallel and perpendicular components of the critical fields (Tanak, Wort1).

When the applied magnetic field is less than the lower critical field H_{C1}, an ideal Type II superconductor acts like a perfect diamagnet, and no magnetic flux penetrates it. When the applied field exceeds the lower critical field, magnetic flux begins to penetrate and the imaginary term χ'' of the susceptibility appears, which gives rise to hysteresis loss.

The ZFC moment of $(La_{0.9}Sr_{0.1})_2CuO_4$ plotted against the applied field on Fig. VIII-13 reaches a maximum at 30 mT, which is approximately the lower critical field H_{C1} (Male1; see Mulle). The upper critical field is well beyond the highest field used, which was 4.5 T, as shown on the inset to the figure. The ratio of the demagnetizing field to the applied field for $H \ll H_{C1}$ was 97.5, 95, and 81.5% for applied fields of 1, 10, and 100 μT, respectively. Corrections of 7–14% were made for demagnetization effects, but no correction was made for the residual field of 0.1 μT. The normalized lower critical field $H_{C1}(T)/H_{C1}(0)$ of LaSr* has the temperature dependence shown in Fig. VIII-14 (Renke). Strong coupling predicts a marked plateau in $H_{C1}(T)$ at low temperatures, which is most pronounced for large electron–phonon coupling (Ramme).

The Pauli spin paramagnetism of the normal state can set a theoretical maximum value for H_{C2} called the Pauli limit which is given by the expression

$$H_{C2} \approx 1.83\,T_c \qquad\qquad \text{(VIII-6)}$$

This limit is 73 T for $T_c = 40$ K and 165 T for $T_c = 90$ K. Plots of H_{C2} versus T generally have initial slopes dH_{C2}/dT near the value -1.8 T/K, and a typical plot of this type shown in Fig. VIII-15 has a slope of -1.3 T/K for data obtained

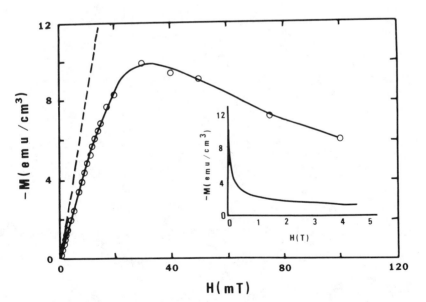

Fig. VIII-13. Zero-field-cooled magnetization of annealed $(La_{0.9}Sr_{0.1})_2CuO_4$ at 5 K, in fields up to 4.5 T. The maximum occurs near the lower critical field H_{c1} of 30 mT. The dashed line indicates complete diamagnetic shielding (Male1).

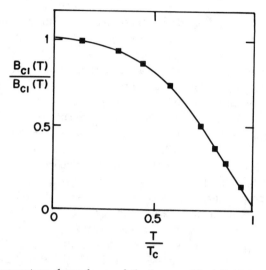

Fig. VIII-14. Temperature dependence of the lower critical field $H_{c1} = B_{c1}$ of LaSr*
plotted as $B_{c1}(T)/B_{c1}(0)$ versus T/T_c where $B_{c1}(0) \cong 18$ mT, $T_c \cong 37$ K (Renke).

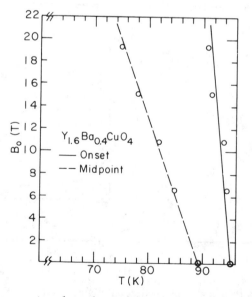

Fig. VIII-15. The temperature dependence of the upper critical field $H_{c2} = B_0(T)$ determined by the onsets (right) and resistance midpoints (left) (Orla1).

at the midpoint of the resistivity transition, which is in the range of typical values shown in Table VIII-2. Data obtained at the onset of the transition give much too high a slope (-5 T/K). Figure I-2 shows a similar plot for several superconductors. Determinations of H_{C2} of LaSr* by magnetoresistance were found to give values about twice as large as those determined by inductive measurements (Renke). The upper critical field was calculated for the LaSrCuO and LaBaCuO systems, and the results may explain the discrepancy between the BCS theory and experimental data (Xuzz1; Xuzz2).

Quoted upper critical fields are usually given for 4.2 K or for extrapolations to 0 K, while the values of technological interest are the 77-K ones. For TmBa$_2$-Cu$_3$O$_{7-\delta}$ the 0-K and 77-K values are 99–175 T and 36 T, respectively (Neume), while the values > 250 T and 80 T, respectively, were reported for YBa* (Ouss2). Calculated upper critical fields for $(La_{0.925}Sr_{0.075})_2CuO_{4-\delta}$ with $T_c = 38$ K ranged from 140 to 21 T at the temperatures 0, 4.2, and 27 K under various assumptions for the electron–phonon coupling constant ($\lambda = 0, 1, 2$) and the spin-orbit-scattering parameter (0 or ∞) (Orla2, Orla3).

Upper critical fields $H_{C2}(T)$ and anisotropy limits were determined for the series of rare-earth-substituted compounds RBa$_2$Cu$_3$O$_{7-\delta}$ where R = Nd, Eu, Gd, Dy, Ho, Er, and Tm (Orlan, Kwok1, Kwok2). We see from Fig. VIII-16 that all of the compounds have H_{C2} values near 160 T, the derivatives $-dH_{C2}/dT$ at $T = T_c$ cluster about 2.8 and the anisotropy $H_{C2\parallel}/H_{C2\perp}$ was less than 12.

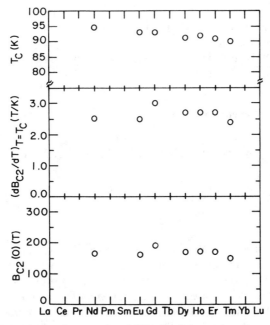

Fig. VIII-16. Dependence of properties of RBa$_2$Cu$_3$O$_{7-\delta}$ specimens versus the rare earth element R. Top, critical temperature T_c; center, $(dB_{c2}/dT)_{T=Tc}$; bottom, calculated $B_{c2}(0)$ based on data above (Orlan).

Table VIII-2 lists the critical fields H_{C1}, H_C, and H_{C2} for various superconductors, and in some cases the derivatives dH_{C1}/dT and dH_{C2}/dT at T_c are given. We see from the table that YBaCuO samples tend to have higher H_{C2} fields than LaSrCuO ones. Critical-field measurements were employed to work out the magnetic phase diagram of the Type II superconductor $YBa_2Cu_3O_{6.9}$.

The magnetic and transport properties of LaSr* at 4.2 K are interpreted as those of a granular Type II superconductor with strongly superconducting grains coupled via weak superconducting links (Senou). Five critical fields were identified: (a) $H_{C1}^W \approx 0.1$ mT and $H_{C1}^G \approx 30$ mT are the lower critical fields for the weak links and grains, where vortex lines begin to penetrate each, respectively; (b) $H^W \approx 1.1$ mT and $H^G \approx 30$ mT correspond to the onset of Abrikosov negative energy behavior in the weak links and grains, respectively; and (c) the usual upper critical field $H_{C2} \approx 50$ T, where the superconductivity of the grain ceases. The field H^G could also be interpreted as the Almeida–Thouless spin-glass line (Mulle). The following regions of reversible (rev) and irreversible (irrev) behavior were found for the M vs H hysteresis loops:

$$0 < \text{rev} < H_{C1}^W < \text{irrev} < H^W < \text{rev} < H_{C1}^G < \text{irrev} < H^G < \text{rev} < H_{C2}$$

Critical-current measurements were correlated with these regions (Senou).

D. TEMPERATURE-DEPENDENT MAGNETISM

Diamagnetism is an intrinsic characteristic of a superconductor (Section VIII-A), but the copper oxide superconductors also exhibit other types of magnetic behavior due to the individual and collective properties of the rare-earth and transition ions that they contain.

1. Temperature Dependence

The susceptibility of a typical oxide superconductor above the transition temperature has a constant contribution χ_0 and a Curie–Weiss term

$$\chi = \chi_0 + N\mu^2/3k_B\,(T + \theta) \qquad \text{(VIII-7)}$$

$$= \chi_0 + C/(T + \theta) \qquad \text{(VIII-8)}$$

where C is the Curie constant and θ is the Curie–Weiss temperature. Often χ_0 may be regarded as Pauli-like, that is, caused by the degenerate free carriers. The positive θ characteristic of oxide superconductors is indicative of antiferromagnetic coupling, while a negative sign would correspond to ferromagnetic coupling. Below T_c the large diamagnetism overwhelms the much smaller terms of Eq. VIII-7 and they become difficult to detect. Nevertheless, some workers have observed the paramagnetism below T_c. For example, in $(La_{0.6}Sr_{0.4})_2CuO_4$

the Curie–Weiss law was found to be obeyed above $T_c = 35$ K and below 7 K where the diamagnetism become saturated (Yoshi).

2. Diamagnetism

An ideal Type I superconductor is a perfect diamagnet with a volume suscepti-bility equal to $-1/4\pi$ (emu/cm³). Experimentally determined susceptibilities must be corrected for the demagnetization factor, which depends on the shape of the specimen.

3. Paramagnetism

Paramagnetism arises from noninteracting or weakly interacting localized mag-netic moments. The temperature dependence is given by the Curie law, which is a special case of the Curie–Weiss behavior of Eq. (VIII-8) with $\theta = 0$.

Oxide materials in which magnetic rare earths replace lanthanum or yttrium provided linear plots of $1/\chi$ versus T above T_c, as shown by the solid curves in Fig. VIII-17, indicating a paramagnetic behavior. For some compounds the temperature-independent term χ_0 of Eq. VIII-7 is zero (Kago1, Tara4). Vacuum annealing the samples destroyed the superconductivity and gave linear Curie–Weiss plots below T_c, shown by the dashed curves on the figure, which provided more precise values of θ. The magnetic moments μ were very close to the values $g(J(J + 1))^{1/2}$ expected for rare-earth ions, where g is the g-factor, J is the total angular momentum quantum number (Mapl1, Tara4), and the sign of θ indi-cates that the rare-earth ions interact antiferromagnetically. These results sug-gest the nearly complete decoupling of the magnetic and superconducting lay-ers, indicating that the superconductivity is anisotropic (Thom2).

The magnetic susceptibility of La_2CuO_4 is weakly paramagnetic ($\approx 10^{-7}$ emu/g) with a cusp at 250 K (Uchi2). The cusp may be indicative of an elec-tronic instability and possibly the formation of a SDW. High magnetic fields (4.5 T) enhanced the cusp.

The series of compounds $YBa_2(Cu_{0.9}M_{0.1})_3O_{8-\delta}$, where M is a first-transition series element, produced the resistivity behaviors of Fig. VII-12 (Xiaoz, Xiao1). The susceptibility above T_c obeyed Eq. (VIII-7) with an effective magnetic mo-ment given by

$$\mu_{eff}^2 = 0.1\,\mu_M^2 + 0.9\,\mu_{Cu}^2 \qquad \text{(VIII-9)}$$

where μ_M and μ_{Cu} are the magnetic moments of the M and Cu atoms, respec-tively. For each sample the Curie constant C and the Curie–Weiss temperature θ were determined from the temperature-dependent data above T_c. We see from Fig. VII-13 that the depression of T_c correlates with the size of the magnetic moment of the substituted transition ion, the larger the moment, the lower the T_c value. Others have reported similar results (Maen2, Osero).

TABLE VIII-2. Lower (H_{c1}), Thermodynamic (H_c), and Upper (H_{c2}) Critical Fields and Their Temperature Derivatives[a]

Material	H_{c1} (mT)	H_c (T)	H_{c2} (T)	$-dH_{c1}/dT$ (mT/K)	$-dH_{c2}/dT$ (T/K)	Ref.
LaSr*	15		≥50		2.13	Aeppl
			64		2.5	Capon
	5.2	0.3	4.3		1.75	Decro
			71		2.7	Finne
					1.5	Kobay
					2.2	Kwokz
	38		58		2.2	Orla2
	20		50	0.8	2.0	Renke
	40	~0.5				Takag
LaSr(0.05)			45		2.2	VanB1
LaSr(0.096)					1.6	Kwokz
LaSr(0.1)	0.06		70		1.8	Nakao
	8–10	0.4				Blaze
			10–15		1.8	Cavaz
	34		47		1.6	Green
			36		1.7	Kobay
						Kwokz
	30	2	10			Male1
	50	0.8	50			Uchid
LaSr(0.3)			23		1.8	Mura1
LaSr(0.3)			76		0.92	Ihar1
LaBa*	1.4	~0.4				Blaze
			≥36			Capon
					1.78	Takag
LaSr* with Ca	40		45		1.8	Kobay
LaSr with Nd					1.3	Kwokz
$(La_{0.89}Ba_{0.11})CuO_{4-\delta}$				43	1.8	Koba1
$La_{1.1}Sr_{0.6}CuO_{2.9}$	~50		>10			Tokum
YBa*	~2	~0.1	~30		4.6	Bezi1
	0.3–2					Blaze

Compound						Reference
	700				~4.5(∥)	Chau1
	40	0.96			~2.9(⊥)	Chau1
	60		100		0.9	Grant
			60			Felic
			80		1.1	Larb1
			250		1.3	Neume
			120			Orla3
		0.8–1.3	200		1.9	Ouss2
						Panso
			280		4.2	VanB1
			140(∥)			VanB1
			29(⊥)		2.3(∥)	Wort1
					0.46(⊥)	Wort1
$YBa_2Cu_3O_{6.9}$	150				0.7–1.0	Zuozi
$YBa_2Cu_3O_{6.9}$	89				1.3	Babic
$Y_{1.2}Ba_{1.8}Cu_3O_{7-\delta}$	50	1		0.7		Cava1
$Y_{1.8}Ba_{1.2}CuO_{7-\delta}$	30					Drumh
$YBa_3Cu_4O_{9-\delta}$			110		1.79	Mura1
YBa*	0.05					Blaze
	0.8					Blaze
	53(∥)					Ding1
	520(⊥)					Ding1
	130(∥)				0.65	Songz
	70(⊥)				~1.0	Songz
$(Y_{0.6}Ba_{0.4})_2CuO_{4-\delta}, \delta \geq 0$			80		1.3	Wuzzz
$(Y_{0.8}Ba_{0.2})_2CuO_{4-\delta}$			80		1.3	Orla1
$Y_{0.9}Ba_{0.1}CuO_{3-\delta}$	10					Taka1
$Y_{0.02}Ba_{0.8}CuO_{4-\delta}$	50					Mehra
$Y_{0.4}Ba_{0.6}CuO_{3-\delta}$	100–200					Toku1
EuBa*			190(∥)			Hikit
			45(⊥)			Hikit
					3.0	Obrad
SmBa*	17				0.7	Neume
TmBa*			≥5.5			
YBa*	150		99–175		2.8	Zuozi

[a] The notation used is: $(La_{1-x}M_{x/2})_2CuO_4 = LaM(x)$; $LaM(0.075) = LaM*$; $YBa_2Cu_3O_{7-\delta} = YBa*$.

157

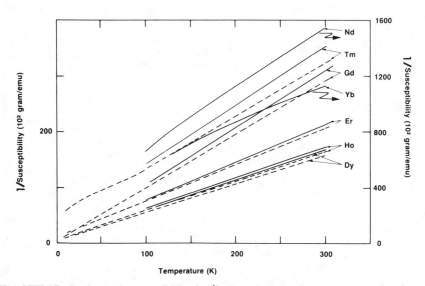

Fig. VIII-17. Reciprocal susceptibility (χ^{-1}) plotted versus the temperature for the superconducting as-prepared (——) and the nonsuperconducting vacuum-annealed (---) rare earth (R) substituted $RBa_2Cu_3O_{7-\delta}$ series over the temperature ranges 100–300 K and 4–300 K, respectively, in a field of 1 T (Tara4).

Data from the substitution of paramagnetic Ni^{2+} and diamagnetic Zn^{2+} ions M for copper ions in $(La_{0.925}Sr_{0.075})_2Cu_{1-x}M_xO_4$ and $YBa_2(Cu_{1-x}M_x)_3O_{7-\delta}$ (Tara3, Tara6, Tara8) are discussed in Section VII-F-4. With $x = 0.008$, Meissner effects of 58 and 42% were observed for Ni and Zn, respectively, supportive of bulk superconductivity. The susceptibility of the Ni compound with $x = 0.1$ fit the Curie–Weiss law with a moment consistent with the valence state Ni^{2+}.

4. Antiferromagnetism

The compound La_2CuO_4 is an antiferromagnet below the Néel temperature of 195–290 K with the copper spins ordered in the CuO_2 planes (Hutir, Malet, Mitsu, Shira), as shown on Fig. VIII-18 (Vakn1). There are 0.4 (Yang4), 0.5 (Vakni, Uemur), or 1.1 (Yamag) Bohr magnetons (μ_B) per copper ion, which is less than the Cu^{2+} spin-only value of 1.73 μ_B. The antiferromagnetic order and Néel temperature are both observed to depend on the heat treatment (Yama1). Vacancy-induced narrow bands near the Fermi level have been suggested as responsible for the antiferromagnetism of nonstoichiometric La_2CuO_4 samples (Kaso1).

Neutron studies show that the compound $GdBa_2Cu_3O_7$ 98.1%, enriched with the low neutron absorbing isotope ^{160}Gd, undergoes antiferromagnetic alignment with the Gd moments ordering below 2.2–2.6 K in the manner shown in Fig. VIII-19. The ordered moment is 7.4 μ_B (Paul1, Farno, Kado1, Linzz,

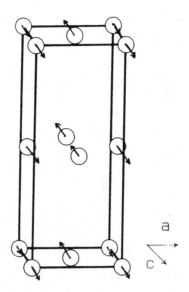

Fig. VIII-18. Proposed magnetic spin structure of antiferromagnetic La_2CuO_4. Only copper atoms are shown. The unit cell of this figure is rotated by 45° around the usual c axis in accordance with Fig. VI-3, and the b axis of this figure is the usual c axis (Vakai).

McGui, Vande). At very low temperatures (below 0.5 K) the Er moments in $ErBa_2Cu_3O_7$ become ordered in a two-dimensional array where chains of spins are aligned ferromagnetically with adjacent chains coupled antiferromagnetically (Lynnz). The compound $DyBa_2Cu_3O_7$ also orders antiferromagnetically below 1 K (Goldm). The reduced compound $YBa_2Cu_3O_6$ exhibits antiferromagnetism up to 500 K with $\mu = 0.48\ \mu_B$ along the tetragonal c axis, and with antiferromagnetic coupling both between two CuO_2 layers of the unit cell and in the basal plane (Rossa). There is some evidence for magnetic ordering in $Y_2Cu_2O_5$ (Trocz). Long-range ordering of the magnetic ions has also been reported in a wide variety of other ceramic samples (Hozzz, Leebw, Park1). Specific heat data support the presence of antiferromagnetic ordering at very low temperatures (cf. Section IX-F-3).

Based on the deGennes model with degenerate orbitals, it has been argued that antiferromagnetism prevails if there is a nonzero crystal field (Ambeg). Band-structure calculations have shown that Fermi surface nesting can be associated with the antiferromagnetic transition (Gorba). The two-dimensional magnetic state has been called a quantum spin fluid because the spins order two dimensionally over distances exceeding 20 nm and the structure factor is dynamical in character (Shira). Neutron studies suggest a spin-fluid configuration at high temperature and an antiferromagnetic state at lower temperatures. The presence of magnetic ordering supports models of superconductivity which invoke magnetic interactions for the pairing of charge carriers (Emer1, Hirsc).

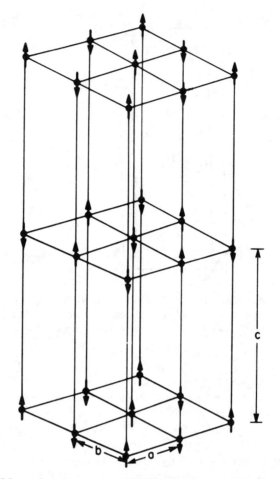

Fig. VIII-19. Magnetic spin structure of $GdBa_2Cu_3O_7$ at 1.5 K. The antiferromagnetically ordered Gd ion spin directions are shown (Paul1).

5. Glassy State

Oxide superconductors exhibit glassy features such as irreversibility, flux trapping, vortex creep, metastability, magnetic and resistive hysteresis, and time-dependent remnant magnetism (Blaze, Crabt, Crone, Datt1, Ebner, Geshk, Giova, Godar, Krusi, McHen, Minam, Tuomi, Xiaoz). Figure VIII-20 shows the regions on the H versus T plane where normal, superconducting, and glass-like behavior are observed (Yeshu), with the superconducting and glass-transition lines shown. The ZFC branch of the susceptibility is metastable while the FC one is more stable. Müller et al. discussed flux trapping and the superconductive glass state of $(La_{1-x}Ba_x)_{2-y}CuO_{4-\delta}$ (Deuts, Mulle, Ravea) in which the ZFC diamagnetism is metastable. They argue that the short coherence length causes

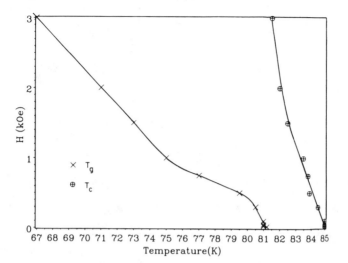

Fig. VIII-20. The field dependence of the glass temperature (T_g) and of the supercon-ducting temperature T_c lines for $YBa_2Cu_3O_7$. The specimen is a superconducting glass on the lower left of the T_g line, a conventional Type II superconductor in the center between the T_g and T_c lines, and a normal metal to the right of the T_c line. The solid lines are a guide for the eye (Yeshu).

weakening of the pair potential at surfaces and interfaces, and this is considered responsible for the existence of internal Josephson junctions at twin boundaries, which are considered the origin of the glassy state.

Relaxation of the isothermal dc magnetization in ZFC LaSrCuO and LaBa-CuO was discussed in terms of spin glasses, with the magnetization decaying logarithmicly as a function of the time, $M(t) \approx \ln(t/t_o)$, over an interval of four decades (Motaz), as shown in Fig. VIII-21. Time-dependent effects have also been studied in YBaCuO. Figure VIII-22 shows the growth of the magnetization with time in the vortex state after ZFC (Drumh). It reaches 93% of its final value in 10 sec and 99% in 1 min. An algebraic expression for the time dependence of the trapped field (i.e., $M \approx t^{-\alpha}$), has been reported (Tjuka) for YBa$_*$ ceramic tubes with $0.005 < \alpha < 0.034$ in the temperature range between 31 and 77 K.

The finite width and the hysteresis of the resistance versus temperature curves have been interpreted as indicative of spin-glass behavior. To show this log R was plotted against $(T - T_{KT})^{-1/2}$ and the result was a straight line, where T_{KT} was defined as the Kosterlitz–Thouless temperature selected to optimize the lin-earity of the plot (Sugah). Only samples with narrow transition regions ($\Delta T \approx$ 1.5 K) exhibit this behavior. In another work (Carol), the temperature $T(0)$ of the onset of hysteretic behavior in zero field (0.02 mT) on the R versus T plot was determined, and then this onset temperature $T(H)$ was measured in magnetic fields between 0.5 mT and 0.25 T. It was observed that $H \approx [(T(H)/T(0))]^\beta$ where β was found to be 1.49, indicative of the existence of a superconducting glass state with an exponent close to the Almeida–Thouless value of 1.5.

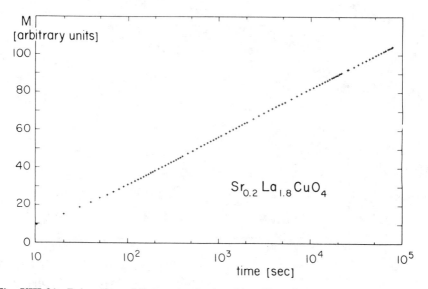

Fig. VIII-21. Relaxation of the magnetization M at $H = 0$ as a function of time for powdered $(La_{0.9}Sr_{0.1})_2CuO_4$ at 4.2 K (Motaz).

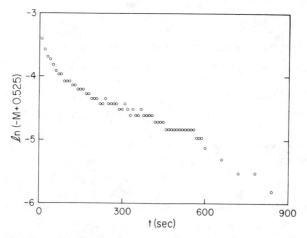

Fig. VIII-22. Semilogarithmic plot of the change in magnetization versus time of YBa*. The temperature was lowered from above T_c in zero field to 41 K, then the magnetic field of 0.5 kG (50 mT) was applied at $t = 0$ (Drumh).

Magnetic glassy behavior was identified in $La_2CuO_{4-\delta}$ by the existence of a difference between the FC and the ZFC susceptibility below 270 K (Toku2). This difference was fairly constant between 115 and 230 K.

We have discussed various properties of superconductors, such as irreversibility, hysteresis, and time-dependent effects, in terms of the presence of a glassy state. Many of these properties have been observed in the older Type II superconductors, and the term "glass" does not even appear in the index of standard works on these subjects (e.g., Parks, Rosei, Tinkh, Vonso). An alternate viewpoint is that what is now called glassy behavior may be explained by pinning effects in ordinary Type II superconductors (Preje).

E. MAGNETIC RESONANCE

Several types of magnetic resonance (Pool1), namely, electron spin resonance (ESR or EPR), nuclear magnetic resonance (NMR), nuclear quadrupole resonance (NQR), Mössbauer resonance, and muon spin resonance (μSR), have been used to study superconductors, and we will discuss some of these results.

Magnetic resonance measurements are made in fairly strong magnetic fields, typically ≈ 0.33 T for ESR and ≈ 10 T for NMR, which are between the lower and upper critical fields H_{C1} and H_{C2}. At these fields most of the external magnetic flux penetrates into the sample, so the average value of H inside is not much less than H outside.

1. Electron Spin Resonance

Electron spin resonance detects unpaired electrons in transition ions, especially those with odd numbers of electrons such as Cu^{2+} ($3d^9$) and Gd^{3+} ($4f^7$). Free radicals such as those associated with defects or radiation damage can also be detected. Oxide superconductors exhibit a conventional $g \approx 2.1$ powder pattern and a low-field nonresonant absorption. The former will be described here and the latter in the next section.

Several groups have observed an ESR powder pattern signal of the type presented in Fig. VIII-23, which is characteristic of Cu^{2+} in $YBa_2Cu_3O_{7-\delta}$. In a typical case g ranged from 2.05 to 2.27 (Osero), the line exhibited no anomaly in passing through T_c (Durny, Oser1), and it decreased (Shalt) or increased (Mehra) in intensity below T_c with a strong overall sample dependence. The signal exhibited a decrease in g-factor from 2.22 to 2.07 between 4 K and $T_c = 90$ K (Benak). Several transitions in the Cu^{2+} ESR signal (96, 78, 30, 10 K) were observed, with the final vanishing of paramagnetism below 10 K, and it was found that $g_\parallel$ (300 K) $-$ $g_\parallel$ (T) varied with the temperature as $(T - T_c)^{-1.1}$ where $g_\parallel$ (300) $= 2.2167$ (Shriv). The perpendicular component $g_\perp$ was also temperature dependent. A related strong pseudocubic ESR signal from Cu^{2+} was observed in unannealed $Y_2Cu_3O_{7-\delta}$ with $T_c \approx 40$ K, which disappears in a sample with bulk superconductivity at $T_c \approx 90$ K (Mehr1).

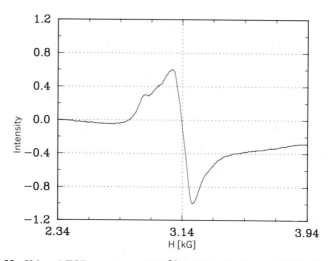

Fig. VIII-23. X-band ESR spectrum of Cu^{2+} in $YBa_2Cu_3O_{7-\delta}$ at 77 K. (Provided by T. Mzoughi and M. Mesa.)

It has been established that the anisotropic Cu^{2+} signal just described does not arise from the superconducting phase $YBa_2Cu_3O_{7-\delta}$, but rather from a non-superconducting fraction (Durny, McKi1, Osero) such as the green phase Y_2BaCuO_5 (Bowde, Jone1, McKi1, McKi2, Owens). Other possibilities are conduction electrons (Shalt), an Anderson lattice with an unusually high Kondo temperature (resistivity minimum associated with localized moments) (Mehra), a hole transferred from copper to oxygen (Benak), copper ions near grain boundaries (Gonca), and Cu^{2+} ions in special coordination (Mehr2).

An anisotropic ESR signal has been reported in $LaSr_2CuO_4$ and $(La_{1-x}Sr_x)_2$ $CuO_{4-\delta}$, and the signal in the latter was attributed to the presence of a trace of the former phase (Davis). The yttrium, bismuth, $(Bi_4Ca_2Sr_4Cu_4O_{16}$, McKi1) and probably thallium $(Tl_2Ca_2BaCu_3O_{9+\delta})$ superconductors are all ESR silent as far as the Cu^{2+} signal is concerned. Antiferromagnetic CuO, which has strong superexchange above $T_N = 230$ K, is also ESR silent, and copper acetate monohydrate exhibits a triplet state spectrum from Cu^{2+} pairs. Superexchange or "resonant valence bonds" could explain the ESR silence of the Cu–O–Cu layered superconductors (McKi1).

The Gd^{3+} in $YBa_2Cu_3O_{7-\delta}$ produced an asymmetric spectrum 130 mT wide with $g = 1.97$ at room temperature, which narrowed somewhat and considerably increased in intensity below T_c (Mehr2). Samples of $EuBa_2Cu_3O_{7-\delta}$ undoped or doped with 3d ions (Cr, Mn, Fe, Ni, Co, Zn) sometimes exhibited the Cu^{2+} powder pattern, but all of them showed the nonresonant absorption described in the next section (Oser1). Spectra from Gd- and Eu-doped YBa∗ have been reported (Osero).

The magnetic field inside a superconducting sample was probed by placing a free radical marker at the face of a specimen normal to the magnetic field direction and another at a face of the specimen that is parallel to the external magnetic field. In the superconducting state the two markers experience different local magnetic fields, so the resonant positions of the lines shift in the manner shown in Fig. VIII-24 (Quagl).

2. Nonresonant Microwave Absorption

The output of an ESR spectrometer has a very high noise level when a superconducting sample is examined below T_c, and in addition a nonresonant microwave absorption signal appears at low fields, far below the $g = 2.0$ point. We show later (Section X-D-2) that irradiating a Josephson junction with microwaves induces an oscillating voltage which depends on the microwave power and frequency (Weng1; see Aberl, Section X-D-2).

A number of workers have observed low-field microwave absorption in the LaSrCuO (Blaze), $YBa_2Cu_3O_{7-\delta}$ (e.g., Bhatz, Cohe2, Durny, Hagen, Khach, Shriv, Rama1, Retto, Sastr, Shriv, Stank), $Bi_4Sr_4Ca_2Cu_4O_{16}$ (McKi1), and TlBaCaCuO (Mzoug) superconductors. The low-field absorption from YBa_2-$Cu_3O_{7-\delta}$, shown in Fig. VIII-25, exhibits superimposed, closely spaced coherent fluctuations below 87 K that were attributed to Josephson oscillations (Bhatz, Stank, Weng1), perhaps involving flux slippage at the Josephson junctions (Rama1). The oscillation level increases with decreasing magnetic field, reaching a maximum near 5 mT (50 G). At low temperatures, large fluctuations occur which decrease with increasing temperature, becoming incoherent at 87 K and disappearing above this temperature.

Figure VIII-26 shows the hysteresis of the low-field absorption in TlBaCa-CuO, with the field-decreasing trace remaining below the field-increasing one,

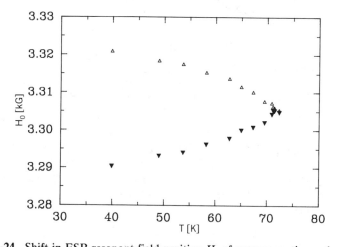

Fig. VIII-24. Shift in ESR resonant field position H_0 of paramagnetic markers located on the side and end of a YBa* sample. The lines split and move apart as the temperature is lowered below T_c. (Provided by T. Mzoughi, M. Mesa, and E. Quagliata.)

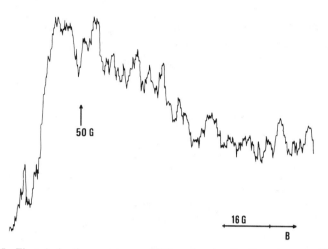

Fig. VIII-25. First derivative spectrum of YBa* showing the Josephson oscillations superimposed on the low-field absorption. The oscillations are evident below 87 K where $T_c = 91.5$ K (Stank).

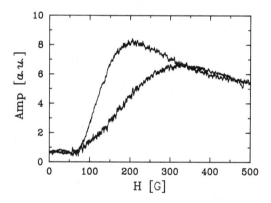

Fig. VIII-26. Low-field microwave absorption for TlBaCaCuO showing ZFC spectrum recorded with increasing field (upper curve) and the reverse scan toward lower fields (lower curve). (Provided by T. Mzoughi, M. Mesa, and E. Quagliata.)

and reaching a peak at a higher field (Mzoug, see also Rama1). The lack of reversibility of the field-increasing and field-decreasing responses was suggested to indicate a superconducting glass state (Blaze). The field H_{MAX} of the low-field absorption peak was used to estimate the average radius r_L of the superconducting loops from the expression

$$\pi r_L^2 = \Phi_0/2H_{MAX} \qquad (VIII\text{-}10)$$

Various samples gave loop radii in the range 0.6–2.5 μm (Blaze).

The absorption of microwaves between 8 MHz and 10 GHz is very sensitive to the temperature, particle size, magnitude of the magnetic field, and the presence of oxygen (Bhatz). The Joule loss appears at T_c and is a small fraction of its high-temperature value at 4.2 K. In general, the surface impedance is observed to be similar to that of weakly linked granular superconductors. The low-field signal was attributed to the transition from the Meissner state to the mixed state (Retto). The behavior for both LaSrCuO- and YBaCuO-type materials has been described by non s-wave BCS in the very dirty limit (Sridh), while others (e.g., Harsh) interpreted their results in terms of s-wave BCS or the possibility of disproportionation of Cu^{3+} to $Cu^{2+} + Cu^{4+}$ in adjacent octrahedra (Sastr).

The surface impedance of $(La_{0.9}Ba_{0.1})_2CuO_4$ at 102 GHz was determined by clamping a polished sample to the bottom of a TE_{011} mode cylindrical resonant cavity and measuring the temperature dependence of the width of the cavity mode. The measurements were repeated with a copper end plate and the surface impedance was reported relative to that of copper. A Lorentzian distribution of energy gaps consistent with the BCS expression $E_g = 3.5 \, kT_c$ was found (Beye2).

The magnetic-field-induced microwave loss occurred within a skin depth equal to the London penetration depth $\Lambda = 21$ nm and was proportional to the surface area rather than to the volume of the sintered samples (Porti). The observed absorption was attributed to microwave conductivity loss brought about by the interaction between the superconducting carriers and damped fluxoids (Porti). Radio frequency loss measurements of YBaCuO-type samples at 2 and 7 GHz provided surface impedance determinations through the transition temperature (Migl1, Reago).

3. Nuclear Magnetic Resonance

Nuclear magnetic resonance (NMR) spectra are sensitive to the immediate chemical environment of the nucleus being observed, to the local magnetic field and field gradient, and in the case of quadrupolar nuclei ($I > \frac{1}{2}$), to the crystal field symmetry. Relaxation time measurements determine the efficiency of spin-energy transfer to the lattice. Relaxation rates have been calculated for various types of superconductivity in two-dimensional systems (Hase2). Several potentially useful nuclei with their spins (I) and natural abundances are: ^{63}Cu ($\frac{3}{2}$, 69.1%); ^{65}Cu ($\frac{3}{2}$, 30.9%); ^{87}Sr ($\frac{9}{2}$, 7.0%); ^{89}Y ($\frac{1}{2}$, 100%); ^{135}Ba ($\frac{3}{2}$, 7.0%); ^{137}Ba ($\frac{3}{2}$, 11.3%); ^{139}La ($\frac{7}{2}$, 99.9%); ^{203}Tl ($\frac{1}{2}$, 29.5%); ^{205}Tl ($\frac{1}{2}$, 70.5%); and ^{209}Bi ($\frac{9}{2}$, 100%). The nucleus ^{16}O (0, 99.8%) has no nuclear spin, and so it cannot be observed.

Pulsed NMR of ^{89}Y nuclei were observed in $YBa_2Cu_3O_{7-\delta}$ at 12.2 MHz and 5.9 T in the temperature range from 59 to 295 K (Marke, Maliz). The value of $T_c = 86$ K at 5.9 T was determined by the onset of increased inhomogeneous broadening from a width of 0.31 mT above T_c to 0.71 mT 10 degrees below T_c. The fraction of ^{89}Y detected decreased from 100% above T_c to about 80% at 59 K owing to incomplete rf penetration in the mixed state. The spin–lattice relaxation time T_1 increased below T_c. In another work the room temperature singlet [89]Y NMR line of YBa* unexpectedly split into a doublet at 100 K, possibly due to the development of a CDW or 1d or 2d nesting at the Fermi surface (Krame). Preparation conditions influence the Y site since different [89]Y chemical shifts were observed in slowly cooled, rapidly cooled, and water-exposed YBa* (Brahm). No NMR signals were detected from the 100% abundant [169]Tm and 14% abundant [171]Yb $I = \frac{1}{2}$ nuclei in the analogous compounds TmBa* and YbBa*.

The spin–lattice relaxation time (T_1) measurements of La nuclei in LaSr* increased with decreasing temperature above $T_c = 34$ K and leveled off in the superconducting region. In addition there was an anomalous minimum at T_c in the plot of T_1 versus T (Butau).

Copper NMR was done in conjunction with copper NQR (nuclear quadrupole resonance) (Fuorz, Lutge, Maliz), as described in the next section. The spectra from the chain Cu(t) exhibited axial symmetry and the Cu(m) in the planes showed evidence for disorder (Walst, Warr1). Loss of oxygen degraded the signal from the copper chains. The NMR of [63]Cu provided the energy gap ratio $E_g/kT_C = 1.3$ in $(La_{0.915}Sr_{0.085})_2CuO_{4-\delta}$ (Leez4).

A [207]Pb NMR study of $BaPbO_3$, and end member of the superconducting oxide series $BaPb_{1-x}Bi_xO_3$, was reported (Tsuda).

4. Quadrupole Resonance

A nucleus with spin $I > \frac{1}{2}$ has an electric quadrupole moment, and a list of several quadrupolar nuclei is given at the beginning of the previous section. The crystalline electric fields at an atomic site with symmetry less than cubic split the spin levels in a manner that depends on the site symmetry, and the spacings between them are determined by nuclear quadrupole resonance (NQR). Table V-14 lists the actual point symmetries for the occupied atomic sites in the stoichiometric La_2CuO_4 and $YBa_2Cu_3O_8$ compounds.

The NQR of [139]La in zero magnetic field of the prototype compound La_2CuO_4 has nine resonant lines from 2.4 to 19.3 MHz with the main splittings arising from the noncubic crystal fields, and additional doublet splittings caused by internal magnetic fields that arise from the magnetic ordering of the copper ions that occurs below 240 K (Kitao, Lutge). The spectra permitted the resolution of this internal field into components parallel and perpendicular to the c axis. The doublet splittings are less well resolved in the barium- and calcium-substituted compounds indicating that the internal magnetic fields decrease with alkaline earth doping. The internal field parallel to c is about 35 mT for low barium contents ($\approx 1\%$) in the superconducting region (Kita2). The electric field gradient at the La site also changes on passing from the normal to the superconducting state (Watan).

The copper NQR spectrum of YBaCuO consists of one doublet centered at 21 MHz arising from the Cu(m)–O planes near Y, and another doublet centered at 30 MHz arising from the Cu(t) coppers in the chain layer or basal plane (Maliz,

Riese, Walst, Warr1). The two isotopes ^{63}Cu and ^{65}Cu produce the doublet splitting. The fact that NQR is observable is evidence for oxygen ordering (Maliz). An anomolous discontinuity in the NQR resonant frequency was observed at T_c (Riese). The linewidth strongly depends on the oxygen content. The spin-lattice relaxation rate drops sharply below T_c for both sites. The NQR measurements on YBa* have provided evidence for two pairing energies (Warre), and there was no indication of magnetic order (Furoz).

The LaSr* exhibited only a weak and broad NMR signal, and none at all appeared in La_2CuO_4 owing to the antiferromagnetism (Lutge).

5. Muon Spin Relaxation

The negative muon μ^- acts in all respects like an electron except that it has a mass 206.77 times larger (Poole). The precession of muons in a solid constitutes a microscopic probe of the distribution of local magnetic fields (Budni), and in particular the width of the muon spin relaxation (μSR) signal from a superconductor provides an estimate of this field distribution (Schne) and the penetration depth Δ (Wappl). The μSR signal represents a simple average over different parts of the sample, so the effect of inhomogeneities should be less serious here than in the case of transport measurements like resistivity (Koss2).

The μSR measurements provided the following values of the penetration depth Λ in LaSrCuO: 2000 Å at 10 K (Kossl); 2500 Å at 6 K (Aeppl); and 2300 Å (Wappl); and in YBaCuO: 1065–1400 Å (Harsh). The temperature dependence of Λ in YBaCuO fits BCS quite well, as shown on Fig. III-6. Random internal fields with components perpendicular to H_0 were induced by an applied external field $H_0 < H_{C1} = 15$ mT. The superconducting carrier density was estimated to be $3 \times 10^{21}/cm^3$ (Aeppl, Kossl). The μSR measurements failed to show a Meissner effect, in contradiction to dc results (Gygax).

Antiferromagnetic ordering of the Cu ions in La_2CuO_4 (Budni, Uemur) and of the Gd ions in $GdBa_2Cu_3O_{7-\delta}$ (Golni) was demonstrated by the observation of a μSR signal in zero-applied field.

6. Mössbauer Resonance

Mössbauer resonance measures gamma rays emitted by recoilless isotopes, and their spectra are sensitive to the valence state and to the strength and symmetry of the local electric field (Pool2).

Measurements of ^{151}Eu in $EuBa_2Cu_3O_{7.1}$ established that the europium is trivalent (Boolc, Eibsc, Freim). The thermal displacements of the Eu site are characterized by a Debye temperature $\theta_D = 280$ K (Boolc). The isomer shifts of ^{155}Gd Mössbauer spectra from $GdBa_2Cu_3O_7$ indicate the absence of conduction electron density within the Gd layers (Smitz).

Mössbauer data from ^{57}Fe substituted for Cu in $YBa_2Cu_3O_{7-\delta}$ show low spin Fe^{2+} or Fe^{3+} in the linear chain layer, with magnetic ordering of the Fe ions considerably below T_c (Gomez, Kimba, Quizz, Zhou2). Similar results were obtained for ^{57}Fe spectra from $GdBa_2Cu_3O_{6.9}$ (Tangz).

Mössbauer measurements of ^{119}Sn replacing Cu in $(La_{0.925}Sr_{0.075})_2CuO_4$ probed the Cu site electronic and vibrational behavior, and revealed complex phonon properties and possible phase instabilities above T_c (Giapi). Mode softening occurred near 75 and 170 K. The 0, 5, 10, and 15% replacement of Cu by Sn^{4+} lowered T_c through the range 35, 30, 27, 25 K, respectively, and changed the average copper charge Q as follows: 2.15, 1.95, 1.75, 1.55. Substituting ^{57}Fe at Cu sites of LaSr* rapidly decreased T_c, and below 15 K magnetic ordering was observed for 5% ^{57}Fe (Matyk). Semiconducting $La_{0.5}Sr_{0.5}Cu_xFe_{1-x}O_y$ was also studied by ^{57}Fe Mössbauer (Errak).

F. MAGNETISM AND SUPERCONDUCTIVITY

Throughout this chapter the motivation has been on explaining how magnetism is involved in superconductivity. This is because the existence of the diamagnetic state is a fundamental property of a superconductor. There are other magnetic properties which are not affected by the transformation to the superconducting state, such as the paramagnetism of ions substituted on La or Y sites where they are essentially uncoupled from the interactions involved in sustaining the superconductivity, as was explained in Section VIII-D-3.

In ordinary or low-temperature superconductors the substitution of magnetic ions generally destroys the superconductivity quite strongly, which suggests that in the LaSrCuO and YBaCuO cases the rare-earth ions substituted for La or Y are far enough from the Cooper pairs to ensure that there is only a small interaction between them (Horzz, Hoso1). On the other hand, the substitution of small percentages of zinc or a 3d transition ion on a copper site substantially degrades or destroys the superconductivity (Haseg), as was shown in Section VII-F-4, suggesting that the superconducting pairs are in the Cu–O planes.

The proposal has been made that the antiferromagnetic coupling discussed in Section VIII-D-4 may provide the interaction mechanism for pairing in the ceramic superconductors (Emer1, Frelt, Hirsc, Shira).

IX

PHONON AND SPECTROSCOPIC PROPERTIES

A. INTRODUCTION

This chapter begins with two topics that are central to superconductivity, namely the energy gap E_g and the density of states (DOS) at the Fermi level. Then it introduces a third subject, the electron–phonon interaction, which is the mechanism responsible for superconductivity in most of the lower-temperature materials, and which may play a role in the newer oxide types. This is followed by a discussion of several spectroscopic techniques that provide information on these and other topics. The chapter concludes with a section on specific heat.

B. ENERGY GAP

Two characteristic properties of a superconductor are its energy gap E_g and the ratio E_g/kT_c of this energy gap to the thermal energy kT_c at the transition temperature. The weak coupling BCS model predicts $E_g/kT_c = 3.53$ (Eq. IV-3).

1. Measurement

Energy gaps can be measured by many methods, such as inelastic neutron scattering (Section IX-E-4), microwave surface impedance (Beye2), NMR (Leez4), susceptibility (Poltu), and ultrasonics (Table IX-2), but by far the most common ways to determine them are by IR spectroscopy or by tunneling experiments. Sometimes both measurements are made on the same sample and the tunneling values tend to be larger (Chau1, Ekino, Kirt1, Maeka, Tsuei), possibly because the far IR beam averages over a volume that is about 10 μm thick and up to ≈ 10

mm^2 in area, as well as over a finite region of k space, whereas tunneling probes a much smaller region in both coordinate and k space that is not necessarily typical (Thom3).

It is beyond the scope of this book to describe the details of these two measurement techniques, but we will comment briefly on the types of experimental data obtained with each. Infrared will be treated in this chapter and tunneling in the next.

2. Gap Values and E_g/kT_c Ratios

Figure IX-1 compares the normal state far IR reflectivity of YB$_a$* above and below T_c. The high reflectivity is indicative of the intrinsic high conductivity of the planes. The frequency for the onset of absorption below T_c provides the energy gap, which was close to the BCS ratio $E_g/kT_c \approx 3.5$ (Thom3).

Table IX-1 lists energy-gap values and gap ratios for various oxide superconductor samples. In most cases the measurement technique is noted. Where feasible, midpoint T_c was used to calculate the E_g/kT_c values listed in the table. When investigators report ranges of possible gap values, sometimes the original range is given, and in other cases average values are listed (Kirt1, Kirt3), especially if the range is small. Some of the gap ratios E_g/kT_c are smaller than the BCS value 3.53, but most are larger, typically about 4.5. Figure IX-2 may be used to find E_g/kT_c when E_g and T_c are known. It was drawn using the standard temperature-to-energy conversion factors

$$100 \text{ K} = 8.6164 \text{ meV} = 69.50 \text{ cm}^{-1} \qquad \text{(IX-1)}$$

which provide the $E_g/T_c = 1$ line on the figure. The BCS line drawn for the ratio $E_g/T_c = 3.53$ corresponds to the conditions

$$100 \text{ K} = 30.42 \text{ meV} = 245 \text{ cm}^{-1} \qquad \text{(IX-2)}$$

Figure IX-3 presents the normalized BCS curve for the temperature dependence of the gap, together with some experimental points (Cromm, Schl2) for YBa$_2$Cu$_3$O$_{7-\delta}$.

Table IX-2 compares the energy-gap ratio E_g/kT_c determined by different methods for some selected old and new superconductors, and Fig. IX-4 shows how E_g depends on T_c for a similar selection of superconductors. When one datum point is plotted on the figure for an element it means that only one point was available, not that the precision is high. These results show that: (a) various investigators measured ranges of gaps for the old and new superconductors so the accuracy of individual determinations can be suspect; (b) the oxide superconductors exhibit a much wider spread in E_g values which probably reflects their great diversity in composition, oxygen content, and grain size distribution; and (c) given these conditions, the agreement of the gap ratio E_g/T_c with the

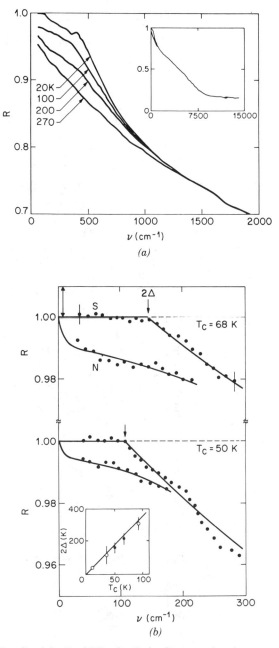

Fig. IX-1. Far IR reflectivity R of $YBa_2Cu_3O_7$ in the normal and superconducting states (a) spectra and (b) replotting of normal (N) and superconducting state (S) data to determine the energy gap. The inset of (b) downpaves data from this work (●) and other workers (○) with the BCS prediction $E_g/\xi T_c = 3.53$ (G. A. Thomas, private communication).

TABLE IX-1. Energy Gap E_g, Energy Gap Ratio E_g/kT_c, and Measurement Method[a]

Material	T_c (K)	E_g (meV)	E_g/kT_c L	Measurement Method	Comments	Ref.
BaPb$_{1-x}$Bi$_x$O$_3$		3.35		Tun		Ekino
LaSr*		10-16	2.9-4.5	IR		Bonnz
	34		0.7-2.7	IR	Mirror reflection	Degio
	36	16-28	≥ 5.2	Tun	PC	Hawle
	36	14	4.5	Tun	STM, polyxtal	Kirt1
			3-6	Tun		Leide
	36	14-18	4.5-5.8	Tun	Rod, BJ	More4
	38.5	24	7	Tun	STM	Panzz
	38.5		2.6	IR	Needle xtals	Suel1
	33		3.5-4	Tun	Pressure varied	VanBe
LaSr(0.1)	~36	8.8	2.8	Tun		Kirt1
	~32	20,30,60		Tun	Film	Naito
	34	52		IR		Ogita
	30	6.5	2.5	IR		Schl1
LaSr(x)	34	14	4.7	Tun		Ekin1

174

	T_c	E_g	E_g/kT_c	Method	Sample	Ref
LaSr(x)	37		5	IR	Powder	Nagas
LaBr*	30	52		IR		Ogita
YBa*	89	27	4.5	Tun	Bulk	Baro2
	90	30–42	3.5	IR		Bonn1
	90	37	4.5	Tun	Epitaxial film	Chau1
	90	30	4.7	IR	Epitaxial film	Colli
	90		3.9	Tun	Polycrys, PC	Cromm
	90		4–5.3	STM	Epitaxial film	Kirt2
		29–43	3.7–5.6		Tun and IR compared	Kirt3
	85	25	3.4	Raman	Thin film	Lyons
	93	39	4.8	Tun	BJ	More5
	34		5	Susc		Poltu
	90	62	8	IR		Schl2
	84	23	3.2	Tun	Pressure varied	VanBe
$YBa_2Cu_3O_{6.9}$	93		2.3–3.5	IR		Thom3
$Y_{0.3}Ba_{0.5}CuO_{4-\delta}$	92	100	13	Tun		Kirkz
$Y_{0.35}Ba_{0.65}CuO_{3-\delta}$	89	18	2.4	Tun	PC	Kochz
$(Y_{0.55}Ba_{0.45})_2CuO_{4-\delta}$	90	31	3.8–4.5	Tun		Ekino

[a] The weak coupling BCS ratio is $E_g/kT_c = 3.53$. The following notation is used: $(La_{1-x}M_x)_2CuO_4 = LaM(x)$; $LaM(0.075) = LaM_*$; $YBa_2Cu_3O_{7-\delta} = YBa_*$; BJ = break junction; PC = point contact; Tun = tunneling; STM = scanning tunneling microscope; Susc = magnetic susceptibility.

175

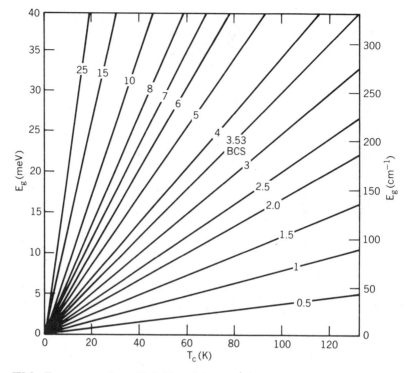

Fig. IX-2. Energy gaps in meV (left) and in cm^{-1} (right) as a function of the critical temperature T_c for various ratios E_g/kT_c (labels of lines) from 0.5 to 25.

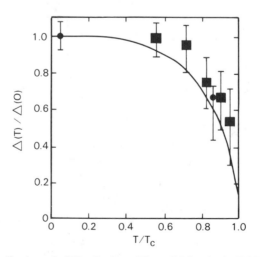

Fig. IX-3. Normalized gap in $YBa_2Cu_3O_{7-\delta}$. The solid line is the BCS gap, and the data shown are from Cromm ($\bullet$) and Schl2 ($\blacksquare$).

TABLE IX-2. Energy Gap Ratio E_g/kT_c for Selection of Superconductors[a]

Material	T_c (K)	IR	Tun	Therm[b]	Ultrasonic
				E_g/kT_c	
Cd	0.56	—	3.2	3.44	
Al	1.2	3.4	2.8–4.2	3.53	
Sn	3.72	3.5	2.8–3.5	3.59	
Ta	4.4	≤3	3.6	3.63	3.5
Pb	7.19	4.2	4.3	3.95	
V	5.4	3.4		3.50	3.1–3.5
Nb	9.26	2.8	3.7	3.65	3.61–3.77
V_3Ge	11	—	3.2		
V_3Si	18	1.0,3.8	1.8,3.8		
Nb_3Sn	18.1	3.8	0.2–3.1		
LaSr*	36	1–4.5	3–6		
YBa*	90	3.5–7	3–5.5		

[a] The BCS value is 3.53 (see Meser, p. 141; Ginsb. p. 216; Glads).
[b] The thermodynamic values are calculated from the expression $E_g/kT_c = [2\pi V_0 H_c^2(0)/3\gamma T_c^2]$ (Meser, p. 141).

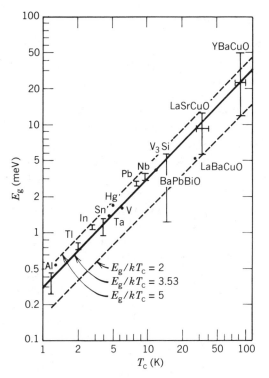

Fig. IX-4. Dependence of the energy gap E_g on the critical temperature T_c for various superconductors. Lines of unit slope are drawn for $E_g/kT_c = 2$ (---), 3.53 (——) and 5 (· · ·) where 3.53 is the BCS value.

weak coupling BCS theory is about as good for the oxide compounds as it is for some of the older types (Thoma, Thom1).

By exerting a large pressure during the tunneling measurement the energy gap of LaSrCuO could be increased from 10–20 meV to 35 meV (Vanbe). If electrons are inelastically scattered at a rate $1/\tau_{IN}$ such that h/τ_{IN} at T_c exceeds kT_c, then E_g/kT_c can exceed the BCS value of 3.5 (Leez1).

3. Infrared versus Tunneling Measurements

There has been a great deal of discussion in the literature about the discrepancies between IR and tunneling-gap determinations, about the significance of the measured gap ratios, and about why many of them do not conform to the BCS prediction. However, these same problems have been plaguing superconductor researchers for many decades, as the data of Table IX-2 and Fig. IX-4 demonstrate. Since the materials listed in this table are considered as BCS types, the data for the oxide superconductors can also be construed as consistent with the BCS mechanism. The discrepancies between far IR and tunneling experiments could arise from anisotropy of the gap (Maeka).

C. DENSITY OF STATES

The density of states $N(E_F)$ at the Fermi level E_F is an important parameter in determining the properties of superconductors. For example, in the BCS theory both the critical temperature and the energy gap have an exponential type of dependence on the DOS (Section IV-B-1). This quantity, $N(E_F)$, has been determined by several techniques. Measurements such as photoemission have provided values of $N(E_F)$ in the units states/eV Cu atom, such as 0.1 (Takag) for $BaPb_{1-x}Bi_xO_3$, 3.6–4.9 (Kita1) for LaSrCuO, 4.5 (Salam), 1.1 (Finn1), 2–2.6 (Tana1) and 6.4 (Zuozz) for $YBa_2Cu_3O_{7-\delta}$. Band structure and other calculations have provided theoretical values of $N(E_F)$, such as 1.6 (tetragonal La_2CuO_4, Bulle), 1.6 (orthorhombic La_2CuO_4, Bulle), 1.3 (($La_{1-x}X_x)_2CuO_4$, Matt5), 1.1 (($La_{1-x}-Ba_x)_2CuO_4$, Yuzzz), 1.5 ($YBa_2Cu_3O_7$, Bulle), 1.05 (BiSrCaCuO, Hyber), and 1.0 (TlBaCaCuO, Yuzz2).

To provide information on the theoretical density of states and the number of valence electrons in the various atomic orbitals of the atoms in the oxide superconductors, self-consistent augmented plane-wave band calculations were carried out for the prototype perovskite-type compound $MCuO_3$ (Papac, Takeg). Table IX-3 lists the theoretical DOS for M = Ba, Cs, La, and Y.

The higher T_c of the YBaCuO system compared with that of the LaSrCuO system has been attributed to an increased density of states at the Fermi level (Vanb1), although there have also been reports that YBaCuO has a lower DOS than LaSrCuO (Tana1).

The technique of positron 2D-ACAR, described in Section IX-E-5, was used to determine the topology of the Fermi surface of $La_2CuO_{4-\delta}$ (Tanig). Contour

TABLE IX-3. Fermi Energy E_F(Ry) and Partial and Total Densities of States (states/Ry) at E_F for Compounds LaCuO$_3$, BaCuO$_3$, CsCuO$_3$, and YCuO$_3$[a]

	M			
	La	Ba	Cs	Y
E_F	0.382	0.280	0.233	0.551
N(M-s)	0.0042	0.0071	0.0645	0.0050
N(M-p)	0.0189	0.4968	15.7506	0.0095
N(M-t_{2g})	0.0442	0.1601	0.3147	0.0307
N(M-e_g)	0.0034	0.0013	0.1767	0.0060
N(M-f)	0.0956	0.0899	0.4303	0.0244
N(Cu-s)	0.1219	0.1205	0.0743	0.1205
N(Cu-p)	0.2607	0.4087	0.2578	0.1923
N(Cu-t_{2g})	0.0754	0.7294	6.6017	0.0156
N(Cu-e_g)	7.3835	4.9116	3.6857	7.0666
N(Cu-f)	0.0181	0.0890	0.0628	0.0119
N(O-s)	0.2175	0.1614	0.1058	0.2733
N(O-p)	9.6388	26.4336	61.3162	6.8178
N(O-t_{2g})	0.0144	0.0135	0.0238	0.0136
N(O-e_g)	0.0868	0.0765	0.0581	0.0982
N(O-f)	0.0229	0.0099	0.0190	0.0277
N(E_F)	19.3697	36.7274	94.9097	15.8271

[a] The calculations were made using the self-consistent APW approximation (Papa1).

plots of the reduced momentum space density and three-dimensional sketches of the first Brillouin zone are presented in this article.

D. BREATHING MODES AND ELECTRON–PHONON INTERACTION

In the BCS theory superconductivity can arise from various interactions which are capable of producing Cooper pairs (Section IV-C). In conventional superconductors phonons mediate pair formation, but other bosons such as excitons can also contribute to the process. A much-debated question concerns the role played by electron–phonon coupling in the new oxide superconductors.

The Cu–O breathing modes, two of which are illustrated in Fig. IX-5, can involve electron–phonon interactions (Sugai; Cavaz, Hase1, Renk1). Sketches of the eigenvectors of other modes are also available (Blume). Raman-active axial (426 cm^{-1}) and in-plane symmetry-breaking (526 cm^{-1}) oxygen stretch modes were observed in La$_2$CuO$_4$ (Kouro). The Raman 664 and 153 cm^{-1} peaks of La$_2$CuO$_4$ are resonantly enhanced at low temperatures, the former being a Cu–O breathing mode (Sugai). This relatively high value of 664 cm^{-1} implies a large phonon interaction (Talia), which favors a high T_c. A formula for T_c involving the breathing modes has been proposed (Guome).

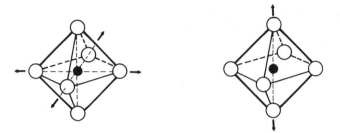

Fig. IX-5. Vibrational breathing modes of the planar (left) and axial (right) types.

Frozen-phonon optical breathing mode calculations indicated a large (≤ 3 eV/Å) electron–phonon interaction matrix element for the dominant $Cu(d_{x^2-y^2}) - O(p_{x,y})$ orbitals (Fuzzz). The 0.4-T thermodynamic critical field H_C in $(La_{0.9}Sr_{0.1})_2CuO_4$ was proposed as indicating the presence of a high DOS and an electron–phonon mechanism (Cavaz). Evidence against an electron–phonon mechanism comes from a model that associates breathing mode phonons with repulsive interactions between electrons on nearest-neighbor unit cells (Ohka1).

E. SPECTROSCOPIC MEASUREMENTS

Standard spectroscopic techniques have been used by many investigators for the study of the oxide superconductors. In this section we will present the results from the viewpoint of what insight they provide on the superconductivity and its mechanism.

Spectroscopists have a propensity for making free use of acronyms, such as BIS, EELS, ESR, EXAFS, IR, NMR, and XANES, and these will be defined in the appropriate sections.

1. Infrared and Raman

Vibrational spectroscopy can clarify the extent to which a strong electron–phonon coupling is involved in the production of high superconducting transition temperatures. It can also provide estimates of the gap energy (Maeka, Ohbay), as was mentioned above in Section IX-B.

Infrared (IR) and Raman studies are complementary to each other because some vibrational transitions are IR-active, others are Raman-active, and some appear in both types of spectra. Infrared spectral lines are due to a change in the electric dipole moment of the molecule and Raman lines appear when there is a change in the polarizability. Of particular interest are the three Raman-active Cu-O stretching modes: A_g involves the symmetric stretching of O(b) in the

O(b)–Cu(t)–O(t′) plane and the B_{2g} and B_{3g} modes involve the stretching of the O(m′) and O(m) oxygens. There are also O-deformation modes in the ab plane and parallel to the c axis (Iqba3).

The room-temperature IR spectra of $(La_{1-x}Ba_x)_2CuO_{4-\delta}$ presented in Fig. IX-6 cover the region from 400 to 4000 cm^{-1} for x between 0 and 0.2 (Ohbay, Ohba1). The spectral lines are centered at ≈ 370, 509 (A_2), ≈ 686 (A_3, for $x \leq 0.04$), and 1430 cm^{-1} (A_4, for $x \geq 0.1$). We see from the figure that in each case the transmittance approaches a constant value I_0 at the highest frequencies. For each value of x a semilogarthmic plot was made of $I(E) - I_0$ versus the energy E, as shown in the inset to Fig. IX-7, and the energy where the extrapolated line falls to $1/e$ of its value at zero energy provided the energy gap ≈ 60 meV (see also Schle). We see from the figure that the gap is a slowly increasing function of x. This gap energy, which was determined from room-temperature spectra, is close to the ones deduced from IR spectra in the superconducting region.

Bare phonon dispersion curves calculated for LaSr* and fitted to four Raman and one IR frequency reproduced the other IR modes. A relatively small number of phonon modes showed on extremely strong interaction with the conduction electrons (Weber). Force or spring constants calculated for the atom pairs in the primitive cell are listed in Table IX-4 (Brunz).

Photoinduced IR-active modes were observed in La_2CuO_4 (Kimzz). This study provided evidence for the importance of the electron–phonon interaction. One Raman line at 1470 cm^{-1} observed at 10 K was attributed to a magnon pair excitation associated with antiferromagnetic ordering. Raman spectra were obtained for Sr_2TiO_4, which has the K_2NiF_4 structure, and for the related compounds $Sr_3Ti_2O_7$, $Sr_4Ti_3O_{10}$ (Burns, Burn3), and $LaNi_2O_4$ (Sule2). They are not superconducting themselves, but their structural similarity to oxide superconductors means that they should have related vibrational spectra.

The weakening and eventual disappearance of the 680 cm^{-1} vibrational line of $(La_{1-x}Sr_x)_2CuO_4$ with increasing x was associated with a CDW suppression (Stavo). The 664 cm^{-1} line was assigned to the planar breathing mode sketched in Fig. IX-5 (Sugai). The Cu–O asymmetric stretching mode at 509 cm^{-1} (Sawad), on the other hand, remains strong under divalent metal substitution. Dimers in LaSr* have a singlet–triplet splitting of 250 cm^{-1} (Marki).

In the case of YBaCuO, except for weak lines at 151 and 310 cm^{-1}, there is no correspondence between IR and Raman frequencies (Macfa). The motion of Cu–O on the conducting planes adjacent to the yttrium atoms is IR- but not Raman-active. The detection of four IR modes in YBa* not observed by reflectance indicates that the microscopic properties of the surface differ from those of the bulk (Talia). Lowering the symmetry of $YBa_2Cu_3O_{7-\delta}$ from tetragonal to orthorhombic did not produce any observable mode splitting of the high-frequency features (Burn1, Burn2). A systematic change in the lattice vibrations due to progressive oxygen deficiency and the related structural phase transition was observed in $YBa_2Cu_3O_{7-\delta}$ (Sait3, Stav1). The softening or frequency decrease associated with the 356 cm^{-1} mode reflects the decrease of the bound charge around the Cu atoms (Suga1).

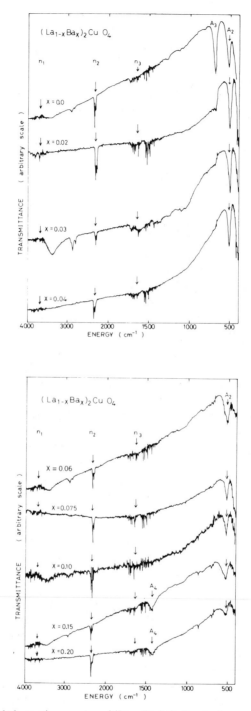

Fig. IX-6. Infrared absorption spectra of $(La_{1-x}Ba_x)_2CuO_{4-\delta}$ in the energy region between 400 and 4000 cm^{-1}. All spectra are measured at room temperature. The symbols n_1, n_2, and n_3 indicate the absorptions due to water, CO_2, and water vapor, respectively (Ohbay).

182

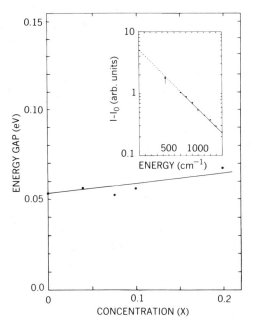

Fig. IX-7. Concentration dependence of energy gap of $(La_{1-x}Ba_x)_2CuO_{4-\delta}$. Inset: The method used to estimate the band gap by plotting the difference $I - I_0$ between the IR intensity I and its high-energy limit I_0 versus the energy on a semilogarithmic scale. The band gap is estimated at the value where the extrapolated line decays to 0.368 $(1/e)$ (Ohbay).

TABLE IX-4. Effective Spring Constants Between Atom Pairs in Unit Cell of LaSr$_*$
(Brunz)

Atom Pairs	Spring Constant (kdyne/cm)
Cu–O(1)	85
Cu–O(2)	20
La^a–O(1)	160
La–O(2)	105
La–O(2)^a	50
La–La^b	30
La–Cu	10
O(1)–O(1)	20
O(2)–O(2)^a	7
O(1)–O(2)	4

^{a} Shifted by $(a/2, a/2, c/2)$.
^{b} Shifted by $(0,0,c)$.

Infrared and Raman spectra of a number of rare-earth-substituted RBa_2-$Cu_3O_{7-\delta}$ specimens (M = Eu, Gd, Ho, Sm, Y) have been obtained (Cardo, Liuz1, Perko, Wittl). The position of the ≈ 500-cm^{-1} line peak assigned to the O(b) symmetric stretch correlated with the Cu(t)–Cu(m) distance in a series of five substituted YBaCuO compounds (Krolz). The 600-cm^{-1} band may be due to a defect-induced Cu–O mode (Krolz). The results from the compounds $Y_{0.9}Ba_{0.6}CuO_{3-\delta}$ (Mingu) and $Y_{1-x}Ba_xCuO_3$ with $0.05 < x < 0.075$ and $0.3 < x < 0.8$ (Daoqi) have also been reported. The effects of K and Ce substitutions for Sr of the LaSrCuO system were studied (Ogita).

Far IR spectra have been interpreted using an anisotropic model of randomly oriented grains highly conducting in two directions and nonconducting in the third (Herrz). A first-principles calculation of the optical properties of YBa* (Zhaoz) appears to support an exciton-enhanced superconducting mechanism and predicts a plasmon energy of 2.8 eV. Experimentally, plasmons at lower frequencies (< 0.1 eV) have been observed (Perko).

Thus far the discussion has involved IR spectra at temperatures above T_c. Reflection and transmission IR spectra of $(La_{1-x}Sr_x)_2CuO_{4-\delta}$ for $x = 0.1$ were recorded in the range from 20 to 4800 cm^{-1} at temperatures from 19 to 298 K (Sawad). Lowering the temperature did not appreciably affect the intensity or the position of the 520-cm^{-1} line from these samples. On the other hand the absorption peak near 255 cm^{-1} increased in intensity, became narrower, and shifted to higher wavenumbers with decreasing temperature. The authors pointed out that this absorption is related to a dynamical (antiferrodistortive) type of ordering in the CuO_2 layers.

Far IR transmission spectra from 20 to 500 cm^{-1} of powdered $(La_{1-x}Sr_x)_{2-y}$-CuO_{4-d} were recorded above and below T_c (Nagas). From these a normalized difference spectrum was constructed and the energy gap was determined. An extra Raman mode appears at 138 cm^{-1} in the vicinity of T_c (Copic).

An analysis of the normal and superconducting IR properties ($La_{0.925}$-$Sr_{0.075})_2CuO_4$ (Schle) suggests that the 60-meV electronic excitation and the phonon in resonance with it might provide the mediating interaction, which gives rise to the superconductivity (Schl3). Figure IX-1 shows how the three sharp lines at 151, 268, and 303 cm^{-1} in the IR spectrum of YB* are enhanced in the superconducting state (Thom3). These lines were assigned to Cu–O, La–O, and Sr–O bond bending in LaSrCuO, and similar modes assignments could apply here also.

In the superconducting state there was no evidence for new features in the Raman spectra of $YBa_2Cu_3O_{7-\delta}$ (Iqbal, Bozov, Sanju). The temperature-dependent shift of one line at 338 cm^{-1} does, however, change sign and exhibits a softening or shift to lower frequencies below T_c, as shown in Fig. IX-8, while the other lines such as the one near 504 cm^{-1} continue increasing in frequency below T_c (Macfa, Masca). Another work assigned the 250 and 338 cm^{-1} lines, which soften somewhat below T_c, to Davydov pairs involving the deformation of the warped Cu–O square frames, and considers them involved in the mechanism of the superconductivity (Wittl).

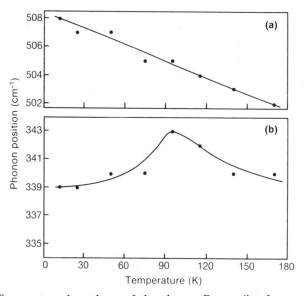

Fig. IX-8. Temperature dependence of the phonon Raman line frequency of (*a*) the 507-cm^{-1} mode, which shows a normal monotonic increase on cooling and (*b*) the 339-cm^{-1} mode, which exhibits anomalous softening or decrease in frequency on cooling the YBa$_*$ sample below T_c (Masca).

    Additional authors have published IR (Iqba1, Geser, Herrz, Ohish, Schl1, Sueli) and Raman (Blume) spectra of the (La$_{1-x}$Sr$_x$)$_2$CuO$_{4-\delta}$ system and IR (Bonn1, Bonn2, Czerw, Genze, Mihai, Sulew, Vuong, Wrobe) and Raman (Hemle, Rosen) spectra of YBaCuO.

2. Photoemission

Photoemission spectroscopy (PES) measures the energy distribution of the emitted photoelectrons from ions in particular charge and energy states. In ultraviolet photoemission spectroscopy (UPS) the exciting energy is high-intensity UV light such as that from the 21.2-eV resonance line (He-I), or the higher frequency 40.8-eV line (He-II) of a helium gas discharge tube. In the X-ray analogue (XPS) the radiation used to excite photoelectrons is obtained from an Mg-K$_\alpha$ (1253.6), Al-K$_\alpha$ (1486.7 eV), or other convenient X-ray source. Monochromatic XPS (MXPS) was used. Inverse photoelectron spectroscopy (IPS), called Bremsstrahlung isochromat spectroscopy (BIS) when UV photons are detected, has also been employed (Gaozz, Gaoz2, Riest, Temme), as well as electron energy-loss spectroscopy (EELS) (Ahnzz, Nucke). Both XPS and BIS give information related to the density of occupied and unoccupied states, respectively. Auger electron spectra have also been studied.

Emitted photoelectrons are at energies characteristic of particular atoms in the solid. The inner core electrons have high binding energies which increase with atomic number, so their emission spectra are at hundreds of electron volts. Figure IX-9b shows core-level spectra from particular levels of each of the atoms Y, Ba, Cu, and O of $YBa_2Cu_3O_7$ (Iqba4, Stein). Of greater interest are the spectra of the outer or valence electrons which occur at lower energies, as shown by the valence band spectra of Fig. IX-9a. One should note the dependence of the valence band spectra on the exciting frequency. Others have published similar results (e.g., Fujim). The O-2p and Cu-3d lines are close together (Yarmo). The positions and widths of the valence bands are of particular significance for elucidating the electronic structure of the oxide superconductors.

Some workers (Ihara, Ihar1, Ihar3) decomposed the Cu-2p $\frac{3}{2}$ line of $(La_{1-x}Sr_x)_2CuO_{4-\delta}$ (shown on Fig. IX-9b) into components arising from the valence states Cu^+, Cu^{2+}, and Cu^{3+}, and they interpreted the narrowing of the line after annealing to the absence of the Cu^{3+} and most of the Cu^{2+} ions. A similar decomposition was done for YBa∗. There is additional PES and X-ray absorption evidence for the presence of Cu^{3+} (Boyce, Liang). Other workers saw little (Iqba4) or no evidence for the valence states Cu^+ (Brow1, Onell, Sarma, Sarm1, Tranq, Yarmo) and Cu^{3+} (Horn1, Kurtz, Stein, Taka3, Tranq, Yarmo). Additional PES results favor all Cu in the 2+ state with one O^- oxygen rather than $2Cu^{2+} + Cu^{3+}$ per unit cell (Tranq, Yarmo, see also Bianc). An Auger spectroscopy study indicated that the covalency of the Cu-O bond increases with T_c (Ramak). The temperature-dependent character of the bond between Cu and O atoms is dominated by the instability of the Cu–O bond in the basal plane (Kohik).

A study of $(La_{1-x}Sr_x)_2CuO_{4-\delta}$ suggests that the density of states near E_F is not very high, shifts of the La, Cu, and O core level lines indicate very little change in charge on these atoms when the Sr content is changed (Nucke), and oxygen vacancies are coordinated to Ba^{2+} (Stein). The DOS of YBaCuO near E_F is also found to be small. Inverse photoelectron studies indicate a small (up to ≈ 3 eV above E_F) density of states above the Fermi level, with larger features occurring higher (Gaozz, Wagen). A number of workers compared their PES results with the density of states calculated from band theory (e.g., Johns, Moogz, Riest, Taka2).

The XPS and UPS photoemission experiments with $YBa_2Cu_3O_7$ show Cu-2p spectra and an indication of large O^{2-} vacancies coordinated with Ba^{2+} sites (Stein). Calculated local density X-ray and photoemission spectra of $YBa_2Cu_3O_{7-\delta}$ indicate that the low-lying O-pσ bonding states show more itinerant character than higher pπ states (Redin). Oxygen contents in the planar and vertical sites of bulk and thin-film LaSrCuO were determined by XPS (Tera1).

Post-annealing experiments revealed two categories of degradation of T_c (Schro), one involving a slight decrease in T_c below 91 K due to the removal of oxygen atoms from the apex of the pyramidal configuration, and the other in which T_c falls below 50 K because of the "removal of oxygens from the linear Cu–O chain in the basal plane." Synchrotron radiation PES data indicate a

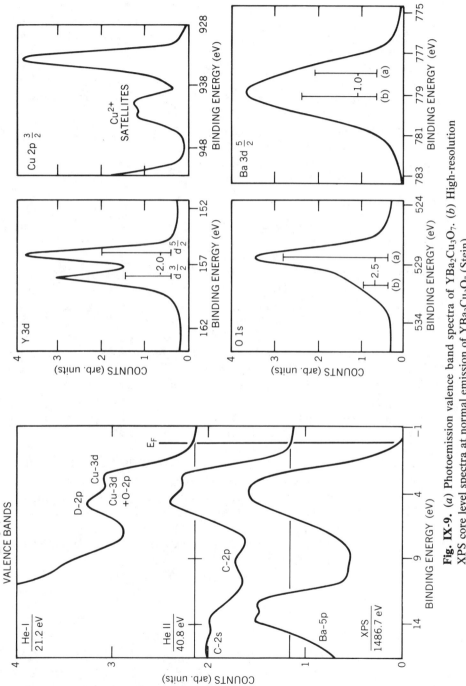

Fig. IX-9. (*a*) Photoemission valence band spectra of $YBa_2Cu_3O_7$. (*b*) High-resolution XPS core level spectra at normal emission of $YBa_2Cu_3O_7$ (Stein).

strong electron correlation role in the electronic structure of YBa* (Taka3). Another synchrotron PES study compared the electronic structure of semiconducting $YBa_2Cu_3O_{6.2}$ with that of superconducting $YBa_2Cu_3O_{6.8}$ (Stoff), and spectra of the green semiconducting phase Y_2BaCuO_5 were compared with those of the orthorhombic and tetragonal variants of $YBa_2Cu_3O_{7-\delta}$ (Fjeil). The PES results revealed that $(La_{1-x}Sr_x)_2CuO_{4-\delta}$ and $YBa_2Cu_3O_{7-\delta}$ have the electronic structures of Mott insulators with their localized Cu-3d electrons and low density of states at the Fermi level (Fujim). Auger spectra of $(La_{1-x}Sr_x)_2CuO_4$ and $YBa_2Cu_3O_7$ were interpreted to be dominated by electron–electron interactions, which may play a significant role in electron pair binding (Ramak).

Other workers have also used PES for the study of oxide superconductors (e.g., Kohik, Kurtz, Mingr, Reihl, Sarma, Sarm1, Schr1, Takah).

3. X-Ray Absorption

Several workers have employed X-ray absorption edge structure (XANES) and extended X-ray absorption fine structure (EXAFS) spectroscopy to obtain information on chemical bonding and valence states.

Broadening of the absorption edge of LaSrCuO was explained by charge fluctuations at the copper sites introduced by substituting Sr^{2+} for La^{3+} (Sonde). The XANES studies of $(La_{1-x}(Ba,Sr)_x)_2CuO_{4-\delta}$ indicate that all of the copper is divalent, independent of x, with doping having the effect of adding holes in the O-2p band which have d symmetry relative to the La and Cu sites (Tranq, Tran1). There is also X-ray absorption support for the presence of O-2p holes in YBaCuO (Horn1, Yarmo). This is at variance with the usual picture of copper valence changes, and lends support to Emery's model (Section IV-E) that superconductivity is due to pairing of O-2p holes.

Oxygen-deficient samples ($\delta = 1$) of $YBa_2Cu_3O_{7-\delta}$ exhibited "significant differences" in the low-energy spectral region that were associated with changes in the oxygen vacancies in the linear chains (Rhyne). The superconductors YBa* and GdBa* were found to be essentially identical by XANES and EXAFS, with a copper valence between +2 and +3 (Boyc1). The near-neighbor vibration frequencies show that the two-dimensional CuO_2 planes and the one-dimensional Cu–O chains form the most rigid part of the structure. These layers are bound more weakly along the c axis by the Ba, O, and Y planes.

Some X-ray absorption results support the presence of Cu^+ in LaSr* (Oyana) and Cu^{3+} in YBa* (Alpzz, Anton, Boyc1, Jeonz), while others do not (Horn1, Yarmo). There are reports of detecting only Cu^{2+} in $(La_{1-x}(Ba,Sr)_x)_2CuO_{4-\delta}$ (Tranq, Tran1).

4. Inelastic Neutron Scattering

Inelastic neutron scattering can provide information on the energy gap and the phonon density of states $D(\omega)$ which determines the electron–phonon coupling constant through the Eliashberg relation (Eq. IV-6) (e.g., see Lynnz, Masak,

Rhyne, Robi1). In LaSr* an enhanced phonon DOS at 10 meV (Ramir) and the absence of any pronounced soft-mode behavior plus indications of coupling between electrons and oxygen breathing modes (Renk1) were observed. The application of this technique to a $YBa_2Cu_3O_{7-\delta}$ sample with oxygen contents $\delta = 0$, 0.02, and 0.38 and respective transition temperatures $T_c = 94$, 91, and 10 K showed no change in the characteristic features of the phonon DOS, and no evidence for soft-phonon mode-driven superconductivity (Miha2).

5. Positron Annihilation

In positron annihilation (PA) spectroscopy the sample is irradiated with positrons from a radioactive source such as $^{22}NaCl$. The various positron lifetimes are determined by the time delay between the 1.28-MeV γ ray emitted simultaneously with the positron and the 0.51-MeV γ rays produced by the annihilation events, and each has a corresponding intensity. The S and W Doppler broadening parameters could also be determined. Oxygen vacancies may be the most likely positron trapping sites.

The lifetimes and intensities exhibited discontinuities at the transition temperature of YBaCuO (Jeanz, Jean1, Tengz), as shown in Fig. IX-10, and the Doppler broadening parameters peaked at the onset and zero resistance temperatures (Zhon2). It was concluded that oxygen vacancies play an important role in the superconductivity and that the electron density at the vacancy sites is higher in the superconducting state (Jeanz).

The PA results obtained with $(La_{1-x}Sr_x)_2CuO_{4-\delta}$ suggest that the oxygen vacancy concentration and hence the Cu^{3+} ion concentration are determined by Sr^{2+} ion clustering on the La lattice (Smeds).

A specialization of the PA technique called 2D-ACAR or two-dimensional angular correlation of annihilation radiations can provide the topology of the Fermi surface, as was mentioned in Section IX-C.

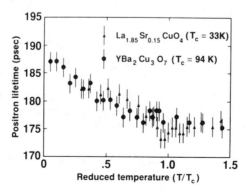

Fig. IX-10. Positron annihilation lifetime versus reduced temperature (T/T_c) for LaSr* and YBa* (Jean1).

F. SPECIFIC HEAT

One of the best ways to check the extent to which a superconductor obeys the BCS theory via a phonon mechanism is through measurements of its low-temperature specific heat, as was explained in Section IV-B-2. Several groups have made such measurements on LaSrCuO (e.g., Maeno, Ramir, Zirng), on YBaCuO (e.g., Junod, Juno1, Mapl2), and on other compounds (e.g., Brown, Hozzz, Ferre, Leez6, Mapl2).

1. Specific Heat above T_c

The specific heat above T_c but considerably below the Debye temperature θ_D is the sum of a linear term arising from the conduction electrons and a cubic term due to the lattice vibrations, so the molar specific heat C_n in the normal state may be written as

$$C_n = \gamma T + AT^3 \qquad (IX\text{-}3)$$

In the free-electron approximation the electronic specific heat coefficient γ, sometimes called the Sommerfeld constant, is

$$\gamma = \tfrac{1}{2}\pi^2 R(k/E_F) \qquad (IX\text{-}4)$$

$$= (\pi^2/3)k^2 N(E_F) \qquad (IX\text{-}5)$$

and in the low-temperature Debye approximation the vibrational factor is

$$A = 12\pi^2 R/5\theta_D^3 \qquad (IX\text{-}6)$$

where the gas constant $R = 8.314$ J/mol K. Using measured values of γ and A from Table IX-5 (Mapl1) on typical samples of $(La_{0.9}Sr_{0.1})_2CuO_{4-\delta}$ and $YBa_2Cu_3O_{7-\delta}$ we find, at T_c

$$\frac{AT_c^3}{\gamma T_c} = \begin{cases} 110 & \text{for LaSr} \\ 450 & \text{for YBa*} \end{cases} \qquad (IX\text{-}7)$$

which means that for oxide superconductors the vibrational term dominates.

The Sommerfeld specific heat factor γ can be estimated from the susceptibility χ using the free-electron conversion (Nevit)

$$\frac{\gamma}{\chi} = \frac{1}{3}\left[\frac{\pi k}{\mu_B}\right]^{1/2} \qquad (IX\text{-}8)$$

The factor γ has been evaluated by other methods such as entropy matching (Kita1) and magnetic critical field methods.

Figure IX-11 shows the relationship between γ and T_c for various types of superconductors. The $Ba(Pb,Bi)O_3$, $(La,Sr)_2CuO_4$, and $YBa_2Cu_3O_7$ types are at the top, and the heavy Fermions are off scale on the lower right (Tana1, Phil1).

A typical value of $\gamma = 17$ mJ/mole $\cdot$ Cu $\cdot$ K for $(La_{0.925}Ca_{0.075})_2CuO_4$ corresponds, with the aid of Eq. (IX-5), to the following density of states $N(E_F)$ at the Fermi level (Kita1)

$$N(E_F) = 3.6 \text{ states/eV} \cdot \text{Cu} \cdot \text{spin} \qquad \text{(IX-9)}$$

which is over five times the calculated value for La_2CuO_4 (Matth). The Debye temperature was estimated from the slope of the normal state C/T versus T^2 curve. The assumption that between two and seven of the seven atoms in the LaSrCuO unit cell contribute to the phonon modes gives the limiting values of $\theta_D = 175$ and 270 K. These values of θ_D correspond to using $\lambda = 2.4$ and $\lambda = 1.5$, respectively, and $\mu_* = 0.12$ in the McMillan expression (IV-7) for the critical temperature.

Calorimetric specific heat measurements of LaSrCuO were fit to the expression

$$C = BT^{-2} + \gamma T + AT^3 \qquad \text{(IX-10)}$$

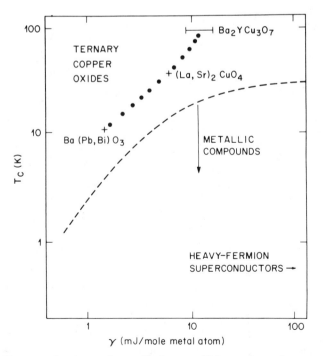

Fig. IX-11. T_c versus the electronic specific heat coefficient γ for various types of superconductors (Phill).

TABLE IX-5. Debye Temperature (θ_D) and Specific Heat Data[a]

Material	T_c (K)	θ_D (K)	γ (mJ/mole CuK²)	$(C_s - C_n)/\gamma T_c$	$(C_s - C_n)/T_c$ (mJ/mole CuK²)	Measurement Method[b]	Ref.
Cu (element)	—	342	0.69	—	—		Glads
Cd (element)	0.55	252	0.67	1.36	0.91		Glads, Meser
Al (element)	1.2	423	1.36	1.45	1.97		Glads, Meser
Sn (element)	3.72	196	1.78	1.60	2.85		Glads, Meser
Pb (element)	7.19	102	3.14	2.71	8.51		Glads, Meser
Nb (element)	9.26	277	7.66	1.93	14.8		Glads, Meser
LaSr*	36	140–270	1.7	1.4	2.4	Susc	Decro
	18		20		32	Cal	Kita1
	31				20	Cal	Kita2
	38		16			CF	Kobay
	38		6			CF	Orla2
	37		7.0	1.43	10	Cal	Ramir
	33		7				VanB1
LaSr(0.1)	34.5		2.6			CF	Kobay
	38		3.4			Cal	Mapl1
(LaSr)$_2$CuO$_4$	28	359	39	0	0	Cal	Reeve
	40		8				Phil3
LaBa*	30	385	4	0	0	Cal	Maeno

LaBa(0.1)	32			0	0	Cal	Reeve
YBa*	28	356	71	0	0	Cal	Wenge
	93	383	5.4	~2	~23	Cal	Ayach
	91.6	460	11		59	Cal	Beckm
	93		41	1.25	10	CF	Bezi1
	91		8			CF	Caval
	92		3–5			CF	Cheon
	90		9	1.2	15	Cal/Susc	Inder
	92.2	400	12	1.4	13	Cal/Susc	Junod
	92		9		11	Cal	Lizzz
	90		7			Cal	Mapl1
	93		8	1.5	18	Cal/CF	Nevit
	~90		12			CF	Panso
	92	374	2–4		—	Cal	Reev1
	91		16			Cal	VanB1
YBa$_2$Cu$_3$O$_{6.9}$	90	340	2.8	3.95			Zhaoj
	91		3–5			CF	Caval
DyBa*	92.5		~30		42	Cal	Sait2
ErBa*	91.2		~32		46	Cal	Sait2
GdBa*	94	374				Cal	Reev1

[a] The dimensionless quantity $(C_s - C_n)/\gamma T_c$ has the value 1.43 in the BCS model.

[b] Cal, calorimetry; CF, critical field measurements; Susc, normal state susceptibility.

obtained by adding the Schottky-type term BT^{-2} to Eq. (IX-3). The former might arise from magnetic impurities or La hyperfine splittings ($I = \frac{7}{2}$) (Wenge).

2. Discontinuity at T_c

In the BCS approximation the electronic specific heat jumps abruptly at T_c from the normal state value γT_c to the superconducting state value C_S with the ratio $(C_S - \gamma T_c)/\gamma T_c = 1.43$ in accordance with Eq. (IV-8). A more sophisticated calculation for a phonon-coupled isotropic Eliashberg superconductor found this ratio to be approximately 3.4 with a slight dependence on the Coulomb pseudopotential $\mu* \approx 0.15$ (Blezi). The abrupt jump of the specific heat at T_c is illustrated in Fig. IX-12. This figure shows the slight dependence of the jump on the heating rate. A similar figure for LaSr$*$ (Phil3) shows that the agreement with the BCS theory is good.

One should note that the range in the ordinate scale of Fig. IX-12 is very small compared with the magnitude of the specific heat. This is because, from Eq. (IX-7), the specific heat jump is superimposed on the much larger T^2 vibrational term (Inder, Tana1). This small change at T_c can be resolved by superimposing curves of C/T versus T^2 obtained in zero field, and in a 5.7-T field which destroys the superconductivity (Kita1).

Other researchers have observed the jump in the specific heat at T_c in the LaSrCuO (Decro, Mapl1, Phil3) and YBaCuO (Mapl1, Nevit) compounds, and data from various investigators are compared in Table IX-5. This table also lists experimental values of T_c, θ_D, and γ and the ratios $(C_S - C_n)/T_c$ and $(C_S - C_n)/\gamma T_c$ for several elements and a number of oxides. Some of the elements are close to the BCS value of 1.43, but the strong coupled ones, Pb and Nb, are higher. Several experimental results for YBa$*$ are quite close to 1.43, as indicated in the table. Some other workers have failed to observe a specific heat discontinuity (Maeno, Reeve, Wenge).

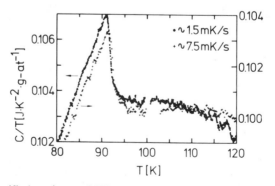

Fig. IX-12. Specific heat jump of YBa$*$ at T_c measured at two heating rates, 0.09 and 0.45 K/min (Juno1).

3. Measurements below T_c

At temperatures far below T_c the vibrational specific heat term AT^3 becomes negligible, and other mechanisms become important. A term linear in T was found to dominate the specific heat of LaSr* and YBa* below 4 K, and was attributed to tunneling (Collo). Such a contribution was observed in TlBaCa-CuO but not in BiSrCaCuO. This linear specific heat term might also arise from the presence of impurities. In LaSr* nuclear hyperfine effects provided the largest contribution in the range from 0.03 to 0.2 K (Gutsm). Antiferromagnetic ordering was also found to contribute strongly below ≈ 2 K, both in $RBa_2Cu_3O_{7-\delta}$, where R = Gd, Er (Rami1, Reev1, Simiz), and in LaSrCuO (Gutsm).

X

TRANSPORT PROPERTIES

A. INTRODUCTION

The principal applications of superconductors are based upon their ability to carry electric current without any loss, and therefore it is important to understand their transport properties. This chapter begins with a discussion of resistivity and critical current flow in the absence of externally applied fields. This is followed by a discussion of several techniques involving applied fields and thermal effects. The chapter concludes with sections on tunneling and the Josephson effect.

B. CURRENT FLOW

Electric currents that flow through a superconductor owing to the action of an external source of potential are called transport currents, and those that arise in an external magnetic field to cancel the magnetic flux inside the superconductor are called diamagnetic screening currents or shielding currents. In magnet applications transport currents are started by an external source and continue to flow (persist) after its removal, while in long-distance electrical transmission applications the source voltage continues to be applied.

Current densities J are intrinsic properties and hence are more useful than currents I for quantitative comparisons between different superconductors. Transport current densities can be comparable in magnitude to shielding current densities.

The velocity of electrons at the Fermi surface v_F was estimated to be 10^7 cm/sec in these materials, which is of the same order as in A-15 compounds, and $\frac{1}{10}$ of the value in aluminum (Garoc).

1. Resistivity

A susceptibility measurement is a better thermodynamic indicator of the superconducting state because magnetization is a thermodynamic state variable. The resistivity, on the other hand, is easier to measure, and can be a better guide for applications. The temperature of zero resistivity shows when continuous superconducting paths are in place between the electrodes. Filamentary paths can produce sharp drops in resistivity at higher temperatures than pronounced onsets of diamagnetism. This can be described in terms of two- and three-dimensional percolation thresholds (see Section III-E).

Many investigators have published figures of resistivity or resistance versus temperature, since this is the most popular way to determine the critical temperature and the sharpness of the transition. It should be remembered that if the specimen is porous, accurate determinations of the resistivity cannot readily be made because of the presence of voids and intergrain problems. In almost all cases T_c determined from the resistivity midpoint is at a higher temperature than its susceptibility counterpart.

Good conductors such as copper and silver have room-temperature resistivities of about 1.5 $\mu\Omega$ cm, and at liquid nitrogen temperature the resistivity typically decreases by a factor of 6–8, as shown by the data in Table X-1. The elemental superconductors, such as Nd, Pb, and Sn, have room-temperature resistivities a factor of 10 greater than good conductors. The other metallic elements present in oxide superconductors, namely, Ba, Bi, Ca, La, Sr, Tl, and Y, have resistivities 10–80 times that of Cu. The copper oxide superconductors have even higher room-temperature resistivities, over three orders of magnitude greater than that of metallic copper, which puts them within a factor of 3 or 4 of the semiconductor range, as shown by the data in Table X-1. The resistivity of these materials above T_c decreases more or less linearly with decreasing temperature (cf. Fig. VII-11) down to the neighborhood of T_c, with a drop by a factor of 2 or 3 from room temperature to this point, as shown by the data in the table.

Needless to say, the concept of resistivity is not a meaningful one to apply to a superconductor below T_c. Nevertheless, it is instructive to study the low-temperature resistance in nonsuperconducting compounds that are closely related to superconductors. For example, in nonsuperconducting crystals of $(La_{1-x}Sr_x)_2(Cu_{1-y}Li_y)O_{4-\delta}$ a variable range type of hopping resistance, that is, $R \approx \exp[(T_0/T)^4]$, was reported. From this it has been argued that the poor conductivity is not due to a large gap, but rather to localization of the states at E_F (Kastn).

The resistivity in the high-temperature ($80 \leq T \leq 1200$ K) region is linear with the temperature for $T < 600$ K and superlinear above 600 K. This linearity has been linked to the two-dimensional character of the electron transport

TABLE X-1. Examples of Resistivities at Room Temperature ρ_{300}, at Low Temperature $\rho(T)$, and Their Ratio $\rho_{300}/\rho(T)^a$

Material	ρ_{300} ($\mu\Omega$ cm)	$\rho(T)$ ($\mu\Omega$ cm)	T (K)	$\rho_{300}/\rho(T)$	Ref.
Cu	1.68	0.18	77	9.3	
Ag	1.60	0.26	77	6.2	
Pt	10.6	1.74	77	6.1	
Sn	12.4				
Nb	12.5				
Tl	18				
Pb	20				
Sr	23				
Ca	53				
Y	57				
La	58				
Ba	60				
Bi	119				
LaSr∗	2200	430	44	5.1	Kobay
	2700	850	50	3.2	Penne
	2300	510	50	4.5	Tara1
LaSr (0.05)		950(∥)	40		Hidak
		19000($\perp$)	40		Hidak
(0.1)			50	3.6	Coppe
		1600	44		Kobay
	1200				Tonou
LaBa(0.1)			50	4.0	Coppe
YBa∗	2000	1000	95	2.0	Bonn1
	650	225	95	2.9	Cava1
	900	470	100	1.9	Mawds
	4000	2000	100	2.0	Neume
		730	120	1.9	Panso
	1350	680	100	2.0	Penne
($\rho\parallel$)	450	200	100	2.3	Tozer
($\rho\perp$)	13000	18000	100	0.7	Tozer
$Y_{0.6}Ba_{0.4}CuO_3$	1.5×10^6				Tonou
DyBa∗				~1.3	Mapl1
EuBa∗	720	410	100	1.8	Hikit
TmBa∗	4600	1900	100	2.4	Neume
YbBa∗				~3.3	Mapl1

^{a}Typical semiconductors have values from 10^4 to 10^{15} $\mu\Omega$ cm and insulators range from 10^{20} to 10^{28} $\mu\Omega$ cm. The notation ∥ and $\perp$ refers to resistivity measurements made parallel to and perpendicular to the Cu–O planes, respectively.

(Micna). Applying the Mott–Ioffe–Regel rule (minimum scattering length < mean free path) to the observed normal state resistivity gave electron–phonon couplings in both LaSr* ($\lambda = 0.1$, $\lambda_{max} = 0.45$) and YBa* ($\lambda = 0.3$, $\lambda_{max} = 1.2$) which were too small to account for the observed T_c (Gurv1, Gurv2).

An exponential dependence of the resistivity of YBa* on the temperature was observed between 80 and 1240 K (Fishe). This may occur via tunneling of electrons through barriers. The value of the exponent was different above and below the temperature T_* (700–750), which is near the tetragonal-to-orthorhombic transition. The temperature dependence of the resistivity appears to result from the loss of oxygen during heating, and the following expression was proposed to reflect this dependence:

$$\rho(T) = \frac{AT}{1 - \delta(T)} \tag{X-1}$$

where A is temperature insensitive and $\delta(T)$ is the oxygen content factor in the formula $YBa_2Cu_3O_{7-\delta}$ (Chaki; see also Fiory). A break in the slope of the logarithmic derivative $(T/\rho)d\rho/dT$ plotted against T occurs at the orthorhombic-to-tetragonal phase transition (Fiory). Evidence for an n-to-p type transition was also reported (Choi1).

The resistivity is much higher when measured perpendicular to the ab planes (i.e., along the c axis) than it is parallel to the planes. Measured ratios $\rho_\perp/\rho_\parallel$ are about 20 for LaSrCuO (Hidak), 50 for YBaCuO (Tozer), and 10^5 for BiSrCa-CuO. We see from Table X-1 that typical measured resistivities, which are on polycrystalline specimens, are much closer to the in-plane values.

Hysteresis effects have been seen in the resistance versus temperature curves, as illustrated on Fig. X-1 for $(Y_{0.875}Ba_{0.125})_2CuO_{4-\delta}$ (Taras). The 2-K shift in T_c for decreasing and increasing temperature measurements is about half of the width of the transition.

2. Critical Current Density

When the current density in a superconductor exceeds a value called the critical current density J_C, the superconductivity is destroyed. This is called the Silsbee effect. The value of $J_C(T)$ increases from zero at $T = T_c$ to a maximum value $J_C(0)$ at 0 K (Leide). Figure X-2 shows the magnetic field dependence of J_C for $YBa_2Cu_3O_{7-\delta}$ from 0.5 to 6 T for several temperatures in the range from 4.2 to 83 K (e.g., Panso; see also Ekinz, Jones).

The value of J_C can be determined directly by the resistivity method by measuring the current at which a small voltage (typically 1 μV) is induced across the sample (≈ 1 cm) in a four-probe resistivity arrangement. An indirect method uses a magnetization versus field hysteresis loop through the expression (Kumak, see also Sunzz, Xiao2).

$$J_C = 30 \, \Delta M/d \qquad A/cm^2 \tag{X-2}$$

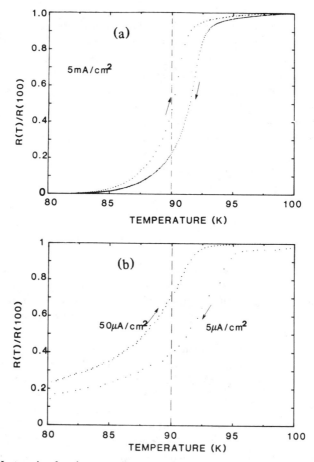

Fig. X-1. Hysteresis of resistance versus temperature for two samples of YBaCuO, for increasing and decreasing temperature, as indicated by the arrows. (*a*) current density 5 mA/cm²; (*b*) current density 50 μA/cm² for increasing temperature and 5 μA/cm² for decreasing temperature (Taras).

where ΔM is the hysteresis of magnetization per unit volume in electromagnetic units per cubic centimeter and d is the size of the sample in centimeters.

The J_C values measured directly are called transport currents and those determined from hysteresis loops are called magnetization currents. Transport currents were found to be smaller than magnetization currents in the LaSrCuO (e.g., Larba) and YBaCuO (e.g., Kuma1, Togan) systems. This could be caused by granularity and intergrain contact, and improving sample quality might bring transport currents closer to their magnetization current counterparts (Wuhlz). In contrast to this, magnetization and transport critical currents of YBa* epitaxial films were reported to be the same (Chaud, Ohzzz).

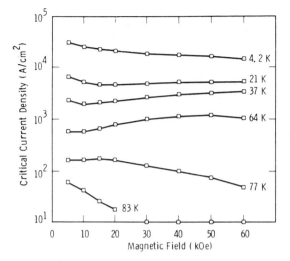

Fig. X-2. Magnetic field dependence of critical current densities of $YBa_2Cu_3O_{7-\delta}$ obtained from hysteresis loops (Panso) (10 kG = 1 T).

Critical current densities for a number of LaSrCuO and YBaCuO materials are listed in Table X-2 (e.g., Camp1, Chau2, Chau3, Daizz, Jinz2, Kagos, Umeza, Xiao2). We see from the table that the YBaCuO compounds tend to have higher values of J_C than LaSrCuO. For the same material J_C is considerably larger at 4.2 K than at 77 K. It is also quite anisotropic, with critical current densities much larger parallel to the Cu–O planes than perpendicular to them. Jin et al. (Palca) employed a technique called melt-textured growth to achieve J_C as high as 10^3 A/cm^2 in a magnetic field of 1 T at 77 K. Values of $J_C = 10^6$ A/cm^2 and greater have been reported in epitaxial thin films (Chaud, Kwoz1, Ohzzz) and single crystals (Crabt, Ding1, Wort1).

Grinding and heat treating samples of LaSrCuO and YBaCuO was found to appreciably increase the critical current density at 4 K (Suena). High critical currents require efficient flux pinning since J_C increases with the pinning force (Huebe, p. 125). Weak pinning leads to flux creep and low critical currents (e.g., Giova).

Very high values of J_C are needed for magnet materials. A niobium–titanium filament has been reported with J_C values as high as 3.7×10^5 A/cm^2 at 5 T (Cheng). The Superconducting Super Collider and Relativisitic Heavy Ion Collider accelerators require 5-μm wire filaments that support J_C of at least 2.8×10^5 A/cm^2 in a 5-T magnetic field (Grego). At present J_C values of technologically suitable oxide materials at 77 K are too small for such high-field magnet applications. Although the oxide superconductors do not yet compete with the old ones in critical currents, they are superior in their critical field (H_{C2}) capability, as shown in Fig. I-2.

TABLE X-2. Critical Current Densities[a]

Compound	J_c (A/cm²)	T (K)	H (T)	Measurement Method	Comments	Ref.
LaSr*	10^5	4		Mg		Suena
LaSr(0.05)	2	4.2	0.025	Tr		Larba
	1	4.2	10^{-3}–3	Tr		Larba
	0.75	4.2	7	Tr		Larba
YBa*	$>10^6$	77		Mg	E-film	Chaud
	$>10^5$	77		Mg	E-film	Chaud
	10^5				E-film	Chau1
	$1.4 \times 10^6 (\perp)$	5	0–1	Mg	M-xt	Crabt
	1.4×10^4	5	0.4	Mg	P-xt	Crabt
	$1.1 \times 10^4 (\perp)$	77	0.1	Mg	M-xt	Crabt
	$4.3 \times 10^3 (\perp)$	77	1	Mg	M-xt	Crabt
	1.2×10^2	77	0.6	Mg	P-xt	Crabt
	4×10^5	4.5		Mg	M-xt	Ding1
	$3 \times 10^6 (\perp)$	4.5		Mg	M-xt	Ding1
	$>10^6$	4.5	>4	Mg	M-xt	Ding1
	1–200	0	0	Tr		Ekinz
	620	77	0	Tr	$J_{c\perp}/J_{c\parallel} \sim 6$	Glowa
	3×10^4	77	1	Mg	sintered rod	Jinzz
	10^6	4.2		Tr	P-xt	Kumak
					E-film, bc plane	Kwoz1

Material	J_c	T	H	Method	Sample	Reference
	10^3	77	1–10	Tr	E-film, bc plane	Kwoz1
	$1.5\text{–}2\times10^4$	4.2	1	Mg		Larb1
	235	77	6	Mg		Larb1
	10	77		Mg		Larb1
	$0.9\text{–}11\times10^5$	4.2	0.3	Mg	E-film	Ohzzz
	$1\text{–}5\times10^4$	78	0	Mg	E-film	Ohzzz
	$0.4\text{–}4\times10^4$	78	0.3	Mg	E-film	Ohzzz
	$1.5\text{–}3\times10^4$	4.2	0.1–6	Mg		Panso
	50–200	77	0.1–6	Mg		Panso
	$9.4\times10^4(\perp)$	5	0.4	Mg	sinter forged	Songz
	$5.9\times10^4(\parallel)$	5	0.4	Mg	sinter forged	Songz
	10^5	4		Mg		Suena
	8×10^3	4.2	0.2	Mg		Togan
	$>10^2$	77	0.1	Mg		Togan
	$3.2\times10^6(\perp)$	4.5	0	Mg	M-xt	Wort1
	$1.6\times10^5(\parallel)$	4.5	0	Mg	M-xt	Wort1
$YBa_2Cu_3O_{6.5}$	168	77		Tr		Capo1
$YBa_2Cu_3O_{6.9}$	>1100	77	0	Tr		Cava1
$Y_{0.8}Ba_{1.2}CuO_x$	14	14	0	Tr		Iguch
$YBaCuO$	2800	4.2		Tr	pulsed	Jones
$YBaCuO$	650	77		Tr	pulsed	Jones

[a] E-film, epitaxial film; Mg, magnetization; M-xt, monocrystal; P-xt, polycrystal; Tr, transport. Some measurements were made in the presence of an applied magnetic field H, and the notation $\parallel$ and $\perp$ refers to currents measured with the field applied parallel and perpendicular to the Cu–O planes, respectively.

3. Persistent Currents

The zero-resistance property of a superconductor implies that an electrical current flowing in a closed path should persist indefinitely. Several investigators have examined this property and set lower limits on the lifetime of the current and upper limits on its associated resistivity. The lifetime of the persistent or superconducting current in a cylindrical LaSrCuO sample is in excess of 3×10^6 sec or more than a month, corresponding to a resistivity of less than $3 \times 10^{-17}\,\Omega$ cm (Wells). Current lifetimes in loops of low-temperature superconductors are much longer and suggest an effective resistivity of $<10^{-23}\,\Omega$ cm (Chand), close to the value of $7 \times 10^{-23}\,\Omega$ cm reported for YBaCuO (Kedve). Other reported minimum resistivity determinations are: $<10^{-9}\,\Omega$ cm (Iguch), $4 \times 10^{-16}\,\Omega$ cm for $(Y_{0.6}Ba_{0.4})_2CuO_4$ (Skoln), $<10^{-16}\,\Omega$ cm in YBa* (Tjuka), and $2 \times 10^{-18}\,\Omega$ cm (Yehzz). Similar results ($10^{-24}\,\Omega$ cm) were obtained for TlBaCaCuO.

The ceramic superconductors are granular and the relaxation of trapped field and critical current loops may be characteristic of glassy structures (cf. Section VIII-D-5). Because of the granularity even small fields can penetrate the materials. This property has been utilized to nondestructively read the supercurrent (Macf1).

C. MISCELLANEOUS TRANSPORT PROPERTIES

In this section we will discuss some transport properties that depend upon the application of electric or magnetic fields, and some that involve thermal effects. Various transport results of YBa*, namely, thermal conductivity, thermopower, Hall constant, and resistivity, were found to be consistent with ordinary metallic behavior with a strong phonon interaction (Uher2). It was concluded that there is no evidence for exotic electronic behavior in YBa*.

1. Magnetoresistance

A number of investigators have studied the resistance versus temperature behavior in low (Hikam), high (Kwokz, Mats1, Mura1, Uher4), and very high (Ouss1) magnetic fields.

Very high field ($H \leq 43$ T) longitudinal and transverse orientation studies of YBa* (Ouss1) show that the magnetoresistance $\Delta\rho(T,H) = \rho(T,H) - \rho(T < 0)$ may be decomposed into three contributions

$$\Delta\rho = \Delta\rho^M + \Delta\rho^N + \Delta\rho^S \tag{X-3}$$

where $\Delta\rho^S$ (>0) is the increase in resistance when the superconducting fluctuations are suppressed by the field, $\Delta\rho^N$ is the normal or Lorentz magnetoresistance, and $\Delta\rho^M$ is an unidentified component that may be associated with magnetic ordering. No change in the superconductivity was observed when $H = 43$ T at 50 K, and only $\approx 50\%$ normal phase resistance was found at 77 K. The upper critical field H_{C2} was estimated as ≈ 125 T.

One of the problems with comparing transport and magnetic T_c data is that resistivity measurements are generally made in zero magnetic field and susceptibility determinations require the presence of a field. The resistivity measurements in magnetic fields (Kobay, Wuzzz) show that for the LaSrCuO and YBaCuO systems, respectively, the transition temperature broadens and shifts downward by perhaps 1 K/T (Ihar1, Nakao), as shown in Fig. X-3. In an ac susceptibility determination a downward shift of T_c by 2 K was reported for an increase in field amplitude from 3.1 to 31 μT (Odazz).

2. Hall Effect

The Hall effect provides information on the sign and the mobility of charge carriers in the normal state, and usually a positive sign indicates that the majority carriers are holes. In the superconducting state the Hall voltage is expected to drop to zero (Hundl, Zhan1).

In this experiment a magnetic field H_0 is applied perpendicular to the direction of the current flow through the sample. The Lorentz force of the magnetic field on the moving charge carriers produces a charge separation which induces an electric field E_x perpendicular to the current and magnetic field directions. The Hall coefficient R_H is the ratio

$$R_H = E_x/JH_0 \tag{X-4}$$

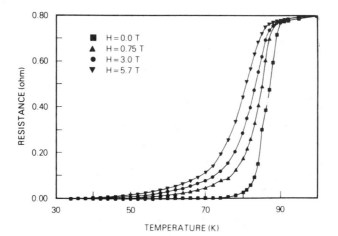

Fig. X-3. Temperature dependence of the resistance of YBaCuO in various magnetic fields from 0 to 5.7 T (Wuzzz).

where J is the current density. When the charge carriers are electrons with the density n per cubic centimeter, the Hall coefficient is negative with the value

$$R_H = -1/nec \tag{X-5}$$

in cgs units. A similar expression with a positive sign applies to hole conduction. Also of interest are the Hall mobility

$$\mu_H = R_H/\rho \tag{X-6}$$

where ρ is the resistivity, and the dimensionless Hall number V_0/R_He, where V_0 is the volume per formula unit:

$$V_0 = 94 \text{ Å}^3 \quad \text{for } (La_{0.925}Sr_{0.075})_2CuO_4 \tag{X-7a}$$

$$V_0 = 174 \text{ Å}^3 \quad \text{for } YBa_2Cu_3O_7 \tag{X-7b}$$

Some authors find a strong temperature dependence of R_H or the Hall number V_0/R_He for LaSrCuO (Tonou) and YBaCuO (Penne, Wangz). Others find a weak dependence for LaSrCuO (Hundl, Penne, Uchi1), and a large anomaly in R_H near T_c has been observed in YBaCuO (Gottw, Yongz). Figure X-4 shows R_H of three substituted YBaCuO compounds increasing strongly with decreasing temperature (Cheon, Tana1). Figure X-5 shows the temperature dependence of the Hall mobility $\mu_H = R_H/\rho$, the Hall number V_0/R_He and the resistivity ρ of an epitaxial film of YBaCuO (Chau1, Penne). Hall-effect data on various compounds are listed in Table X-3.

Hall-effect data on LaSrCuO provided the room-temperature electron concentration $n = 1.5 \times 10^{21}/cm^3$ and the room-temperature mobility $\mu_H = 4.17$

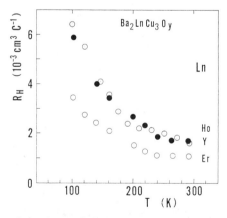

Fig. X-4. Temperature dependence of the Hall coefficient R_H of $LnBa_2Cu_3O_{7-\delta}$ for Ln = Y, Ho, and Er (Tana1).

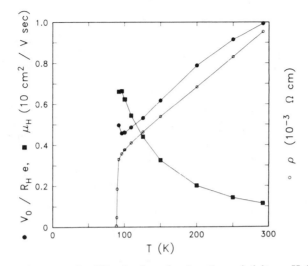

Fig. X-5. Transport data for $YBa_2Cu_3O_{7-\delta}$ showing the resistivity ρ, Hall number V_0/R_He, and Hall mobility $\mu_H = R_H/\rho$ versus temperature (Chau1).

cm^2/V sec (Tonou). The electron concentration decreases with increasing temperature, probably owing to significant capture of thermally excited electrons at deep traps in grain boundaries surrounding the conductive channels. The Hall coefficient measurement of $(La_{1-x}Sr_x)_2CuO_4$ single-crystal films grown epitaxially on $SrTiO_3$ provided a room-temperature carrier density $n = 6.8 \times 10^{21}/cm^3$ which decreased notably with falling temperature (Suzuk).

Hall voltage measurements in LaSr* indicated the presence of granular or inhomogeneous superconductivity (Hundl). The temperature dependence of the Hall coefficient of $(La_{1-x}Sr_x)_2CuO_{4-\delta}$ was interpreted in terms of a relatively large phonon coupling and conduction via both electron and hole bands (Uher1, Uher3). The chemically determined electron deficiency or hole concentration in this compound exhibited a direct correlation with T_c for $x \leq 0.15$, suggestive of single-band transport and supportive of an all electronic mechanism for superconductivity (Shafe, Shaf1).

Hall-effect measurements on $YBa_2Cu_3O_{8-\delta}$ single crystals with the magnetic field on ± 1 T in the ab plane yielded a negative Hall constant in the range -0.75 to -9×10^{-10} m^3/C corresponding to 1.2–1.5 electrons per formula unit (Tozer). A large temperature dependence in R_H of YBa* was attributed to multiband conduction (Hongm) and to the temperature dependence of the 2-D and 1-D chain mobility ratio (Wangz). The Hall coefficient was inversely dependent on the temperature in the compounds YBa* and GdBa*, which are judged as moderately heavy Fermion-like, with a Coulomb correlation energy comparable to or larger than the bandwidths (Cheon). Around the transition a larger increase was observed in R_H for both YBa* and DyBa* (Yongz, Zhan1). In DyBa* the peak in R_H corresponds to $n \approx 4 \times 10^{19}$ cm^{-3}, and was interpreted as a grain

TABLE X-3. Hall Effect Data[a]

Material	R_H (cm³/°C)	V_0/R_He[b]	n (cm⁻³ × 10²¹)	μ (cm²/Vsec)	ρ (μΩ cm)	T (K)	Ref.
LaSr*	+0.001		6.0			300	Hund1
	+0.004		2.1	4.3	600	77	Orngzz
		0.3		+0.75	2600	300	Penne
LaSr(0.1)	−0.005		1.5	−4.17	1200	300	Tonou
LaSrCuO			6.8	1.2	710	300	Suzuk
La₂CuO₄	0.11			1.0	10⁵	300	Uchi2
YBa*	0.1	1.0		+1.0	1000	300	Chau1
		1.8		±0.5	1300	300	Penne
		1.4		±0.8	950	300	Penne
	0.0005−0.002		1.4			290	Wangz

[a] Hall coefficient R_H, Hall number V_0/R_He, carrier density n, mobility μ, resistivity ρ, and measurement temperature T. The notation used is: $(La_{1-x}Sr_x)_2CuO_4$ = LaSr(x), LaSr(0.075) = LaSr*, $YBa_2Cu_3O_{7-\delta}$ = YBa*.

[b] Hole concentration per formula unit.

boundary effect. In the case of YBa*, both the intrinsic quality and the grain boundary origin were mentioned (Yongz). The Hall coefficient R_H decreases with decreasing oxygen content in $YBa_2Cu_3O_{7-\delta}$, with a plateau of very little change from $\delta = 0.1$ to $\delta = 0.4$ (Ongz1, Wangz).

Several papers (Allen, Alle1) have discussed the transport coefficients in the relaxation time approximation. They concluded that the Hall tensor tends to be holelike in the *ab* plane and electronlike in the *ac* and *bc* planes, which explains why sign differences have been reported in the literature.

3. Thermoelectric Effects

A conductor with a temperature gradient and no electric current develops a steady-state electrostatic potential difference between the high and low temperature regions; this phenomenon is called the thermoelectric, thermopower, or Seebeck effect. This and other related effects such as the Peltier effect vanish in the superconducting state (Hundl).

Figure X-6 shows the temperature dependence of the thermoelectric power or Seebeck coefficient S of La_2CuO_4 (Gran1) and $(La_{1-x}Ba_x)_2CuO_4$ for several values of x (Alle3, Hundl, John3, Marcu, Uchi1, Uher1, Uher2, Yanz1). The coefficient decreases with increasing $x = 0.025$, 0.05, and 0.075, and the latter two compounds exhibit a rapid decrease to zero near the phase transition. No drop was observed in S at low temperatures for $x = 0$. In another study the same result was found for $x = 0.075$, but S for the $x = 0$ sample decreased dramatically to a very low value below 90 K. This was attributed to a changeover from an activated type of semiconductor transport at high temperatures to a variable range-hopping type at low temperatures (Maeno).

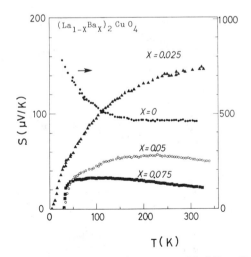

Fig. X-6. Temperature dependence of the thermopower S of $(La_{1-x}Ba_x)_2CuO_4$ for several values of x (Uchi1).

The thermopower of polycrystalline granular $(La_{1-x}M_x)_2CuO_{4-\delta}$ with M = Ba or Sr was calculated in the effective medium approximation and the results for the x dependence of $S(T)$ are in good agreement with experimental values (Xiazz). The room-temperature thermopower data of $(La_{1-x}M_x)_2CuO_4$ fit the Heikes expression (Coope, Heike)

$$S = -\frac{k}{e}\left[2\ln 2 + \ln\left(\frac{2x}{1-2x}\right)\right] \qquad (X\text{-}8)$$

where $1 - 2x$ is the number of electrons per copper site. Note that this expression has no adjustable parameters.

There are reports that above T_c the thermopower in YBaCuO and its rare-earth-substituted analogues is positive (Mawds, Mitra), and also is negative (Khimz, Yaozh, Yuzzz), and examples of both cases are shown in Fig. X-7. An electronic or negative $S(T)$ was observed in O_2-heated superconducting samples, and holelike behavior with positive $S(T)$ was seen in air-heated nonsupercon-ducting ones (Raych). A sample of $YBa_2Cu_3O_{6.9}$ exhibited a negative thermo-power between 300 and 125 K attributed to a diffusion mechanism, a positive $S(T)$ from 125 to 90 K ascribed to phonon drag effects, with $S = 0$ below $T_c = 90$ K (Yaozh; see also Khimz).

A large peak in the thermopower of YBa∗, called a precursor effect, was ob-served just above T_c (Mawds, Uher2). It was suggested that high-T_c materials be used for thermopower test leads for the absolute determination of $S(T)$ above the temperature range covered by the present standard Nb–Ti leads (Uherz).

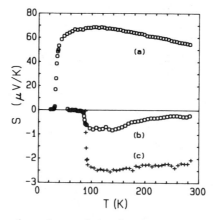

Fig. X-7. Temperature dependence of the thermopower S of (a) a ceramic sample of $La_{1.85}Sr_{0.15}CuO_{4-\delta}$, ($b$) a ceramic sample of $YBa_2Cu_3O_{7-\delta}$, and (c) the ab plane of a $YBa_2Cu_3O_{7-\delta}$ single-crystal sample. Note the different scales on the positive and negative parts of the y axis (Yuzz3).

4. Photoconductivity

Photoconductivity studies of LaCuO, YCuO, and YBaCuO samples suggest that polarons and excitons play a substantial role in the mechanisms of superconductivity (Masum, Masu1, Robas).

5. Thermal Conductivity

The thermal conductivity $K(T)$ is helpful in determining the fraction of the thermal energy that is transported by charge carriers and the amount carried by the lattice (phonons). It can provide information about the electron–phonon interaction, mean-free path, carrier density, and other physical properties. The thermal conductivity is not necessarily divergent or zero in the superconducting state. Thermal conductivity measurements have been reported on both polycrystalline and single-crystal samples (e.g., Bayot, More6, Uher2), and an example of the temperature dependence of $K(T)$ in polycrystalline YBa$_*$ is shown in Fig. X-8.

A detailed study of the transport properties of YBa$_*$ (Gottw) included an analysis of the low-temperature $(0.1 < T < 2$ K$)$ behavior of $K(T)$ in terms of a phonon or lattice contribution K_{ph} and an electronic contribution K_{el}

$$K(T) = K_{ph} + K_{el} \qquad \text{(X-9)}$$

where

$$K_{ph} = aT, \qquad K_{el} = bT^3 \qquad \text{(X-10)}$$

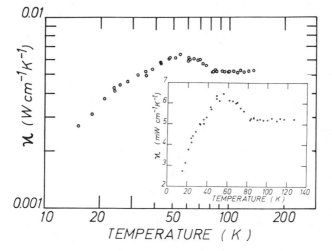

Fig. X-8. Log–log plot of the thermal conductivity of YBa$_*$ versus the temperature. The inset shows the data replotted on a linear scale (More6).

with $a = 16$ $\mu W/K^2$ cm and $b = 47$ $\mu W/K^4$ cm. Hence even at ≈ 0.1 K large portions of the thermal carriers appear to be in the normal state. An observed increase in $K(T)$ below T_c indicated large electron–phonon scattering. The authors estimated $v_F \approx 10^7$ cm/sec, $N_{eff} \approx 10^{22}/cm^3$, a small carrier mean-free-path, and pointed out the similarity with the heavy Fermion system $CeCu_2S_4$.

There was a report that $K_{el} \ll K_{ph}$ at 300 K and $n \approx 0.13$ carriers/Cu atom (Mori6). The observed increase in $K(T)$ below T_c may be due to freer phonon flow, and suggests that a strong electron–phonon coupling is present in YBa*. A T^3 dependence of $K(T)$ in YBa* below 5 K provided evidence for a temperature-independent lifetime (Herem). At 50 K the Lorentz number $L = K\rho/T$ was estimated to be 26×10^{-8} $W\Omega/K^2$, a factor of 20 larger than the Wiedermann–Franz value, and $K_{el} \approx 0.1$ K_{ph} in polycrystalline $(La_{0.9}Sr_{0.1})_2CuO_4$ (Bartk). The thermal conductivity of sintered YBa* in the temperature range from 0.1 to 10 K was linear in T at the lowest temperatures, and had a T^3 dependence at the highest temperatures, in accordance with Eq. (X-9) (Graeb). This was claimed to be consistent with strong Rayleigh scattering of the phonons from the granular particles. In the ab plane single crystals of microtwinned HoBa* obeyed a power law $K(T) = aT^n$ with $n \approx 2$, a behavior similar to that observed in glasses (Graeb).

D. TUNNELING PROPERTIES

Tunneling can be carried out through an insulating layer, I, between two superconductors (S–I–S), between a superconductor and a normal material (S–I–N), and between two normal materials (N–I–N) such as two semiconductors. The dc and ac Josephson effects involve particular types of tunneling phenomena across a barrier between two superconductors. The SQUID is an application of Josephson tunneling that involves macroscopic quantum phenomena. These topics will be discussed in the following three sections.

One of the preferred ways to measure an energy gap is through tunneling experiments, and many values of the gap energy determined by this technique are recorded in Table IX-1.

1. Tunneling Measurements

A tunneling study of LaSrCuO thin films was made using both the sandwich-type and the point-contact-type tunneling techniques (Naito). Sandwich-type junctions were prepared by forming a small window ≈ 150 μm square and depositing a Pb counter electrode, which formed a natural high-resistance tunnel barrier, typically 1 $M\Omega$. Point contact tunneling was done with the aid of a scanning electron microscope (SEM) using a tungsten probe with a tip radius less than 100 nm. Continuous scanning was not possible, but several points on the same specimen within a 1×1 μm^2 window could be sampled. The curves of differential tunneling conductances dI/dV as a function of the bias voltage presented in Fig.

X-9 have ordinate scales proportional to the density of states, and they provide gap energies.

In a low-temperature SEM tunneling study of $(La_{0.9}Sr_{0.1})_2CuO_{4-\delta}$ at 5 K the electrons tunneled from the tip into many superconducting grains with inhomogeneities small compared to the grain sizes (Kirtl). The data were analyzed in terms of the model of Zeller and Giaever (Zelle), and gave E_g/kT_c in the range from 3.5 to 6.3. Another experiment (Tsue1) was consistent with tunneling into grain sizes ≤ 1 nm which are much smaller than the apparent crystal sizes. A study of $(Y_{0.55}Ba_{0.45})_2CuO_{4-\delta}$ using point-contact tunneling at 4.2 and 27.4 K provided gap ratios E_g/kT_c in the range 3.8–4.2, somewhat larger than the BCS ratio (Ekino). Figure X-10 shows an example of the way I versus V SEM electron tunneling curves give a range of gap values (5 meV $\leq E_g \leq$ 190 meV) for a niobium tip located at various points of an aluminum-doped YBa* sample (Gall1).

Electron tunneling between the grains of oxygen-poor YBa* exhibited a differential resistance dV/dI which appears to have an asymptotic peak that did not appear in homogeneous oxygenated samples (Escud). The gap signature was

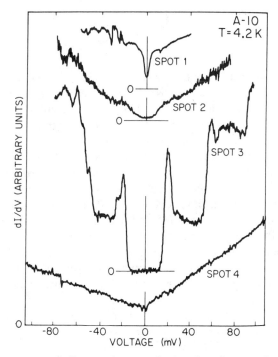

Fig. X-9. Point-contact tunneling conductance for four locations on a LaSrCuO thin film at 4.2 K (Naito).

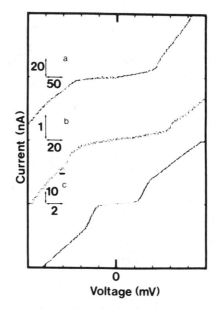

Fig. X-10. Current versus voltage plots obtained using a niobium tip at different positions on the surface of aluminum-doped $YBa_2Cu_3O_{7-\delta}$. The difference scales used for each curve are indicated. The spectra from top to bottom show jumps in current at 95, 30, and 2.5 meV, respectively (Gal11).

not very well resolved and $E_g/kT_c \approx$ 7–13. Tunneling studies of YBa* thin films (Ogale) exhibited a dependence of the normalized critical current I_C/I_{C0} on T/T_c, which differs from some recent S–I–S junction calculations (Ambeg). The deGennes model was reported to be better over a small temperature range, but the predicted curvature was larger than that of the data.

In the break-junction tunneling technique (More3, More5) a small piece of bulk material is electromechanically broken under liquid helium, and the freshly fractured surfaces are adjusted to form a tunneling barrier with the liquid helium acting as the insulator (More4). The most common current–voltage curves obtained by this technique (More4) closely followed a quadratic current dependence on the voltage. Some curves have derivatives that exhibit structure indicative of the presence of gap sum voltages, and others have shapes that suggest quasi-particle tunneling between superconducting electrodes (Morel). Related behaviors are observed, for example, for tunneling from a PtIr tip into LaSrCuO (Tsuei, Zelle).

In the absence of high-quality "sandwich" junctions these break-junction results are interesting. However, the effects of the strain and elastic failure on the electronic properties of the specimens and junctions is not clear. This could be of concern, particularly in the case of ceramic superconductors that are sensitive to sample treatment.

2. Josephson Effect

When two superconductors are separated by a thin layer of insulating material, electron pairs can tunnel through the insulator from one superconductor to the other. There are three effects of pair tunneling, namely:

1. The dc Josephson effect, which is the flow of a dc current $J = J_0 \sin \delta$ across the junction in the absence of an applied electric or magnetic field, where δ is a phase factor and J_0 is the maximum zero voltage current.

2. The ac Josephson effect relates to the flow of a sinusoidal current $J = J_0 \sin[\delta - (4\pi e Vt/h)]$ across a junction with an applied voltage V where $\nu = 2eV/h$ is the frequency of oscillation.

3. Macroscopic quantum interference effects involving a tunneling current J with an oscillatory dependence on the applied field, given by

$$J = J_0 \frac{\sin \pi \Phi/\Phi_0}{\pi \Phi/\Phi_0} \tag{X-11}$$

where the magnetic flux Φ may be approximated as the product of the average magnetic field strength times the cross-sectional area, and $\Phi_0 = hc/2e$ is the fluxoid or quantum of magnetic flux.

In the reverse ac Josephson experiment, dc voltages are induced across an unbiased junction by introducing an rf current into the junction, or by radiatively coupling an rf signal through a coil surrounding the sample (Chenz, Weng1). It was suggested that this result supports the existence of granular superconductivity in YBaCuO at 240 K (Weng1).

Anomalous voltage excursions as a function of temperature and magnetic field strength were reported above 100 K (Caizz). The onset of the excursions was 20 mT, they reached a maximum at about 33 mT, and disappeared for fields above 56 mT. The voltage jumps were different for cooling and heating, and they were more frequent for larger samples. These transient voltages were attributed to flux jumps of granular superconductors. It was proposed that some grains have transition temperatures as high as 160 K.

Superconducting oxide materials are porous with chains of grains measuring a few microns in size. Owing to the inverse Josephson effect an applied rf current could cause individual Josephson junctions in these materials to develop quantum voltages given by $V_j = nh\nu/2e$, which is on the order of nanovolts. Such junctions could even be inside the grains themselves (Blaze). Thermal smearing can prevent the detection of individual quantum voltages, but observable dc voltages in the millivolt range can result from the summation of thousands of junctions with n values as large as 100.

Josephson junctions in thin films of LaSr(0.1) are believed to form at grain boundaries. In one experiment the Josephson current was found to be proportional to $[1 - (T/T_c)]^2$. In contrast, samples of YBa* (Cuizz) and BaPb$_{1-x}$Bi$_x$O$_3$ (Suzu2) produced current proportional to $[1 - (T/T_c)]^{3/2}$, which may suggest proximity effects (Fuku3, Kobes, Kres3, Lynto) with comparatively long-range leakage of Cooper pairs (Moriw). The proximity effect can cause two superconductors with different T_c values in contact with each other to exhibit one intermediate T_c, and it can cause a nonsuperconductor–superconductor pair to act like a superconductor with a lower T_c (Kres3).

A LaSr* sample mounted in a point contact current–voltage probe with a conducting tip produced a hysteretic I versus V characteristic of the type shown in Fig. X-11 (Estev, Tsaiz). Microwave irradiation produced the Shapiro steps (Baron), which result from the beating of the oscillating Josephson supercurrent with the microwaves. The separation in voltage between these steps is proportional to the microwave frequency, and their amplitude is Bessel-like. The Josephson junction characteristics are observed even when the metal tip is nonsuperconducting, which indicates that the superconducting junction is inside the material under the tip (Estev). The absence of voltage steps during the microwave irradiation of a Nb–YBaCuO point-contact junction suggests that the coupling between the superconducting regions along the percolative path is non-Josephson (Tsaiz). Other workers studying Nb–YBa* point-contact junctions observed Shapiro steps and an unusual noise behavior along the I–V characteristic (Kuzni), and an estimate was made of $\Delta \approx 19.5 \pm 20$ mV and $2\Delta/kT_c \approx 4.8$ (Baro1). Very clear steps were observed in a weak link fabricated by carving a bridge (0.1×0.2 mm^2) in a $10 \times 2 \times 2$ mm^3 YBa* sample. Both harmonic ($n\Phi_0$) and subharmonic ($n\Phi_0/m$) flux quantum steps were reported at 77 K (Chan2). Other microwave radiation experiments have also been carried out (McGra, Nieme, Mengz).

Theoretically microwave absorption by S–I–S junctions with square-well potentials at low temperatures is calculated to occur in sharp steps whenever the microwave energy is an integral multiple of the minimum energy needed to excite quasi particles from the ground state to bound Andreev excited states (Aberl, Andre). Andreev reflections at an Ag–YBa* interface on a thin film have been reported. It was argued to provide evidence for Cooper or zero-momentum carrier pairs. The gap as determined from the reflection data was 12.5 meV, compared with 14 meV deduced from tunneling on the same film. These results are consistent with a BCS picture (Hoeve).

3. Macroscopic Quantum Phenomena

Macroscopic quantum phenomena were reported in Sn–YBaCuO and YBa-CuO–YBaCuO point contacts. The critical current and the voltage are periodic in the magnetic field, and for each voltage there is a minimum and a maximum value of the current between which the oscillations take place when the magnetic field is varied (DeWae, DeWa1). This dc SQUID (Finkz, Jakle, Soule) behavior has been observed up to 40 K (Kawab), 66 K (DeWae, DeWa1), and 90 K (Tsai1).

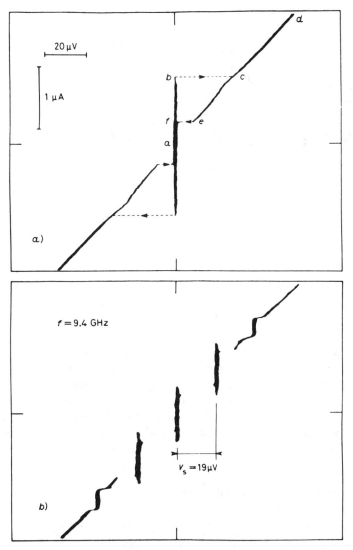

Fig. X-11. (*a*) Oscilloscope trace of a current–voltage characteristic obtained at 4.2 K with an aluminum tip on a LaSr* sample. Letters *a* through *f* indicate the sense of the trace. Dashed lines have been added to indicate the switching between the two branches. (*b*) Steps induced by microwave irradiation at the frequency $f = 9.4$ GHz (Estev).

SQUIDs have been fabricated from $YBa_2Cu_3O_{7-\delta}$ (Colcl, Kawab, Zimme). The flux quantum Φ_0 obtained from these devices was the expected $hc/2e$ (Gough, Koch1, Koch2). For a YBa* specimen to exhibit SQUID behavior the sample need not necessarily be multiply connected. For such a YBa* SQUID operating in the magnetometer mode the field spectral density was 5.8×10^{-10} $T/\sqrt{H_C}$ at 4.2 K and 1.5×10^{-10} $T/\sqrt{H_C}$ at 77 K (Pegru). This value is better than that of a good flux-gate magnetometer.

XI

CONCLUDING REMARKS

A. INTRODUCTION

The worldwide growth in this area of research over the past 2 years has been phenomenal. We believe that the relative ease with which some of the copper oxide superconductors can be produced, together with the ready availability of the starting materials and the liquid nitrogen cryogenics, are to a great extent responsible for this rapid growth in research. Other factors are the excitement of being the first to discover new properties and provide meaningful explanations, and of course the intrinsic competitiveness of the whole adventure.

B. GROWTH OF THE LITERATURE

We estimate that the reprint and preprint literature has reached several thousand articles. There has been an enormous amount of duplication; some of us who have surveyed other areas in condensed-matter physics are amazed by the repetitions that we found. Not only have different groups that are oceans apart reported analogous results, but sometimes individual groups publish somewhat lengthy reports of minor extensions of earlier work. During the course of our literature search some figures became so familiar to us that we would identify the authors of some articles just by looking at figures present therein. Unquestionably this is partly because many journals adapted innovative editorial and reviewing policies for this field, but the overall effect of these policies has surely been beneficial.

Aside from editorial concerns, there are several positive results of this duplication. Some which come to mind are:

1. The observation of above liquid nitrogen temperature superconductivity by hundreds of researchers in specimens produced from different sources under different conditions is a confirmation of some underlying commonality of these materials. It implies that one does not really have to fine tune the system to achieve 100 K, in some cases just putting the chemicals together in the correct proportions appears to be adequate.

2. There is a thrill of discovery and satisfaction even when one is reproducing a result mentioned in the *New York Times* or just heard over the grapevine.

3. Perhaps a field cannot grow as fast as that of high-temperature superconductivity without this type of corroboration. Even in the established disciplines classic measurements such as the Etövös experiment and the like are duplicated. Hence as a literature grows there is a proportionate growth in the "repeat" work.

Despite our intentions we have been unable to cite many of the papers. Most of our discussion concerned articles in print, but some had been received earlier as preprints. The overseas preprints were received in fewer numbers, and may not be representative of their true volume. It is unfortunate that we were unable to cover the literature from laboratories in China, India, Japan, the Soviet Union, and other countries as much as we would have liked. Nevertheless, it is clear that large numbers of condensed-matter researchers in these parts of the globe have done much productive research in superconductivity.

C. THE PRESENT SITUATION

In this book we have summarized the present experimental knowledge regarding copper oxide superconductors and compared their properties with those of the different types of previously known superconductors. We have also introduced and discussed the principal theoretical explanations to provide some perspective for the experimental results. The conventional BCS theory has been treated as the benchmark, and comparisons of the data with this and with other models have been made throughout the text.

A general consensus has been emerging concerning the physical properties of these copper oxide materials. They exhibit the two hallmarks of a superconductor, namely, zero resistivity and perfect diamagnetism, although most specimens do not achieve the latter. Other common properties are electron pairing, the existence of an energy gap, and flux quantization.

The copper oxide materials are Type II superconductors with strong anisotropies, a granular (ceramic) composition, and a relatively high room-temperature resistivity. The upper critical field is quite high, and the highest critical currents have been reported in single crystals and epitaxial films. Their crystallographic structures contain (oxygen-deficient) copper oxide layers which appear to be responsible for the superconducting properties. Atomic substitutions on the copper oxide layers seriously degrade or destroy the superconductivity, while atomic substitutions on the other sites tend to have very little effect on it.

D. COMPARISON WITH PREVIOUS SUPERCONDUCTORS

A comparison with the transition ion superconducting compounds shows many commonalities and many marked differences. Commonalities include (a) the widespread presence of a group-VI ion (O, S, Se, Te), the new materials having oxygen and the older ones generally another member of this group, (b) perhaps a large inferred electron–phonon coupling constant, and (c) similar quantitative agreements with the BCS theory.

Differences between the old and new superconductors include

(a) A much greater adherence to stoichiometry among the transition ion materials.

(b) The newer superconducting compounds tend to have more atoms per formula unit, namely, 7 for the prototype La_2CuO_4, 13 for $YBa_2Cu_3O_7$ and 19 for $Tl_2Ca_2Ba_2Cu_3O_{10+\delta}$, in contrast to, for example, 3 for the laves phases (e.g., $LaOs_2$), 4 for the A-15 compounds (e.g., Nb_3Ge), and 5 for some heavy Fermions (e.g., $CeCu_2Si_2$). There are exceptions such as Chevrel phases (e.g., $NaMo_6S_8$) and some heavy Fermions (e.g., UBe_{13}).

(c) The crystallographic symmetry tends to be higher among the old superconductors, which are ordinarily cubic or hexagonal, than among the newer oxides, which are tetragonal or orthorhombic.

(d) The newer oxides have smaller coherence lengths, larger penetration depths, and larger Ginzburg–Landau κ parameters.

(e) Among the copper oxide types there is a separation of the lattice into regions (Cu–O layers) believed to be involved in the superconductivity, and regions not so involved; there was no such clear subdivision among the earlier transition ion types.

(f) The occasional presence of layered structures among some older superconductor types and its widespread occurrance among the newer ones.

(g) The newer types have much higher transition temperatures!

E. MECHANISMS AND MODELS

The theoretical literature is not nearly as voluminous as the experimental one. However, it shows evident signs of dynamism and vitality, and is full of many creative ideas about high-temperature superconductivity. Many theoretical workers have favored the archetypal phonon-mediated BCS model, and others favor BCS with different interactions involving, for example, excitons. Possible alternatives to the phonon-mediated BCS theory include antiferromagnetic models, and some theoretical approaches involve such factors as fractons, solitons, spin bags, and real-space bosons.

Before discussing the new models it may be fair to say that most of the empirical facts about these copper oxide superconductors are not inconsistent with

Type II superconductivity. The BCS theory has been highly successful in explaining the behaviors of most previously known superconductors, and indications are that it may also be adequate for these new high-temperature materials.

The newer superconductors exhibit the following evidence for being BCS types: (a) many estimated ratios of the band gap to kT_c are close to the BCS value of 3.53, (b) the BCS type specific heat anomaly at T_c has been observed, (c) charge pairs are involved in the superconductivity, (d) there is evidence for the presence of electron–phonon coupling and the electron–phonon coupling constants have been determined, and (e) the isotope effect has been observed in some materials.

Nonphonon mechanisms for superconductivity within the framework of the BCS theory have also been reported. These models introduce other boson excitations such as excitons and plasmons, but they are open to criticisms similar to those applied to the phonon model. Evidence for many of these excitations has been reported in studies involving IR and optical spectra, photoemission, neutron scattering, tunneling experiments, and so on, but there is no consensus on the extent to which these excitations are important.

There are several other theoretical questions. For example, is there Fermi surface nesting of the 2-D half-filled bands as the band structure calculations require? The antiferromagnetism of La_2CuO_4 appears to introduce a new aspect to some of the theoretical procedures. How should the antiferromagnetic correlation effects be introduced? This could draw attention to the "magnetic" (Hubbard) models.

There are broadly two classes of antiferromagnetic models, the resonant and nonresonant type. In the resonant "magnetic" models such as RVB the major difficulties are: (a) the ground state in La_2CuO_4 is observed to be closer to the conventional antiferromagnetic Néel state than it is to a resonant state of dimers; (b) the predicted gap is small and calculations seems to indicate $(E_g/kT_c)_{RVB} \ll (E_g/kT_c)_{EXP}$; (c) a great deal of elaborate analysis appears needed to calculate the measurable superconducting properties with this model.

A typical nonresonant magnetic approach such as the antiferromagnetic formalism of Emery and others may be a possible option. In this model there are indications that superconductivity may be absent in a single-band model, and phonons may be required to produce superconductivity. Also, evidence for the presence of real-space paired bosons above T_c is lacking. It is not yet clear to what extent magnetic interactions are central or peripheral in these new superconductors.

F. VERY HIGH T_c

It may be possible that superconductivity above 200 K, or even above room temperature, can exist. However, as of this writing the reports of very high T_c values, well above 130 K, have not been confirmed. So many really high T_c "discoveries" have been released to the general press that were never reproduced by others that now new reports are met with skepticism, and the emphasis is on confirmation by several independent groups. This is what happened with the bismuth and thallium compounds.

G. OPEN QUESTIONS

Many properties of the copper oxide superconductors have been reported as being glasslike; for example, they show unconventional irreversibility and slow relaxation. Interestingly, this type of behavior is observed even in single crystals. It is possible that these properties are a consequence of Type II superconductivity.

Other structural questions are the importance of 1-D chains, 2-D planes, and 3-D couplings, the role of orthorhombicity and $a-b$ anisotropy, the significance of antiferromagnetism, and whether symmetries and observed commonalities are essential for high T_c. Another question is the role of alkaline earth–oxygen layers, alkaline earth layers without oxygen, and perhaps no intervening layer at all between copper oxide planes. Many authors have addressed these points, but a great deal remains to be done before satisfactory answers are forthcoming.

The possibility of totally replacing copper and/or oxygen without degrading superconductivity remains a challenge. There was a report of 50-K superconductivity in a silver oxide compound, and partial replacement of oxygen by fluorine and sulphur without a serious degradation of the superconductivity has also been achieved, but more work remains to be done.

Two persistent questions are (1) how do we raise T_c and (2) what is the mechanism responsible for high temperature superconductivity.

H. CONCLUSION

In summary, the superconducting properties of copper oxide materials are, in many ways, similar to those of conventional Type II superconductors. However there are still some uncertainties concerning the nature of the mechanism of the superconductivity. The BCS phonon-based model appears to be consistent with most empirical observations. Several other microscopic explanations for this behavior are also available, but it is not obvious if any of them is superior to BCS and provides the correct model.

It is also possible that an adaptation of the BCS theory with an eclectic admixture of several mechanisms such as phonon, plasmon, exciton, and others will be required to account for all observed behavior. It may be that copper oxides have everything just right. Or a totally new theoretical approach may eventually provide the ultimate explanation.

At the end of the first chapter we drew attention to the recent very rapid increase in the transition temperature over the past 2 years, and we mentioned the possibility that $T_c > 300$ K could be reached in the not too distant future. It will be an exciting challenge to strive for the goal of room-temperature superconductivity, and perhaps enter the new century along with it. We hope that this book will be helpful to the many scientists, young and mature, who have embarked on this merry chase.

APPENDIX

Various authors have used different units for some quantities such as energy, magnetic field, and pressure, and the following conversion factors and constants might be useful to the reader.

BCS gap ratio: $E_g/kT_c = 3.53$

BCS specific heat jump: $(C_S - C_n)/C_n = 1.43$

BCS isotope effect exponent $(T_c = M^{-\alpha})$ for pure elements: $\alpha = \frac{1}{2}$

Energy: 1 cm^{-1} = 0.124 meV; 1 meV = 8.066 cm^{-1}

Energy gap: $E_g = 0.08616 T_c (E_g/kT_c)$ meV

Flux quantum or fluxoid $\Phi_0 = hc/2e = 2.0679 \times 10^{-7}$ G cm^2

Josephson frequency $\omega = 2\pi (2\ eV/h)$; $[1\ \mu V = 483.6$ MHz$]$

Length: 1 Å $= 10^{-10}$ m $= 10^{-8}$ cm $= 10^{-4}\ \mu$m $= 0.1$ nm

Magnetic field: 1 tesla (T) $= 10^4$ gauss (G); 1 mT $= 10$ G

Pressure: 0.987×10^4 atm ≈ 10 kbar $= 1$ GPa $= 7.5 \times 10^6$ torr

Temperature: 1 K $= 0.6950$ cm^{-1} $= 0.08616$ meV

The molecular weight (MW), cell volume per formula unit (V_0), and typical lattice constants (a,b,c) for LaSr$_*$ = $(La_{0.925}Sr_{0.075})_2CuO_4$, YBa$_*$ = $YBa_2Cu_3O_7$, BiSr$_*$ = $Bi_2Sr_2CaCu_2O_8$ and TlBa$_*$ = $Tl_2Ba_2CaCu_2O_8$ are:

LaSr$_*$: MW $= 397.7$; $V_0 = 94$ Å^3; $a = b = 3.77$ Å; $c = 13.2$ Å.

YBa$_*$: MW $= 666.2$; $V_0 = 174$ Å^3; $a = 3.83$ Å; $b = 3.89$ Å; $c = 11.68$ Å.

BiSr$_*$: MW $= 888.4$; $V_0 = 223$ Å^3; $a = b = 3.82$ Å; $c = 30.6$ Å.

TlBa$_*$: MW $= 978.6$; $V_0 = 218$ Å^3; $a = b = 3.86$ Å; $c = 29.3$ Å.

Note that V_0 is the unit cell volume (abc) for YBa$_*$ and half of the unit cell volume $(a^2c/2)$ for LaSr$_*$, BiSr$_*$ and TlBa$_*$.

REFERENCES

ACS Symp. (1987) denotes: *Chemistry of High-Temperature Superconductors*, D. L. Nelson, M. S. Whittingham, and T. F. Georgs (Eds.), ACS Symposium Series No. 351, American Chemical Society, Washington, D.C., 1987.

MRS Anaheim Symp. (1987) denotes: *High Temperature Superconductors, Proceedings of Symposium*. S. 1987 Spring Meeting of the Materials Research Society, April 1987, Anaheim, California, D. U. Gubser and M. Schluter (Eds.), Materials Research Society, Pittsburgh, PA, 1987.

MRS Boston Symp. (1987) denotes: *High Temperature Superconductors, Proceedings of Symposium*, Vol. 99, Fall Meeting of the Materials Research Society, Nov.-Dec., 1987, Boston, MA. M. B. Brodsky, R. C. Dynes, K. Kitazawa, and H. L. Tuller (Eds.), Materials Research Society, Pittsburgh, PA, 1988.

Novel SC (1987) denotes: *Novel Superconductivity*, International Workshop, Berkeley, California, June, 1987. S. A. Wolf and V. Z. Kresin (Eds.), Plenum, New York, 1987.

Aberl   H. Aberle and R. Kummel, Phys. Rev. Lett. *57*, 3206 (1986).

Abrik   A. A. Abrikosov and L. P. Gor'kov, Sov. Phys. JETP *12*, 1243 (1961).

Adach   H. Adachi, K. Setsune, and K. Wasa, Phys. Rev. B *35*, 8824 (1987).

Aeppl   G. Aeppli, R. J. Cava, E. J. Ansaldo, J. H. Brewer, S. R. Kreitzman, G. M. Luke, D. R. Noakes, and R. F. Kiefl, Phys. Rev. B *35*, 7129 (1987).

Aharo   A. Aharony, S. Alexander, O. Entin-Wohlmann, and R. Orbach, Phys. Rev. B *31*, 2565 (1985).

Ahar1   A. Aharony, R. J. Birgeneau, A. Coniglio, M. A. Kastner, and H. E. Stanley, Phys. Rev. Lett. *60*, 1330 (1988).

Ahnzz   C. C. Ahn, D. S. Lee, and K. Samwer, MRS Boston Symp. (1987), p. 793.

Akimi   J. Akimitsu, T. Ekino, H. Sawa, K. Tomimoto, T. Nakamichi, M. Oshiro, Y. Matsubara, H. Fujiki, and N. Kitamura, Jpn. J. Appl. Phys. *26*, L449 (1987).

Aleks   A. S. Aleksandrov, V. N. Grebenev, and E. A. Mazur, Pis'ma Zh. Eksp. Teor. Fiz. *45*, 357 (1987).

Alexa   A. S. Alexandrov, J. Ranninger, and S. Robaskiewicz, Phys. Rev. B *33*, 4526 (1986).

Allen   P. B. Allen, W. E. Pickett, and H. Krakauer, Phys. Rev. B *36*, 3926 (1987).

Alle1   P. B. Allen, W. E. Pickett, and H. Krakauer, MRS Boston Symp., p. 183 (1987).

Alle2   D. Allender, J. Bray, and J. Bardeen, Phys. Rev. B *7*, 1020 (1973).

Alle3   P. B. Allen, W. E. Pickett, and H. Krakauer, Novel SC, 489 (1987).

Allge   C. Allgeier, J. S. Schilling, and E. Amberger, Phys. Rev. B *35*, 8791 (1987).

Allg1   C. Allgeier, J. S. Schilling, H. C. Ku, P. Klavins, and R. N. Shelton, Sol. St. Comm. *64*, 227 (1987).

Almas   C. Almasan, J. Estrada, C. P. Poole, Jr., T. Datta, H. A. Farach, D. U. Gubser, S. A. Wolf, and L. E. Toth, MRS Boston Symp., p. 451 (1987).

Almon   D. P. Almond, E. Lambson, G. A. Saunders, and W. Hong, J. Phys. F: Met. Phys. *17* L221 (1987).

Almo1   D. P. Almond, E. F. Lambson, G. A. Saunders, and W. Hong, J. Phys. F: Met. Phys. *17*, L261 (1987).

Alpzz   E. E. Alp, G. K. Shenoy, D. G. Hinks, D. W. Capone, II, L. Soderholm, H. B. Schuttler, J. Guo, D. E. Ellis, P. A. Montano, and M. Ramanathan, Phys. Rev. B *35*, 7199 (1987).

Ambeg   V. Ambegaokar, P. G. de Gennes, and A. M. Sequin-Tremblay, C. R. Acad. Sci. Paris, *305*, 757 (1987).

Ander   P. W. Anderson, Science *235*, 1196 (1987); see also Ande5 and Ande6.

Ande1   P. W. Anderson, G. Baskaran, Z. Zou, and T. Hsu, Phys. Rev. Lett. *58*, 2790 (1987).

Ande2   P. W. Anderson, "It's Not Over 'Till the Fat Lady Sings." Am. Phys. Soc. Meeting, New York, March (1987).

Ande3   P. W. Anderson, Phys. Rev. *86*, 694 (1952).

Ande4   P. W. Anderson, Mater. Res. Bull. *8*, 153 (1973).

Ande5   P. W. Anderson, Novel SC 295 (1987).

Ande6   P. W. Anderson, Phys. Rev. Lett. *59*, 2497 (1987); comment on Ander.

Andre   A. F. Andreev, Sov. Phys. JETP *19*, 1823 (1964); *22* 455 (1966).

Andr1   J. P. Andreeta, H. C. Basso, E. E. Castellano, J. N. H. Gallo, A. A. Martin, and O. E. Piro, Phys. Rev. B *36*, 5588 (1987).

Anton   G. M. Antonini, C. Calandra, F. Corni, F. C. Matacotta, and M. Sacchi, Europhys. Lett. *4*, 851 (1987).

Antso   O. K. Antson, P. E. Hiismäki, H. O. Pöyry, A. T. Tiitta, K. M. Ullakko, V. A. Trunov, and V. A. Ul'yanov, Sol. St. Comm. *64*, 757 (1987).

Aokiz   H. Aoki and H. Kamimura, Sol. St. Comm. *63*, 665 (1987).

Arend   R. H. Arendt, A. R. Gaddipati, M. F. Garbauskas, E. L. Hall, H. R. Hart, Jr., K. W. Lay, J. D. Livingston, F. E. Luborsky, and L. L. Schilling, MRS Boston Symp., 203 (1987).

Asela   T. L. Aselage, B. C. Bunker, D. H. Doughty, M. O. Eatough, W. F. Hammetter, K. D. Keefer, R. E. Loehman, B. Morosin, E. L. Venturini, and J. A. Voigt, MRS Anaheim Symp., 157 (1987).

Ashau   B. Ashauer, W. Lee, and J. Rammer, Z. Phys. B Cond. Mat. *67*, 147 (1987).

Atobe   K. Atobe and H. Yoshida, Phys. Rev. B *36*, 7194 (1987).

Ayach   C. Ayache, B. Barbara, E. Bonjour, R. Calemczuk, M. Couach, J. H. Henry, and J. Rossat-Mignod, Sol. St. Comm. *64*, 247 (1987).

Ayyub   P. Ayyub, P. Guptasarma, A. K. Rajarajan, L. C. Gupta, R. Vijayaraghavan, and M. S. Multani, Z. Phys. C: Sol. St. Phys. *20* L673 (1987).

Babic   E. Babić, Z. Marohnić, M. Prester, and N. Brnicević, Phil. Mag. Lett. *56*, 91 (1987).

Bagle   B. G. Bagley, L. H. Greene, J. M. Tarascon, and G. W. Hull, Appl. Phys. Lett. *51*, 622 (1987).

Bakke   H. Bakker, D. O. Welch, and O. W. Lazareth, Jr., Sol. St. Comm. *64*, 237 (1987).

Baozz   Z. L. Bao, F. R. Wang, Q. D. Jiang, S. Z. Wang, Z. Y. Ye, K. Wu, C. Y. Li, and D. L. Yin, Appl. Phys. Lett. *51* 946 (1987).

Barat   A. Baratoff and G. Binnig, Physica *188B*, 1335 (1981).

Barde   J. Bardeen, L. N. Cooper, and J. R. Schrieffer, Phys. Rev. *108*, 1175 (1957).

Bard1   J. Bardeen, D. M. Ginsburg, and M. B. Salamon, Novel SC, 333 (1987); J. Bardeen, MRS Boston Symp., 27 (1987).

Baris   S. Barisić, I. Batistić, and J. Friedel, Europhys. Lett. *3*, 1231 (1987).

Barns   R. L. Barns and R. A. Laudise, Appl. Phys. Lett. *51*, 1373 (1987).

Baron   A. Barone and G. Paterno, *Physics and Applications of the Josephson Effect*, Wiley, New York, 1982.

Baro1   A. Barone, A. Di Chiara, G. Peluso, U. Scotti di Uccio, A. M. Cucolo, R. Vaglio, F. C. Matacotta, and E. Olzi, Phys. Rev. B *36*, 7121 (1987).

Baro2   A. Barone, A. Di Chiara, G. Peluso, G. Pepe, M. Russo, and U. Scotti di Uccio, MRS Boston Symp., 869 (1987).

Barri   R. A. Barrio, C. Wang, J. Tagüena-Martinez, D. Rios-Jara, T. Akachi, and R. Escudero, MRS Boston Symp., 801 (1987).

Barso   M. Barsoum, D. Patten, and S. Tyagi, Appl. Phys. Lett. *51*, 1954 (1987).

Bartk   K. Bartkowski, R. Horyń, A. J. Zaleski, Z. Bukowski, M. Horobiowski, C. Marucha, J. Rafalowicz, K. Rogacki, A. Stepień-Damm, C. Sulkowski, E. Trojnar, and J. Klamut, Phys. Stat. Sol. *103*, K37 (1987).

Baska   G. Baskaran, Z. Zou, and P. W. Anderson, Sol. St. Comm., *63*, 973 (1987).

Baszy   J. Baszyński, Phys. Lett. A *123*, 31 (1987).

Batlo   B. Batlogg, R. J. Cava, A. Jayaraman, R. B. van Dover, G. A. Kourouklis, S. Sunshine, D. W. Murphy, L. W. Rupp, H. S. Chen, A. White, K. T. Short, A. M. Mujsce, and E. A. Rietman, Phys. Rev. Lett. *58*, 2333 (1987).

Batl1   B. Batlogg, G. Kourouklis, W. Weber, R. J. Cava, A. Jayaraman, A. E. White, K. T. Short, L. W. Rupp, and E. A. Rietman, Phys. Rev. Lett. *59*, 912 (1987).

Batl2   B. Batlogg, R. J. Cava, C. H. Chen, G. Kourouklis, W. Weber, A. Jayaraman, A. E. White, K. T. Short, E. A. Rietman, L. W. Rupp, D. Werder, and S. M. Zahurak, Novel SC, 653 (1987).

Batl3   B. Batlogg, R. J. Cava, and R. B. van Dover, Phys. Rev. Lett. *59*, 2616 (1987); reply to Salam comment on Cava1.

Bayot   V. Bayot, F. Delannay, C. Dewitte, J. P. Erauw, X. Gonze, J. P. Issi, A. Jonas, M. Kinany-Alaoui, M. Lambricht, J. P. Michenaud, J. P. Minet, and L. Piraux, Sol. St. Comm. *63*, 983 (1987).

Bayo1   V. Bayot, C. Dewitte, J. P. Erauw, X. Gonze, M. Lambricht, and J. P. Michenaud, Sol. St. Comm. *64*, 327 (1987).

Beckm   O. Beckman, L. Lundgren, P. Nordblad, L. Sandlund, P. Svedlindh, T. Lundström, and S. Rundqvist, Phys. Lett. A *125*, 425 (1987).

Bedno   J. G. Bednorz and K. A. Müller, Z. Phys. B *64*, 189 (1986).

Bedn1   J. B. Bednorz, M. Takashige, and K. A. Müller, Europhys. Lett. *3*, 379 (1987).

Bedn2   J. G. Bednorz, K. A. Müller, and M. Takashige, Science *236*, 73 (1987).

Bedn3   J. G. Bednorz, M. Takashige, and K. A. Müller, "Preparation and Characterization of Alkaline–Earth Substituted Superconducting La_2CuO_4," IBM Zürich preprint (1987).

Bedn4   J. G. Bednorz, K. A. Müller, and A. Reller, MRS Anaheim Symp., 1 (1987).

Beech   F. Beech, S. Miraglia, A. Santoro, and R. S. Roth, Phys. Rev. B *35*, 8778 (1987).

Beill   J. Beille, R. Cabanel, C. Chaillout, B. Chevalier, G. Demazeau, F. Deslandes, J. Étourneau, P. Lejay, C. Michel, J. Provost, B. Raveau, A. Sulpice, J. L. Tholence, and R. Tournier, C. R. Acad. Sci. Paris, *304*, 1097 (1987).

Belit   D. Belitz, Phys. Rev. B *36*, 47 (1987).

Benak   S. Benakki, E. Christoffel, A. Goltzene, B. Meyer, C. Schwab, M. J. Besnus, A. Meyer, S. Vilminot, and M. Drillon, J. Mater. Res. *2*, 765 (1987).

Benoz   M. A. Beno, L. Soderholm, D. W. Capone, II, D. G. Hinks, J. D. Jorgensen, J. D. Grace, and I. K. Schuller, Appl. Phys. Lett. *51*, 57 (1987).

Berni   P. Berning, J. Collazo, A. Mashayekhi, and R. E. Benenson, MRS Boston Symp., 939 (1987).

Beyer   R. Beyers, G. Lim, E. M. Engler, V. Y. Lee, M. L. Ramirez, R. J. Savoy, R. D. Jacowitz, T. M. Shaw, S. La Placa, R. Boehme, C. C. Tsuei, S. I. Park, M. W. Shafer, W. J. Gallagher, Appl. Phys. Lett. *51*, 614 (1987).

Beye1   R. Beyers, G. Lim, E. M. Engler, R. J. Savoy, T. M. Shaw, T. R. Dinger, W. J. Gallagher, R. L. Sandstrom, Appl. Phys. Lett. *50*, 1918 (1987).

Beye2   W. P. Beyermann, B. Alavi, and G. Grüner, Phys. Rev. B *35*, 8826 (1987).

Beye3   R. Beyers, G. Lim, E. M. Engler, V. Y. Lee, M. L. Ramirez, R. J. Savoy, R. D. Jacowitz, T. M. Shaw, K. G. Frase, E. G. Liniger, D. R. Clarke, S. La Placa, R. Boehme, C. C. Tsuei, S. I. Park, M. W. Shafer, W. J. Gallagher, and G. V. Chandrashekhar, MRS Anaheim Symp., 149 (1987).

Bezi1   A. Bezinge, J. L. Jorda, A. Junod, and J. Muller, Sol. St. Comm. *64*, 79 (1987).

Bharg   R. N. Bhargara, S. P. Herko, and W. N. Osborne, Phys. Rev. Lett. *59*, 1468 (1987).

Bhatt   A. Bhattacharya, P. G. McQueen, C. S. Wang, and T. L. Einstein, MRS Boston Symp., Abstract AA7.57 (1987).

Bhatz   S. V. Bhat, P. Ganguly, T. V. Ramakrishnan, and C. N. R. Rao, J. Phys. C: Sol. St. Phys. *20* L559 (1987).

Bianc   A. Bianconi, A. C. Castellano, M. De Santis, P. Rudolf, P. Lagarde, A. M. Flank, and A. Marcelli, Sol. St. Comm. *63*, 1009 (1987).

Binni   B. Binnig, A. Baratoff, H. E. Hönig, and J. G. Bednorz, Phys. Rev. Lett. *45*, 1352 (1980).

Birge   R. J. Birgeneau, C. Y. Chen, D. R. Gabbe, H. P. Jenssen, M. A. Kastner, C. J. Peters, P. J. Picone, T. Thio, T. R. Thurston, H. L. Tuller, J. D. Axe, P. Böni, and G. Shirane, Phys. Rev. Lett. *59*, 132 (1987).

Bisho   D. J. Bishop, P. L. Gammel, A. P. Ramirez, R. J. Cava, B. Batlogg, and E. A. Rietman, Phys. Rev. B *35*, 8788 (1987).

Bish1   D. J. Bishop, A. P. Ramirez, P. L. Gammel, B. Batlogg, R. J. Cava, E. A. Rietman, and L. F. Schneemeyer, MRS Anaheim Symp., 111 (1987); see Bish3.

Bish2   D. J. Bishop, A. P. Ramirez, P. L. Gammel, B. Batlogg, E. A. Rietman, R. J. Cava, and A. J. Millis, Phys. Rev. B *36*, 2408 (1987).

Bish3   D. J. Bishop, P. L. Gemmel, A. P. Ramirez, B. Batlogg, R. J. Cava, and A. J. Millis, Novel SC, 659 (1987).

Blami   M. G. Blamire, G. W. Morris, R. E. Somekh, and J. E. Evetts, J. Phys. D Appl. Phys. *20*, 1330 (1987).

Blaze   K. W. Blazey, K. A. Müller, J. G. Bednorz, W. Berlinger, G. Amoretti, E. Buluggiu, A. Vera, and F. C. Matacotta, Phys. Rev. B *36*, 7241 (1987).

Blend   J. E. Blendell, C. K. Chiang, D. C. Cranmer, S. W. Freiman, E. R. Fuller, Jr., E. Drescher-Krasicka, W. L. Johnson, H. M. Ledbetter, L. H. Bennett, L. J. Swartzendruber, R. B. Marinenko, R. L. Myklebust, D. S. Bright, and D. E. Newbury, ACS Symp., 240 (1987).

Blezi   J. Blezius and J. P. Carbotte, Phys. Rev. B, *36*, 3622 (1987).

Blume   S. Blumenroeder, E. Zirngiebl, J. D. Thompson, P. Killough, J. L. Smith, and Z. Fisk, Phys. Rev. B *35*, 8840 (1987).

Bonnz   D. A. Bonn, J. E. Greedan, C. V. Stager, T. Timusk, M. G. Doss, S. L. Heer, K. Kamarás, C. D. Porter, D. B. Tanner, J. M. Tarascon, W. R. McKinnon, and L. H. Greene, Phys. Rev. B *35*, 8843 (1987).

Bonn1   D. A. Bonn, J. E. Greedan, C. V. Stager, T. Timusk, M. G. Doss, S. L. Herr, K. Kamarás, and D. B. Tanner, Phys. Rev. Lett. *58*, 2249 (1987).

Bonn2   D. A. Bonn, J. E. Greedan, C. V. Stager, T. Timusk, M. Doss, S. Herr, K. Kamarás, C. Porter, D. B. Tanner, J. M. Tarascon, W. R. McKinnon, and L. H. Greene, MRS Anaheim Symp., 107 (1987).

Boolc   P. Boolchand, R. N. Enzweiler, I. Zitkovsky, R. L. Meng, P. H. Hor, C. W. Chu, and C. Y. Huang, Sol. St. Comm. *63*, 521 (1987).

Borde   P. Bordet, C. Chaillout, J. J. Capponi, J. Chenavas, and M. Marezio, Nature *327*, 687 (1987).

Borge   H. A. Borges, R. Kwok, J. D. Thompson, G. L. Wells, J. L. Smith, Z. Fisk, and D. E. Peterson, Phys. Rev. B *36*, 2404 (1987).

Bourn   L. C. Bourne, M. F. Crommie, A. Zettl, H. C. zur Loye, S. W. Keller, K. L. Leary, A. M. Stacy, K. J. Chang, M. L. Cohen, and D. E. Morris, Phys. Rev. Lett. *58*, 2337 (1987).

Bour1   L. C. Bourne, M. L. Cohen, W. N. Creager, M. F. Crommie, A. M. Stacy, and A. Zettl, Phys. Lett. A *120*, 494 (1987).

Bour2   L. C. Bourne, A. Zettl, K. J. Chang, M. L. Cohen, A. M. Stacy, and W. K. Ham, Phys. Rev. B *35*, 8785 (1987).

Bour3   L. C. Bourne, M. L. Cohen, W. N. Creager, M. F. Crommie, and A. Zettl, Phys. Lett. A *123*, 34 (1987).

Bour4   L. C. Bourne, A. Zettl, T. W. Barbee, III, and M. L. Cohen, Phys. Rev. B *36*, 3990 (1987).

Bowde   G. J. Bowden, P. R. Elliston, K. T. Wan, S. X. Dou, K. E. Easterling, A. Bourdillon, C. C. Sorrell, B. A. Cornell, and F. Separovic, J. Phys. C. Solid State *20*, L545 (1987).

Boyce   J. B. Boyce, F. Bridges, T. Claeson, T. H. Geballe, C. W. Chu, and J. M. Tarascon, Phys. Rev. B *35*, 7203 (1987).

Boyc1   J. B. Boyce, F. Bridges, T. Claeson, R. S. Howland, and T. H. Geballe, Phys. Rev. B *36*, 5251 (1987).

Bozov   I. Bozović, D. Mitzi, M. Beasley, A. Kapitulnik, T. Geballe, S. Perkowitz, G. L. Carr, B. Lou, R. Sudharsanan, and S. S. Yom, Phys. Rev. B *36*, 4000 (1987).

Bozo1   I. Bozović, D. Kirillov, A. Kapitulnik, K. Char, M. R. Hahn, M. R. Beasley, T. H. Geballe, Y. H. Kim, and A. J. Heeger, Phys. Rev. Lett. *59*, 2219 (1987).

Brahm   U. S. Brahme, G. Kordas, and R. J. Kirkpatrick, MRS Boston Symp., 809 (1987).

Brewe   J. H. Brewer, E. J. Ansaldo, J. F. Carolan, A. C. D. Chaklader, W. N. Hardy, D. R. Harshman, M. E. Hayden, M. Ishikawa, N. Kaplan, R. Keitel, J. Kempton, R. F. Kiefl, W. J. Kossler, S. R. Kreitzman, A. Kulpa, Y. Kuno, G. M. Luke, H. Miyatake, K. Nagamine, Y. Nakasawa, N. Nishida, K. Nishiyama, S. Ohkuma, T. M. Riseman, G. Roehmer, P. Schleger, D. Shimada, C. E. Stronach, T. Takabatake, Y. J. Uemura, Y. Watanabe, D. Ll. Williams, T. Yamazaki, and B. Yang, Phys. Rev. Lett. *60*, 1073 (1988).

Brokm   A. Brokman, Sol. St. Comm. *64*, 257 (1987).

Brown   S. E. Brown, J. D. Thompson, J. O. Willis, R. M. Aikin, E. Zirngiebl, J. L. Smith, Z. Fisk, and R. B. Schwarz, Phys. Rev. B *36*, 2298 (1987).

Brow1   F. C. Brown, T. C. Chiang, T. A. Friedmann, D. M. Ginsberg, G. N. Kwawer, T. Miller, and M. G. Mason, J. Low Temp. Phys. *69*, 151 (1987).

Bruck   T. Brückel, H. Capellmann, W. Just, O. Schärpf, S. Kemmler-Sack, R. Kiemel, and W. Schaefer, Europhys. Lett. *4*, 1189 (1987).

Brunz   T. Brun, M. Grimsditch, K. E. Gray, R. Bhadra, V. Maroni, and C. K. Loong, Phys. Rev. B *35*, 8837 (1987).

Budha   R. C. Budhani, S. M. H. Tzeng, H. J. Doerr, and R. F. Bunshah, Appl. Phys. Lett. *51*, 1277 (1987).

Budni   J. I. Budnick, A. Golnik, C. Niedermayer, E. Recknagel, M. Rossmanith, A. Weidinger, B. Chamberland, M. Filipkowski, and D. P. Yang, Phys. Lett. A *124*, 103 (1987).

Bulae   L. N. Bulaevskii, V. B. Ginodman, and A. V. Gdenko, Sov. Phys. JETP Lett. *45*, 451 (1987); L. N. Bulaevskii and O. V. Dolgov, ibid., p. 526.

Bulle   D. W. Bullett, and W. G. Dawson, J. Phys. C: Sol. St. Phys. *20*, L853 (1987).

Burbi D. S. Burbidge, S. K. Dew, B. T. Sullivan, N. Foruer, R. R. Parsons, P. J. Mulhern, J. F. Carolan, and A. Chaklader, Sol. St. Comm. *64*, 749 (1987).

Burge J. P. Burger, L. Lesueur, M. Nicolas, J. N. Daou, L. Dumoulin, and P. Vajda, J. Phys. *48*, 1419 (1987).

Burns G. Burns, F. H. Dacol, and M. W. Shafer, Sol. St. Comm. *62*, 687 (1987).

Burn1 G. Burns, F. H. Dacol, P. Freitas, T. S. Plaskett, and W. König, Sol. St. Comm. *64*, 471 (1987).

Burn2 G. Burns, F. H. Dacol, P. Freitas, W. König, and T. S. Plaskett, MRS Boston Symp., 813 (1987).

Burn3 G. Burns, F. H. Dacol, G. Kliche, W. König, and M. W. Shafer, MRS Boston Symp., 817 (1987).

Butau P. Butaud, P. Ségransan, C. Berthier, Y. Berthier, C. Paulsen, J. L. Tholence, and P. Lejay, Phys. Rev. B *36*, 5702 (1987).

Buttn H. Büttner and A. Blumen, Nature *329*, 700 (1987).

Caizz X. Cai, R. Joynt, and D. C. Larbalestier, Phys. Rev. Lett. *58*, 2798 (1987).

Cales G. Calestani and C. Rizzoli, Nature *328*, 606 (1987).

Calla J. Callaway, D. G. Kanhere, and P. K. Misra, Phys. Rev. B *36*, 7141 (1987).

Camps R. A. Camps, J. E. Evetts, B. A. Glowacki, S. B. Newcomb, and W. M. Stobbs, J. Mater. Res. *2*, 750 (1987).

Camp1 R. A. Camps, J. E. Evetts, B. A. Glowacki, S. B. Newcomb, R. E. Somekh, and W. M. Stobbs, Nature *329*, 229 (1987).

Capli D. Caplin, Nature *328*, 376 (1987).

Capon D. W. Capone, II, D. G. Hinks, J. D. Jorgensen, and K. Zhang, Appl. Phys. Lett. *50*, 543 (1987).

Capo1 D. W. Capone, II, G. D. Hinks, L. Soderholm, M. Beno, J. D. Jorgensen, I. K. Schuller, C. U. Segre, K. Zhang, and J. D. Grace, MRS Anaheim Symp., 45 (1987).

Capo2 D. W. Capone, II and B. Flandermeyer, MRS Anaheim Symp., 181 (1987).

Cappo J. J. Capponi, C. Chaillout, A. W. Hewat, P. Lejay, M. Marezio, N. Nguyen, B. Raveau, J. L. Soubeyroux, J. L. Tholence, and R. Tournier, Europhys. Lett. *3*, 1301 (1987).

Cardo M. Cardona, L. Genzel, R. Liu, A. Wittlin, Hj. Mattausch, F. García-Alvarada, and E. García-González, Sol. St. Comm. *64*, 727 (1987).

Carol J. F. Carolan, W. N. Hardy, R. Krahn, J. H. Brewer, R. C. Thompson and A. C. D. Chaklader, Sol. St. Comm. *64*, 717 (1987).

Cavaz R. J. Cava, R. B. van Dover, B. Batlogg, and E. A. Rietman, Phys. Rev. Lett. *58*, 408 (1987).

Cava1 R. J. Cava, B. Batlogg, R. B. van Dover, D. W. Murphy, S. Sunshine, T. Siegrist, J. P. Remeika, E. A. Rietman, S. Zahurak, and G. P. Espinosa, Phys. Rev. Lett. *58*, 1676 (1987); Batl3 is reply to Salam comment on this article.

Cava2 R. J. Cava, B. Batlogg, C. H. Chen, E. A. Rietman, S. M. Zahurak, and D. Werder, Nature *329*, 423 (1987).

Cava3 R. J. Cava, B. Batlogg, C. H. Chen, E. A. Rietman, S. M. Zahurak, and D. Werder, Phys. Rev. B *36*, 5719 (1987).

Chail   C. Chaillout, M. A. Alario-Franco, J. J. Capponi, J. Chenavas, J. L. Hodeau, and M. Marezio, Phys. Rev. B *36*, 7118 (1987).

Chaki   T. K. Chaki and M. Rubinstein, Phys. Rev. B *36*, 7259 (1987).

Chakr   S. Chakravarty, S. Kivelson, G. T. Zimanyi, and B. I. Halperin, Phys. Rev. B *35*, 7256 (1987).

Chand   B. S. Chandrasekhar, Chapter 1 in Parks.

Chang   C. Chang-feng, Y. Dao-le, and H. Ru-shan, Sol. St. Comm. *63*, 411 (1987).

Chan1   Y. Chang, M. Onellion, D. W. Niles, R. Joynt, G. Margaritondo, N. G. Stoffel, and J. M. Tarascon, Sol. St. Comm. *63*, 717 (1987).

Chan2   F. Changxin, S. Lin, M. Bocal, and L. Jun, Sol. St. Comm. *64*, 689 (1987).

Chan3   C. Changgeng, H. Yeye, G. Shuquan, L. Shanlin, C. Lie, L. Jinxiang, and C. Liquan, Cryogenics *27*, 481 (1987).

Charz   K. Char, A. D. Kent, A. Kapitulnik, M. R. Beasley, and T. H. Geballe, Appl. Phys. Lett. *51*, 1370 (1987).

Chaud   P. Chaudhari, R. H. Koch, R. B. Laibowitz, T. R. McGuire, and R. J. Gambino, Phys. Rev. Lett. *58*, 2684 (1987).

Chau1   P. Chaudhari, R. T. Collins, P. Freitas, R. J. Gambino, J. R. Kirtley, R. H. Koch, R. B. Laibowitz, F. K. LeGoues, T. R. McGuire, T. Penney, Z. Schlesinger, A. P. Segmuller, S. Foner, and E. J. McNiff, Jr., Phys. Rev. B *36*, 8903 (1987).

Chau2   P. Chaudhari, F. K. LeGoues, and A. Segmüller, Science *238*, 342 (1987).

Chau3   P. Chaudhari, R. H. Koch, R. B. Laibowitz, T. R. McGuire, and R. J. Gambino, Jpn. J. Appl. Phys. *26*, 2684 (1987).

Cheng   L. Chengren and D. C. Larbalestier, Cryogenics *27*, 171 (1987).

Chenz   J. T. Chen, L. E. Wenger, C. J. McEwan, and E. M. Logothetis, Phys. Rev. Lett. *58*, 1972 (1987).

Chen1   C. H. Chen, D. J. Werder, S. H. Liou, J. R. Kwo, and M. Hong, Phys. Rev. B *35*, 8767 (1987).

Chen2   I. W. Chen, S. J. Keating, C. Y. Keating, X. Wu, J. Xu, P. E. Reyes-Morel, and T. Y. Tien, Sol. St. Comm. *63*, 997 (1987).

Chen3   C. X. Chen, R. B. Goldfarb, J. Nogués, and K. V. Rao, J. Appl. Phys. *63*, 980 (1988).

Cheon   S. W. Cheong, S. E. Brown, Z. Fisk, R. S. Kwok, J. D. Thompson, E. Zirngiebl, G. Gruner, D. E. Peterson, G. L. Wells, R. B. Schwarz, and J. R. Cooper, Phys. Rev. B *36*, 3913 (1987).

Cheva   B. Chevalier, B. Buffat, G. Demazeau, B. Lloret, J. Etourneau, M. Hervieu, C. Michel, B. Raveau, and R. Tournier, J. Phys. *48*, 1619 (1987).

Ching   W. Y. Ching, Y. Xu, G. L. Zhou, K. W. Wong, and F. Zandiehnadem, Phys. Rev. Lett. *59*, 1333 (1987).

Choiz   C. H. Choi and P. Muzikar, Phys. Rev. B *36*, 54 (1987).

Choi1   G. M. Choi, H. L. Tuller, and M. J. Tsai, MRS Boston Symp., 141 (1987).

Chuzz   C. W. Chu, P. H. Hor, R. L. Meng, L. Gao, and Z. J. Huang, Science *235*, 567 (1987).

Chuz1   C. W. Chu, P. H. Hor, R. L. Meng, L. Gao, Z. J. Huang, and Y. Q. Wang, Phys. Rev. Lett. *58*, 405 (1987).

Chuz2   C. W. Chu, P. H. Hor, R. L. Meng, L. Gao, Z. J. Huang, J. Bechtold, M. K. Wu, and C. Y. Huang, MRS Anaheim Symp., 15 (1987).

Chuz3   C. W. Chu, S. Huang, and A. W. Sleight, Sol. St. Comm. *18*, 977 (1976).

Chuz4   C. W. Chu, P. H. Hor, R. L. Meng, L. Gao, Z. J. Huang, Y. Q. Wang, and J. Bechtold, Novel SC, 581 (1987).

Chuz5   C. W. Chu, J. Bechtold, L. Gao, P. H. Hor, Z. J. Huang, R. L. Meng, Y. Y. Sun, Y. Q. Wang, and Y. Y. Xue, Phys. Rev. Lett. *60*, 941 (1988).

Cimaz   M. J. Cima, R. Chiu, and W. R. Rhine, MRS Boston Symp., 241 (1987).

Clark   G. J. Clark, A. D. Marwick, R. H. Koch, and R. B. Laibowitz, Appl. Phys. Lett. *51*, 139 (1987).

Clar1   G. J. Clark, F. K. LeGoues, A. D. Marwick, R. B. Laibowitz, and R. Koch, Appl. Phys. Lett. *51*, 1462 (1987).

Clar2   G. J. Clark, F. K. LeGoues, A. D. Marwick, R. B. Laibowitz, and R. Koch, MRS Boston Symp., 127 (1987).

Clar3   J. Clark, Nature *333*, 29 (1988).

Cohen   M. L. Cohen, A. M. Stacy, and A. Zettl, MRS Anaheim Symp., 93 (1987).

Cohe1   M. L. Cohen, D. E. Morris, A. Stacy, and A. Zettl, Novel SC (1987), p. 733.

Cohe2   L. Cohen, I. R. Gray, A. Porch, and J. R. Waldram, J. Phys. F: Met. Phys. *17*, L179 (1987).

Colcl   M. S. Colclough, C. E. Gough, M. Keene, C. M. Muirhead, N. Thomas, J. S. Abell, and S. Sutton, Nature, *328* 47 (1987).

Coles   B. R. Coles, Cont. Phys. *28*, 143 (1987).

Colli   R. T. Collins, Z. Schlesinger, R. H. Koch, R. B. Laibowitz, T. S. Plaskett, P. Freitas, W. J. Gallagher, R. L. Sandstrom, and T. R. Dinger, Phys. Rev. Lett. *59*, 704 (1987).

Collo   S. J. Collocott, G. K. White, S. X. Dou, and R. K. Williams, Phys. Rev. B, *36*, 5684 (1987).

Coll1   G. Collin and R. Comes, C. R. Acad. Sci. Paris, *304*, 1159 (1987).

Compa   A. Compaan and H. Z. Cummins, Phys. Rev. *86*, 4753 (1972).

Cooke   D. W. Cooke, H. Rempp, Z. Fisk, J. L. Smith, and M. S. Jahan, Phys. Rev. B *36*, 2287 (1987).

Cookz   R. F. Cook, T. R. Dinger, and D. R. Clarke, Appl. Phys. Lett. *51*, 454 (1987).

Coope   J. R. Cooper, B. Alavi, L. W. Zhou, W. P. Beyermann, and G. Grüner, Phys. Rev. B. *35*, 8794 (1987).

Coop1   J. R. Cooper, L. W. Zhou, B. Dunn, C. T. Chu, B. Alavi, and G. Grüner, Sol. St. Comm. *64*, 253 (1987).

Coop2   E. I. Cooper, M. A. Frisch, E. A. Giess, A. Gupta, E. J. M. O'Sullivan, S. I. Raider, and G. J. Scilla, MRS Boston Symp., 165 (1987).

Copic   M. Copic, D. Mihailovic, M. Zgonik, M. Prester, K. Biljakovic, B. Orel, and N. Brnicevic, Sol. St. Comm. *64*, 297 (1987).

Coxzz   D. E. Cox, A. R. Moodenbaugh, J. J. Hurst, and R. H. Jones, J. Phys. Chem. Sol. *49*, 47 (1988); see also Novel SC, 746 (1987).

Coxz1   D. E. Cox and A. W. Sleight, Sol. St. Comm. *19*, 969 (1976).

Coxz2   D. E. Cox and A. W. Sleight, Acta Cryst. B *35*, 1 (1979).

Crabt   G. W. Crabtree, J. Z. Liu, A. Umezawa, W. K. Kwok, C. H. Sowers, S. K. Malik, B. W. Veal, D. J. Lam, M. B. Brodsky, and J. W. Downey, Phys. Rev. B 36, 4021 (1987).

Crab1   G. W. Crabtree, W. K. Kwok, A. Umezawa, L. Soderholm, L. Morss, and E. E. Alp, Phys. Rev. B 36, 5258 (1987).

Crisa   M. Crisan, Phys. Lett. A 124, 195 (1987).

Cromm   M. F. Crommie, L. C. Bourne, A. Zettl, M. L. Cohen, and A. Stacy, Phys. Rev. B, 35, 8853 (1987).

Crone   D. C. Cronemeyer and A. P. Malozemoff, MRS Boston Symp., 837 (1987).

Cross   L. Eric Cross, private communication.

Csach   K. Csach, P. Diko, V. Kavecanský, J. Miskuf, M. Reifers, and P. Batko, Czech. J. Phys. B 37, 1207 (1987).

Cuizz   G. J. Cui, X. F. Meng, S. G. Wang, K. Shao, H. M. Jiang, Y. D. Dai, Y. Zhang, R. P. Peng, Z. L. Bao, F. R. Wang, S. Z. Wang, Z. Y. Ye, C. Y. Li, K. Wu, and D. L. Yin, Sol. St. Comm. 64, 321 (1987).

Cyrot   M. Cyrot, Sol. St. Comm. 63, 1015 (1987).

Czerw   E. Czerwosz, J. Bukowski, P. Przyslupski, J. Igalson, A. Pajaczkowska, and J. Rauluszkiewicz, Sol. St. Comm. 64, 535 (1987).

Daizz   U. Dai, G. Deutscher, and R. Rosenbaum, Appl. Phys. Lett. 51, 460 (1987).

Dalic   Y. Dalichaouch, M. S. Torikachvili, E. A. Early, B. W. Lee, C. L. Seaman, K. N. Yang, H. Zhou, and M. B. Maple, Sol. St. Comm. 65, 1001 (1988).

Damen   M. A. Damento, K. A. Gschneidner, Jr., and R. W. McCallum, Appl. Phys. Lett. 51, 690 (1987).

Daole   Y. Daole and H. Rushan, Sol. St. Comm. 63, 645 (1987).

Daoqi   Y. Daoqi, F. Rangchuan, Z. Qirui, Z. Yong, H. Zhenhui, Z. Minjian, X. Cuenyi, Z. Han, C. Zuyao, and Q. Yitai, Sol. St. Comm. 64, 877 (1987).

Datta   T. Datta, C. P. Poole, Jr., H. A. Farach, C. Almasan, J. Estrada, D. U. Gubser, and S. A. Wolf, Phys. Rev. B 37, 7843 (1988).

Datt1   T. Datta, C. Almasan, D. U. Gubser, S. A. Wolf, M. Osofsky, and L. E. Toth, Novel SC, 817 (1987).

Datt2   T. Datta, C. Almasan, J. Estrada, and C. E. Violet, J. Appl. Phys. 63, 4204 (1988).

Datt3   T. Datta, H. M. Ledbetter, C. E. Violet, C. Almason, and J. Estrada, Phys. Rev. B 37, 7502 (1988).

David   A. Davidson, A. Palevski, M. J. Brady, R. B. Laibowitz, R. Koch, M. Scheuermann, and C. C. Chi, "In-situ Resistance of $Y_1Ba_2Cu_3O_x$ Films During Anneal," IBM T. J. Watson preprint (1987).

Davie   A. H. Davies and R. J. D. Tilley, Nature, 326, 859 (1987).

Davis   S. Davison, K. Smith, Y. C. Yang, J. H. Liu, R. Kershaw, K. Dwight, P. H. Rieger, and W. Wold, ACS Symp., 65 (1987).

Davi1   W. I. F. David, W. T. A. Harrison, R. M. Ibberson, M. T. Weller, J. R. Grasmeder, and P. Lanchester, Nature 328, 328 (1987).

Dayzz   P. Day, M. Rosseinsky, K. Prassides, W. I. F. David, O. Moze, and A. Soper, J. Phys. C: Sol. St. Phys. 20, L429 (1987).

Decro   M. Decroux, A. Junod, A. Bezinge, D. Cattani, J. Cors, J. L. Jorda, A. Stett-
         ler, M. Francois, K. Yvon, O. Fischer, and J. Muller, Europhys. Lett. *3*,
         1035 (1987).

Defon   D. de Fontaine, L. T. Wille, and S. C. Moss, Phys. Rev. B *36*, 5709 (1987).

Degio   L. Degiorgi, E. Kaldis, and P. Wachter, Sol. St. Comm. *64*, 873 (1987).

Degro   P. A. J. deGroot, B. D. Rainford, D. McK. Paul, P. C. Lanchester, M. T.
         Weller, G. Balakrishnan, and J. Grasmeder, J. Phys. F: Met. Phys. *17*,
         L185 (1987).

DeLee   D. M. deLeeuw, C. A. H. A. Mutsaers, C. Langereis, H. C. A. Smoorenburg,
         and P. J. Rommers, Physica C *152*, 39 (1988).

Delim   O. F. de Lima, J. Mattson, C. H. Sowers, and M. B. Brodsky, Appl. Phys.
         Lett. *51* 369 (1987).

Deuts   G. Deutscher and K. A. Müller, Phys. Rev. Lett. *59*, 1745 (1987).

DeWae   A. Th. A.M. de Waele, R. T. M. Smokers, R. W. van der Heijden, K. Kado-
         waki, Y. K. Huang, M. van Sprang, and A. A. Menovsky, Phys. Rev. B *35*,
         8858 (1987).

DeWa1   A. Th. A.M. de Waele, R. T. M. Smokers, R. W. van der Heijden, K. Kado-
         waki, Y. K. Huang, M. van Sprang, and A. A. Menovsky, MRS Anaheim
         Symp., 215 (1987).

DewHu   D. Dew-Hughes, Cryogenics *26*, 660 (1986).

Dharz   S. K. Dhar, P. L. Paulose, A. K. Grover, E. V. Sampathkumaran, and
         V. Nagarajan, J. Phys. F: Met. Phys. *17*, L105 (1987).

Dijkk   D. Dijkkamp, T. Venkatesan, X. D. Wu, S. A. Shaheen, N. Jisrawi, Y. H.
         Min-Lee, W. L. McLean, and M. Croft, Appl. Phys. Lett. *51*, 619 (1987).

Dikoz   P. Diko, M. Reiffers, I. Batko, K. Csach, O. Hudák, V. Kavecanský, J. Mis-
         kuf, M. Timko, and A. Zentko, Czech. J. Phys. B *37*, 1085 (1987).

Ding1   T. R. Dinger, T. K. Worthington, W. J. Gallagher, and R. L. Sandstrom,
         Phys. Rev. Lett. *58*, 2687 (1987).

DiSal   F. J. DiSalvo, ACS Symp., 49 (1987).

Djure   D. Djurek, M. Prester, S. Knezović, N. Brnicević, Z. Medunić, and T. Vu-
         kelja, Phys. Lett. A *122*, 443 (1987).

Djur1   D. Djurek, S. Knezović, N. Brnicević, Z. Medunić, and T. Vukelja, Europhys.
         Lett. *4*, 1195 (1987).

Domen   B. Domenges, M. Hervieu, C. Michel, and B. Raveau, Europhys. Lett. *4*, 211
         (1987).

Douzz   S. X. Dou, A. J. Bourdillon, C. C. Sorrell, S. P. Ringer, K. E. Easterling,
         N. Savvides, J. B. Dunlop, and R. B. Roberts, Appl. Phys. Lett. *51*, 535
         (1987).

Dries   A. Driessen, R. Griessen, N. Koeman, E. Salomons, R. Brouwer, D. G. de
         Groot, K. Heeck, H. Hemmes, and J. Rector, Phys. Rev. B *36*, 5602 (1987).

Drumh   J. E. Drumheller, G. V. Rubenacker, W. K. Ford, J. Anderson, M. Hong,
         S. H. Liou, J. Kwo, and C. T. Chen, Solid St. Comm. *64*, 509 (1987).

Durny   R. Durný, J. Hautala, S. Ducharme, B. Lee, O. G. Symko, P. C. Taylor, D. J.
         Zheng, and J. A. Xu, Phys. Rev. B *36*, 2361 (1987).

Dvora   V. Dvorák, Phys. Stat. Sol. *143*, K15 (1987).

Eagle   D. J. Eaglesham, C. J. Humphreys, N. McN. Alford, W. J. Clegg, M. A. Harmer, and J. D. Birchall, Appl. Phys. Lett. *51*, 457 (1987).

Early   E. A. Early, C. L. Seaman, K. N. Yang, and M. B. Maple, Am. J. Phys., July (1988).

Eatou   M. O. Eatough, D. S. Ginley, B. Morosin, and E. L. Venturini, Appl. Phys. Lett. *51*, 367 (1987).

Ebner   C. Ebner and D. Stroub, Phys. Rev. B *31*, 165 (1985).

Edels   E. A. Edelsack, D. U. Gubser, and S. A. Wolf, Novel SC, 1 (1987).

Egami   T. Egami, Sol. St. Comm. *63*, 1019 (1987).

Eibsc   M. Eibschutz, D. W. Murphy, S. Sunshine, L. G. Van Uitert, S. M. Zahurak, and W. H. Grodkiewicz, Phy. Rev. B *35*, 8714 (1987).

Ekino   T. Ekino and J. Akimitsu, Jpn. J. Appl. Phys. *26*, L452 (1987); see also Novel SC, 927 (1987).

Ekinz   J. W. Ekin, A. J. Panson, A. I. Braginski, M. A. Janocko, M. Hong, J. Kwo, S. H. Liou, D. W. Capone, II, and B. Flandermeyer, MRS Anaheim Symp., 223 (1987).

Elias   G. M. Eliashberg, Zhur. Eksp. Teor. Fiz. *38*, 966 (1960).

Elia1   G. M. Eliashberg, Zhur. Eksp. Teor. Fiz. *39*, 1437 (1960).

Emery   V. J. Emery, Phys. Rev. Lett. *58*, 2794 (1987).

Emer1   V. J. Emery, Nature *328*, 750 (1987).

Engle   E. M. Engler, V. Y. Lee, A. I. Nazzal, R. B. Beyers, G. Lim, P. M. Grant, S. S. P. Parkin, M. L. Ramirez, J. E. Vazquez, and R. J. Savoy, J. Am. Chem. Soc. *109*, 2848 (1987).

Engl1   E. M. Engler, Chemtech. *17*, 542 (1987).

Engl2   E. M. Engler, R. B. Beyers, V. Y. Lee, A. I. Nazzal, G. Lim, S. S. P. Parkin, P. M. Grant, J. E. Vazquez, M. I. Ramirez, and R. D. Jacowitz, ACS Symp., 266 (1987).

Errak   L. Er-Rakho, C. Michel, F. Studer, and B. Raveau, J. Phys. Chem. Solids *48*, 377 (1987).

Erski   D. Erskine, E. Hess, P. Y. Yu, and A. M. Stacy, J. Mater. Res. *2*, 783 (1987).

Eschr   H. Eschrig and G. Seifert, Sol. St. Comm. *64*, 521 (1987).

Escud   R. Escudero, L. Rendón, T. Akachi, R. A. Barrio, and J. Tagüena–Martínez, Phys. Rev. B *36*, 3910 (1987); see also R. Escudero, T. Akachi, R. A. Barrio, and T. Tangüena–Martínez, Novel SC, 1011 (1987).

Escu1   R. Escudero, L. E. Rendon-Diazmirón, T. Akachi, J. Heiras, C. Vázquez, L. Banos, F. Estrada, and G. González, Jpn. J. Appl. Phys. *26*, L1019 (1987).

Escu2   R. Escudero, T. Akachi, R. Barrio, L. E. Rendon-Diazmirón, C. Vázquez, L. Banos, G. González, and F. Estrada, Sol. St. Comm. *64*, 235 (1987).

Espar   D. A. Esparza, C. A. D'Ovidio, J. Guimpel, E. Osquiguil, L. Civale, and F. de la Cruz, Sol. St. Comm. *63*, 137 (1987).

Esqui   P. Esquinazi, J. Luzuriaga, C. Duran, D. A. Esparza, and C. D'Ovidio, Phys. Rev. B *36*, 2316 (1987).

236 REFERENCES

Estev   D. Estève, J. M. Martinis, C. Urbina, M. H. Devoret, G. Collin, P. Monod, M. Ribault, and A. Revcolevschi, Europhys. Lett. *3*, 1237 (1987).

Evett   J. E. Evetts, R. E. Somekh, M. G. Blamire, Z. H. Barber, K. Butler, J. H. James, G. W. Morris, E. J. Tomlinson, A. P. Schwarzenberger, and W. M. Stobbs, MRS Anaheim Symp., 227 (1987).

Falte   T. A. Faltens, W. K. Ham, S. W. Keller, K. J. Leary, J. N. Michaels, A. M. Stacy, H. C. zur Loye, D. E. Morris, T W. Barbee, III, L. C. Bourne, M. L. Cohen, S. Hoen, and A. Zettl, Phys. Rev. Lett. *59*, 915 (1987).

Farno   B. Farnoux, R. Kahn, A. Brulet, G. Collin, and J. P. Pouget, J. Phys. *48*, 1623 (1987).

Farre   D. E. Farrell, M. R. DeGuire, B. S. Chandrasekhar, S. A. Alterovitz, P. R. Aron, and R. L. Fagaly, Phys. Rev. B *35*, 8797 (1987).

Farr1   D. E. Farrell, B. S. Chandrasekhar, M. R. DeGuire, M. M. Fang, V. G. Kogan, J. R. Clem, and D. K. Finnemore, Phys. Rev. B *36*, 4025 (1987).

Felic   R. Felici, J. Penfold, R. C. Ward, E. Olsi, and C. Matacotta, Nature *329*, 523 (1987).

Felne   I. Felner, I. Nowik, and Y. Yeshurun, Phys. Rev. B *36*, 3923 (1987).

Feln1   I. Felner and B. Barbara, Phys. Rev. B *37*, 5820 (1988).

Fento   E. W. Fenton and G. C. Aers, Sol. St. Comm. *63*, 993 (1987).

Fent1   E. W. Fenton, Novel SC, 421 (1987).

Ferre   J. M. Ferreira, B. W. Lee, Y. Dalichaouch, M. S. Torikachvili, K. N. Yang, and M. B. Maple, Phys. Rev. B *37*, 1580 (1988).

Finez   S. M. Fine, M. Greenblatt, S. Simizu, and S. A. Friedberg, Phys. Rev. B *36*, 5716 (1987).

Fine1   S. M. Fine, M. Nagano, S. Li, H. T. Shih, K. V. Ramanujachary, and M. Greenblatt, MRS Boston Symp., abstract AA4.2 (1987).

Fine2   S. M. Fine, M. Greenblatt, S. Shimizu, and S. A. Friedberg, ACS Symp., 95 (1987).

Finkz   H. J. Fink, V. Grünfeld, and S. M. Roberts, Phys. Rev. B *36*, 74 (1987).

Finne   D. K. Finnemore, R. N. Shelton, J. R. Clem, R. W. McCallum, H. C. Ku, R. E. McCarley, S. C. Chen, P. Klavins, and V. Kogan, Phys. Rev. B *35*, 5319 (1987).

Finn1   D. K. Finnemore, M. M. Fang, J. R. Clem, R. W. McCallum, J. E. Ostenson, L. Ji, and P. Klavins, Novel SC, 627 (1987).

Fiory   A. T. Fiory, M. Gurvitch, R. J. Cava, and G. P. Espinosa, Phys. Rev. B *36*, 7262 (1987).

Fisch   O. Fischer, Appl. Phys. *16*, 1 (1978).

Fishe   B. Fisher, E. Polturak, G. Koren, A. Kessel, R. Fischer, and L. Harel, Sol. St. Comm. *64*, 87 (1987).

Fiskz   Z. Fisk, J. D. Thompson, E. Zirngiebl, J. L. Smith, and S. W. Cheong, Solid State Comm. *62*, 743 (1987).

Fjeil   H. Fjeilvag, P. Karen, A. Kjekshus, and J. K. Grepstad, Sol. St. Comm. *64*, 917 (1987).

Flemi   R. M. Fleming, B. Batlogg, R. J. Cava, and E. A. Rietman, Phys. Rev. B *35*, 7191 (1987).

Forga   T. Forgan, Nature *329*, 483 (1987).

Fossh   K. Fossheim, T. Laegreid, E. Sandvold, F. Vassenden, K. A. Müller, and J. G. Bednorz, Sol. St. Comm. *63*, 531 (1987).

Franc   J. P. Franck, J. Jung, and M. A. K. Mohamed, Phys. Rev. B *36*, 2308 (1987).

Frase   K. G. Frase, E. G. Liniger, and D. R. Clarke, J. Am. Ceram. Soc. *70*, C204 (1987).

Freem   A. J. Freeman, C. L. Fu, D. D. Koelling, S. Massidda, T. J. Watson-Yang, J. Yu, and J. H. Xu, MRS Anaheim Symp., 27 (1987).

Free1   A. J. Freeman, J. Yu, and C. L. Fu, Phys. Rev. B *36*, 7111 (1987).

Freim   A. Freimuth, S. Blumenröder, G. Jackel, H. Kierspel, J. Langen, G. Buth, A. Nowack, H. Schmidt, W. Schlabitz, and E. Zirngiebl, Z. Phys. B Cond. Matt. *68*, 433 (1987).

Freit   P. P. Freitas, C. C. Tsuei, and T. S. Plaskett, Phys. Rev. B *36*, 833 (1987).

Frei1   P. P. Freitas and T. S. Plaskett, Phys. Rev. B *36*, 5723 (1987).

Fried   J. Friedel, C. R. Acad. Sci. Paris, *305*, 543 (1987).

Frohl   H. Frohlich, Phys. Rev. *79*, 845 (1950); see also Novel SC, 473 (1987).

Fueki   K. Fueki, K. Kitazama, K. Kishio, T. Hasegawa, S. I. Uchida, H. Takagi, and S. Tanaka, ACS Symp., 38 (1987).

Fujim   A. Fujimori, E. Takayama-Muromachi, Y. Uchida, and B. Okai, Phys. Rev. B *35*, 8814 (1987).

Fujit   T. Fujita, Y. Aoki, Y. Maeno, J. Sakurai, H. Fukuba, and H. Fujii, Jpn. J. Appl. Phys. *26*, L368 (1987).

Fuji1   A. Fujimori, E. Takayama-Muromachi, and Y. Uchida, Sol. St. Comm. *63*, 857 (1987).

Fukuy   H. Fukuyama and K. Yosida, Jpn. J. Appl. Phys. *26*, L371 (1987); see Fuku2.

Fuku1   H. Fukuyama and Y. Hasegawa, J. Phys. Soc. Jpn. *56*, 1312 (1987).

Fuku2   H. Fukuyama, Y. Hasegawa, and K. Yosida, Novel SC, 407 (1987).

Fuku3   H. Fukuyama, Novel SC, 51 (1987).

Furoz   I. Furo, A. Jánossy, L. Mihály, P. Bánki, I. Pócsik, I. Bakonyi, I. Heinmaa, E. Joon, and E. Lippmaa, Phys. Rev. B *36*, 5690 (1987).

Fuzzz   C. L. Fu and A. J. Freeman, Phys. Rev. B *35* 8861 (1987).

Fuzz1   W. S. Fu, D. S. Ginley, M. A. Mitchell, E. L. Venturini, B. Morosin, J. F. Kwak, and R. J. Baughman, MRS Boston Symp., 669 (1987).

Gagli   E. R. Gagliano, A. G. Rojo, C. A. Balseiro, and B. Alascio, Sol. St. Comm. *64*, 901 (1987).

Galla   W. J. Gallagher, R. L. Sandstrom, T. R. Dinger, T. M. Shaw, and D. A. Chance, Sol. St. Comm. *63*, 147 (1987).

Gall1   M. C. Gallagher, J. A. Adler, J. Jung, and J. P. Franck, Phys. Rev. B *37*, 7846 (1988).

Gallo   C. F. Gallo, L. R. Whitney, and P. J. Walsh, MRS Anaheim Symp., 165 (1987); Novel SC, 385 (1987).

Gamme   P. L. Gammel, D. J. Bishop, G. J. Dolan, J. R. Kwo, C. A. Murray, L. F. Schneemeyer, and J. V. Waszczak, Phys. Rev. Lett. *59*, 2592 (1987).

Ganzz   Z. Gan, Y. Dai, and R. Han, Sol. St. Comm. *64*, 679 (1987).

Gaozz   Y. Gao, T. J. Wagener, J. H. Weaver, A. J. Arko, B. Flandermeyer, and D. W. Capone, II, Phys. Rev. B *36*, 3971 (1987).

Gaoz1   Y. Gao, T. J. Wagener, J. H. Weaver, B. Flandermeyer, and D. W. Capone, II, Appl. Phys. Lett. *51*, 1032 (1987).

Gaoz2   L. Gao, Z. J. Huang, R. L. Meng, P. H. Hor, J. Bechtold, Y. Y. Sun, C. W. Chu, Z. Z. Sheng, and A. M. Hermann, Nature *332*, 623 (1988).

Gaoz3   Y. Gao, T. J. Wagener, D. M. Hill, H. M. Meyer, III, J. H. Weaver, A. J. Arko, B. K. Flandermeyer, and D. W. Capone, II, ACS Symp., 212 (1987).

Garci   F. Garcïa-Alvarado, E. Moran, M. Vallet, J. M. González-Calbet, M. A. Alario, M. T. Pérez-Frías, J. L. Vicent, S. Ferrer, E. García-Michel, and M. C. Asensio, Sol. St. Comm. *63*, 507 (1987).

Garc1   J. Garcia, C. Rillo, F. Lera, J. Bartolomé, R. Navarro, D. H. A. Blank, and J. Flokstra, J. Magn. Mat. *69*, L225 (1987).

Garla   M. M. Garland, Appl. Phys. Lett. *51*, 1030 (1987).

Garoc   P. Garoche and C. Noguera, MRS Anaheim Symp., 243 (1987).

Garwi   L. Garwin, Nature *327*, 101 (1987).

Gebal   T. H. Geballe, A. Kapitulnik, M. R. Beasley, R. H. Hammond, D. J. Webb, D. B. Mitzi, J. Z. Sun, A. D. Kent, J. W. P. Hsu, S. Arnason, M. I Gusman, D. L. Hildenbrand, S. M. Johnson, M. A. Quinlan, and D. J. Rowcliffe, MRS Anaheim Symp., 59 (1987).

Geise   U. Geiser, M. A. Beno, A. J. Schultz, H. H. Wang, T. J. Allen, M. R. Monaghan, and J. M. Williams, Phys. Rev. B *35*, 6721 (1987).

Genze   L. Genzel, A. Wittlin, J. Kuhl, H. Mattausch, W. Bauhofer, and A. Simon, Sol. St. Comm. *63*, 843 (1987).

Geser   H. P. Geserich, G. Scheiber, and B. Renker, Sol. St. Comm. *63*, 657 (1987).

Geshk   V. B. Geshkenbein, A. I. Larkin, and A. Barone, Phys. Rev. B *36*, 235 (1987).

Giapi   J. Giapintzakis, J. M. Matykiewicz, C. W. Kimball, A. E. Dwight, B. D. Dunlap, M. Slaski, and F. Y. Fradin, Phys. Rev. Lett. A *121*, 307 (1987).

Gilbe   L. R. Gilbert, R. Messier, and R. Roy, Thin Solid Films *54*, 129 (1978).

Ginle   D. S. Ginley, E. L. Venturini, J. F. Kwak, R. J. Baughman, B. Morosin, and J. E. Schirber, Phys. Rev. B *36*, 829 (1987).

Ginl1   D. S. Ginley, E. L. Venturini, C. H. Seager, W. K. Schubert, R. J. Baughman, J. F. Kwak, J. E. Schirber, and B. Morosin, MRS Anaheim Symp., 201 (1987); see also J. Mater. Res. *2*, 732 (1987).

Gins1   D. M. Ginsberg and L. C. Hebel, Chapter 4 in Parks.

Ginzb   V. L. Ginsburg and D. A. Kirzhnits, High Temperature Superconductivity, Consultants Bureau, New York, 1982 (translated from Russian, published by Nauka, Moscow, 1977).

Giova   C. Giovannella, G. Collin, P. Rouault, and I. A. Campbell, Europhys. Lett. *4*, 109 (1987).

Glads   G. Gladstone, M. A. Jensen, and J. R. Schrieffer, Chapter 13 in Parks.

Glowa   B. A. Glowacki and J. E. Evetts, MRS Boston Symp. (1987).

Godar   C. Godart and L. C. Gupta, Phys. Lett. A *120*, 427 (1987).

Golbe   J. P. Golben, S. I. Lee, S. Y. Lee, Y. Song, T. W. Noh, X. D. Chen, J. R. Gaines, and R. T. Tettenhorst, Phys. Rev. B *35*, 8705 (1987).

Golb1   J. P. Golben, S. I. Lee, Y. Song, X. D. Chen, R. D. McMichael, J. R. Gaines, and D. L. Cox, ACS Symp., 192 (1987).

Goldf   R. B. Goldfarb, A. F. Clark, A. J. Panson, and A. I. Braginski, MRS Anaheim Symp., 261 (1987); see Gold1.

Goldi   B. Goldling, N. O. Birge, W. H. Haemmerle, R. J. Cava, and E. Rietman, Phys. Rev. B 36, 5606 (1987).

Goldm   A. I. Goldman, B. X. Yang, J. Tranquada, J. E. Crow, and C. S. Jee, Phys. Rev. B 36, 7234 (1987).

Gold1   R. B. Goldfarb, A. F. Clark, A. I. Braginski, and A. J. Panson, Cryogenics 27, 475 (1987).

Golni   A. Golnik, C. Niedermayer, E. Recknagel, M. Rossmanith, A. Weidinger, J. I. Budnick, B. Chamberland, M. Filipkowsky, Y. Zhang, D. P. Yang, L. L. Lynds, F. A. Otter, and C. Baines, Phys. Lett. A 125, 71 (1987).

Gomez   R. Gómez, S. Aburto, M. L. Marquina, M. Jiménez, V. Marquina, C. Quintanar, T. Akachi, R. Escudero, R. A. Barrio, and D. Rios-Jara, Phys. Rev. B 36, 7226 (1987).

Gonca   A. M. Ponte Goncalves, C. Jee, D. Nichols, J. E. Crow, G. N. Myer, R. E. Salomon, and P. Schlottmann, MRS Boston Symp., 583 (1987).

Gopal   I. K. Gopalakrishnan, J. V. Yakhmi, M. A. Vaidya, and R. M. Iyer, Appl. Phys. Lett. 51, 1367 (1987).

Gorba   A. A. Gorbatsevich, V. Ph. Elesin and Yu. V. Kopaev, Novel SC, 429 (1987).

Gottw   U. Gottwick, R. Held, G. Sparn, F. Steglich, H. Rietschel, D. Ewert, B. Renker, W. Bauhofer, S. von Molnar, M. Wilhelm, and H. E. Hoenig, Europhys. Lett. 4, 1183 (1987).

Gough   C. E. Gough, M. S. Colclough, E. M. Forgan, R. G. Jordan, M. Keene, C. M. Muirhead, A. I. M. Rae, N. Thomas, J. S. Abell, and S. Sutton, Nature 326, 855 (1987).

Graeb   J. E. Graebner, L. F. Schneemeyer, R. J. Cava, J. V. Waszczak, and A. E. Rietman, MRS Boston Symp., 745 (1987).

Graha   R. A. Graham, E. L. Venturini, B. Morosin, and D. S. Ginley, Phys. Lett. A 123, 87 (1987).

Grant   P. M. Grant, R. B. Beyers, E. M. Engler, G. Lim, S. S. P. Parkin, M. L. Ramirez, V. Y. Lee, A. Nazzal, J. E. Vazquez, and R. J. Savoy, Phys. Rev. B 35, 7242 (1987).

Gran1   P. M. Grant, S. S. P. Parkin, V. Y. Lee, E. M. Engler, M. L. Ramirez, J. E. Vasquez, G. Lim, R. D. Jacowitz, and R. L. Greene, Phys. Rev. Lett. 58, 2482, (1987).

Gran2   P. Grant, New Scientist, July 30, 1987, 36.

Greav   C. Greaves and T. Forgan, Nature 332, 305 (1988).

Greed   J. E. Greedan, A. H. O'Reilly, and C. V. Stager, Phys. Rev. B 35, 8770 (1987).

Green   R. L. Greene, H. Maletta, T. S. Plaskett, J. G. Bednorz, and K. A. Müller, Sol. St. Comm. 63, 379 (1987).

Gree1   J. E. Greedan, A. H. O'Reilly, C. V. Stager, F. Razavi, and W. Abriel, MRS Boston Symp. 749 (1987).

Grego   E. Gregory, T. S. Kreilick, J. Wong, A. K. Ghosh, and W. B. Sampson, Cryogenics 27, 178 (1987).

240   REFERENCES

Gries   R. Griessen, Phys. Rev. B *36*, 5284 (1987).

Grosz   C. Gros, R. Joynt, and T. M. Rice, Z. Phys. B Cond. Matt. *68*, 425 (1987).

Grove   A. K. Grover, S. K. Dhar, P. L. Paulose, V. Nagarajan, E. V. Sampathku-
         maran, and R. Nagarajan, Sol. St. Comm. *63*, 1003 (1987).

Gubse   D. U. Gubser, R. A. Hein, S. H. Lawrence, M. S. Osofsky, D. J. Schrodt,
         L. E. Toth, and S. A. Wolf, Phys. Rev. B *35*, 5350 (1987).

Guome   Z. Guomeng, D. Zhanguo, C. Ning, and L. Hongcheng, Sol. St. Comm. *63*,
         151 (1987).

Gupta   A. Gupta, G. Koren, E. A. Giess, N. R. Moore, E. J. M. O'Sullivan, and E. I.
         Cooper, "$Y_1Ba_2Cu_3O_{7-\delta}$ Thin Films Grown by a Simple Spray Deposition,"
         IBM, T. J. Watson preprint (1987).

Gurvi   M. Gurvitch and A. T. Fiory, Appl. Phys. Lett. *51*, 1027 (1987).

Gurv1   M. Gurvitch and A. T. Fiory, Phys. Rev. Lett. *59*, 1337 (1987).

Gurv2   M. Gurvitch and A. T. Fiory, Novel SC, 663 (1987).

Gutfr   H. Gutfreund and W. A. Little, Chap. 7. "The Prospects of Excitonic Super-
         conductivity," in *Highly Conducting One-Dimensional Solids*, J. T.
         Devresse, R. P. Evrard, and N. E. van Doren (Eds.), Plenum, New York,
         1979; see also Gutf1.

Gutf1   H. Gutfreund, Novel SC, 465 (1987).

Gutsm   P. Gutsmiedl, G. Wolff, and K. Andres, Phys. Rev. B *36*, 4043 (1987).

Gygax   F. N. Gygax, B. Hitti, E. Lippelt, A. Schenck, D. Cattani, J. Cors, M. De-
         croux, O. Fischer, and S. Barth, Europhys. Lett. *4*, 473 (1987).

Hagen   M. Hagen, M. Hein, N. Klein, A. Michalke, G. Müller, H. Piel, R. W. Röth,
         F. M. Mueller, H. Sheinberg, and J. L. Smith, J. Magn. Mater. *68*, L1
         (1987).

Hage1   M. Hagen, T. W. Jing, Z. Z. Wang, J. Horvath, and N. P. Ong, Phys. Rev. B
         37, 7928 (1988).

Hajko   V. Hajko, Jr., P. Diko, K. Csach, and S. Molokác, Czech. J. Phys. B *37*, 1205
         (1987).

Hajk1   V. Hajko, Jr., P. Diko, K. Csach, and S. Molokác, Czech. J. Phys. B *37*, 1205
         (1987).

Halda   R. Haldar, Y. Z. Lu, and B. C. Giessen, Appl. Phys. Lett. *51*, 538 (1987).

Halle   I. Haller, M. W. Shafer, R. Figat, and D. B. Goland, "Kinetics of Oxygen
         Uptake in $YBa_2Cu_3O_x$," IBM, T. J. Watson preprint (1987).

Hami1   D. C. Hamilton, Ph.D. Thesis, University of California, La Jolla, 1964 (un-
         published).

Hammo   R. H. Hammond, M. Naito, B. Oh, M. Hahn, P. Rosenthal, A. Marshall,
         N. Missert, M. R. Beasley, A. Kapitulnik, and T. H. Geballe, MRS Ana-
         heim Symp., 169 (1987).

Haned   H. Haneda, M. Isobe, S. Hishita, Y. Ishizawa, S. Shirasaki, T. Yamamoto,
         and T. Yanagitani, Appl. Phys. Lett. *51*, 1848 (1987).

Hangs   W. Hang-sheng and W. Zheng-yu, Solid St. Comm. *63*, 269 (1987).

Hardy   J. R. Hardy and J. W. Flocken, Phys. Rev. Lett. *60*, 2191 (1988).

Harsh   D. R. Harshman, G. Aeppli, E. J. Ansaldo, B. Batlogg, J. H. Brewer, J. F.
         Carolan, R. J. Cava, and M. Celio, Phys. Rev. B, *36*, 2386 (1987).

Haseg   T. Hasegawa, K. Kishio, M. Aoki, N. Ooba, K. Kitazawa, K. Fueki, S. I. Uchida, and S. Tanaka, Jpn. J. Appl. Phys. *26*, 337, (1987).

Hase1   Y. Hasegawa and H. Fukuyama, Jpn. J. Appl. Phys. *26*, L322 (1987); see also Novel SC, 401, 407 (1987).

Hase2   Y. Hasegawa and H. Fukuyama, J. Phys. Soc. Jpn. *56*, 2619 (1987).

Hatan   T. Hatano, A. Matsushita, K. Nakamura, K. Honda, T. Matsumoto, and K. Ogawa, Jpn. J. Appl. Phys. *26*, L374 (1987).

Hatch   D. M. Hatch and H. T. Stokes, Phys. Rev. B *35*, 8509 (1987).

Hatfi   Proceedings of the Symposium on High Temperature Superconducting Materials, University of North Carolina, Chapel Hill, N.C., W. E. Hatfield and J. H. Miller, Jr., (Eds.) in press M. Dekker, New York (1988).

Hauck   J. Hauck, K. Bickmann, and F. Zucht, J. Mater. Res. *2*, 762 (1987).

Hause   B. Häuser and H. Rogalla, Novel SC, 951 (1987).

Hawle   M. E. Hawley, K. E. Gray, D. W. Capone, II, and D. G. Hinks, Phys. Rev. B, *35*, 7224 (1987).

Hayri   E. A. Hayri, K. V. Ramanujschary, S. Li, M. Greenblatt, S. Simizu, and S. A. Friedberg, Sol. St. Comm. *64*, 217 (1987).

Hazen   R. M. Hazen, L. W. Finger, R. J. Angel, C. T. Prewitt, N. L. Ross, H. K. Mao, C. G. Hadidiacos, P. H. Hor, R. L. Meng, and C. W. Chu, Phys. Rev. B *35*, 7238 (1987); Erratum B *36*, 3966 (1987).

Haze1   R. M. Hazen, L. W. Finger, R. J. Angel, C. T. Prewitt, R. L. Ross, C. G. Hadidiacos, P. J. Heaney, D. R. Veblen, Z. Z. Sheng, A. El. Ali, and A. M. Hermann, Phys. Rev. Lett. *60*, 1657 (1988).

Haze2   R. M. Hazen, Sci. Am. June, 74 (1988).

Heike   R. Heikes, in Buhl International Conference on Materials, E. R. Shatz (Ed.), Gordon R. Breach, New York (1979).

Heldz   G. A. Held, P. M. Horn, C. C. Tsuei, S. J. LaPlaca, J. G. Bednorz, and K. A. Müller, Sol. St. Comm. *64*, 75 (1987).

Hemle   R. J. Hemley and H. K. Mao, Phys. Rev. Lett. *58*, 2340 (1987).

Herem   J. Heremans, and D. T. Morelli, MRS Boston Symp. (1987), p. 761.

Herma   A. M. Hermann and Z. Z. Sheng, Appl. Phys. Lett. *51*, 1854 (1987).

Herm1   F. Herman, R. V. Kasowski, and W. Y. Hsu, Phys. Rev. B *36*, 6904 (1987).

Herrm   R. Herrmann, N. Kubicki, T. Schurig, H. Dwelk, U. Preppernau, A. Krapf, H. Krüger, H. U. Müller, L. Rothkirch, W. Kraak, W. Braune, N. Pruss, G. Nachtwei, F. Ludwig, and E. Kemnitz, Phys. Stat. Sol. b *142*, K53 (1987).

Herrz   S. L. Herr, K. Kamarás, C. D. Porter, M. G. Doss, D. B. Tanner, D. A. Bonn, J. E. Greedan, C. V. Stager, and T. Timusk, Phys. Rev. B *36*, 733 (1987).

Hervi   M. Hervieu, B. Domenges, C. Michel, and B. Raveau, Europhys. Lett. *4*, 205 (1987).

Herv1   M. Hervieu, B. Domenges, C. Michel, G. Heger, J. Provost, and B. Raveau, Phys. Rev. B *36*, 3920 (1987).

Hewat   E. A. Hewat, M. Dupuy, A. Bourret, J. J. Capponi, and M. Marezio, Sol. St. Comm. *64*, 517 (1987).

Hewa1   A. W. Hewat, J. J. Capponi, C. Chaillout, M. Marezio, and E. A. Hewat, Sol. St. Comm. *64*, 301 (1987).

Hidak   Y. Hidaka, Y. Enomoto, M. Suzuki, M. Oda, and T. Murakami, Jpn. J. Appl. Phys. *26*, L377 (1987).

Hikam   S. Hikami, S. Kagoshima, S. Komiyama, T. Hirai, H. Minami, and T. Masumi, Jpn. J. Appl. Phys. *26*, L347 (1987).

Hikat   T. Hikata, Y. Kasatani, R. Yoshizaki, H. Uwe, T. Sakudo, and T. Suzuki, Phys. Rev. B *36*, 5578 (1987).

Hika1   S. Hikama, T. Hirai, and S. Kagoshima, Jpn. J. Appl. Phys. *26*, 314 (1987).

Hikit   M. Hikita, Y. Tajima, A. Katsui, Y. Hidaka, S. Iwata, and S. Tsurumi, Phys. Rev. B *36*, 7199 (1987).

Hillz   D. M. Hill, H. M. Meyer, III, J. H. Weaver, B. Flandermeyer, and D. W. Capone, II, Phys. Rev. B *36*, 3979 (1987).

Hirab   M. Hirabayashi, H. Ihara, N. Terada, K. Senzaki, K. Hayashi, S. Waki, K. Murata, M. Tokumoto, and Y. Kimura, Jpn. J. Appl. Phys. *26*, L454 (1987).

Hirot   Y. Hirotsu, S. Nagakura, Y. Morata, T. Nishihara, M. Takata, and T. Yamashita, Jpn. J. Appl. Phys. *26*, L380 (1987).

Hirsc   J. E. Hirsch, Phys. Rev. Lett. *59*, 228 (1987); ibid. *59*, 2616 (1987).

Hirs1   J. E. Hirsch, Phys. Rev. B *31*, 4403 (1985); Phys. Rev. Lett. *54*, 1317 (1985).

Hoeve   H. F. C. Hoevers, P. J. M. Van Bentum, L. E. C. Van De Leemput, H. van Kempen, A. J. C. Schellinger-hout and D. Van Der Marel, Physica C *152*, 105 (1988).

Holla   C. F. Holland, R. L. Hoskins, M. A. Dixon, P. D. VerNooy, H. C. zur Loye, G. Brimhall, D. Sullivan, R. Comia, H. W. Zandbergen, R. Gronsky, and A. M. Stacy, ACS Symp., 102 (1987).

Holtz   F. Holtzberg, D. L. Kaiser, B. A. Scott, T. R. McGuire, T. N. Jackson, A. Kleinsasser, and S. Tozer, ACS Symp., 79 (1987).

Hongm   D. Hong-min, L. Li, W. Xie-mei, L. Shu-yuan, and Z. Dian-lin, Sol. St. Comm. *64*, 489 (1987).

Hongz   M. Hong, S. H. Liou, J. Kwo, and B. A. Davidson, Appl. Phys. Lett. *51*, 694 (1987).

Horie   Y. Horie, T. Fukami, and S. Mase, Sol. St. Comm. *63*, 653 (1987).

Hori1   Y. Horie, Y. Terashi, H. Fukuda, T. Fukami, and S. Mase, Sol. St. Comm. *64*, 501 (1987).

Hornz   P. M. Horn, D. T. Keane, G. A. Held, J. L. Jordan-Sweet, D. L. Kaiser, F. Holtzberg, and T. M. Rice, Phys. Rev. Lett. *59*, 2772 (1987).

Horn1   S. Horn, J. Cai, S. A. Shaheen, Y. Jeon, M. Croft, C. L. Chang, and M. L. denBoer, Phys. Rev. B *36*, 3895 (1987).

Horzz   P. H. Hor, R. L. Meng, Y. Q. Wang, L. Gao, Z. J. Huang, J. Bechtold, K. Forster, and C. W. Chu, Phys. Rev. Lett. *58*, 1891 (1987).

Horz1   P. H. Hor, L. Gao, R. L. Meng, Z. J. Huang, Y. Q. Wang, K. Forster, J. Vassilious, C. W. Chu, M. K. Wu, J. R. Ashburn, and C. J. Torng, Phys. Rev. Lett. *58*, 911 (1987).

Hosoy   S. Hosoya, S. I. Shamoto, M. Onoda, and M. Sato, Jpn. J. Appl. Phys. *26*, L325 (1987).

Hoso1   S. Hosoya, S. Shamoto, M. Onoda, and M. Sato, Jpn. J. Appl. Phys. *26*, L456 (1987); see Hoso2

Hoso2   S. Hosoya, S. Shamoto, M. Onoda, and M. Sato, Novel SC, 909 (1987).

Hozzz   J. C. Ho, P. H. Hor, R. L. Meng, C. W. Chu, and C. Y. Huang, Sol. St. Comm. *63*, 711 (1987).

Hsuzz   W. H. Hsu and R. V. Kasowski, Novel SC, 373 (1987).

Huang   C. Y. Huang, L. J. Dries, P. H. Hor, R. L. Meng, C. W. Chu, and R. B. Frankel, Nature *328*, 403 (1987).

Huan1   Y. K. Huang, K. Kadowaki, M. van Sprang, E. Salomons, A. C. Moleman, and A. A. Menovsky, MRS Anaheim Symp., 137 (1987).

Huebe   R. P. Huebener, Magnetic Flux Structures in Superconductors, Springer Verlag, New York, 1979.

Huimi   S. Huimin, W. Yening, Z. Zhifang, Z. Shiyuan, and S. Linhai, J. Phys. C: Sol. St. Phys. *20*, L889 (1987).

Hundl   M. F. Hundley, A. Zettl, A. Stacy, and M. L. Cohen, Phys. Rev. B *35*, 8800 (1987).

Hutir   G. Y. Hutiray, A. Jánossy, G. Kriza, L. Mihály, S. Pekker, P. Ségransan, Z. Szôkefalvi-Nagy, and E. Zsoldos, Sol. St. Comm. *63*, 907 (1987).

Huxfo   N. P. Huxford, D. J. Eaglesham, and C. J. Humphreys, Nature *329*, 812 (1987).

Hyber   M. S. Hybersten and L. F. Mattheiss, Phys. Rev. Lett. *60*, 1661 (1988).

Iguch   I. Iguchi, H. Watanabe, Y. Kasai, T. Mochiku, A. Sugishita, S. I. Narumi, and E. Yamaka, Jpn. J. Appl. Phys. *26*, L327 (1987).

Ihara   H. Ihara, M. Hirabayashi, N. Terada, Y. Kimura, K. Senzaki, and M. Tokumoto, Jpn. J. Appl. Phys. *26*, L463 (1987).

Ihar1   H. Ihara, M. Hirabayashi, N. Terada, K. Bushida, M. Akimoto, N. Kobayashi, N. Toyota, and Y. Muto, Jpn. J. Appl. Phys. *26*, L458 (1987).

Ihar2   H. Ihara, M. Hirabayashi, N. Terada, Y. Kimura, K. Senzaki, M. Akimoto, K. Bushida, F. Kawashima, and R. Uzuka, Jpn. J. Appl. Phys. *26*, L460 (1987).

Ihar3   H. Ihara, M. Hirabayashi, N. Terada, Y. Kimura, and H. Oyanagi, MRS Anaheim Symp., 129 (1987).

Inamz   A. Inam, X. D. Wu, T. Venkatesan, S. B. Ogale, C. C. Chang, and D. Dijkkamp, Appl. Phys. Lett. *51*, 1112 (1987).

Inder   S. E. Inderhees, M. B. Salamon, T. A. Friedmann, and D. M. Ginsberg, Phys. Rev. B *36*, 2401 (1987); see also Phys. Rev. Lett. *60*, 1178 (1988).

Inoue   M. Inoue, T. Takemori, K. Ohtaka, R. Yoshizaki, and T. Sakudo, Sol. St. Comm. *63*, 201 (1987).

Iqba1   Z. Iqbal, S. W. Steinhauser, H. Eckhardt, and L. W. Shacklette, MRS Anaheim Symp., 231 (1987).

Iqba2   Z. Iqbal, S. W. Steinhauser, A. Bose, N. Cipollini, and H. Eckhardt, Phys. Rev. B *36*, 2283 (1987).

Iqba3   Z. Iqbal, S. W. Steinhauser, A. Bose, N. Cipollini, F. Reidinger, and H. Eckhart, SPIE Raman Lumin. Spectros. Technol., *822*, 100 (1987).

Iqba4   Z. Iqbal, E. Leone, R. Chin, A. J. Signorelli, A. Bose, and H. Eckhardt, J. Mater. Res. *2*, 768 (1987).

Iqba5   Z. Iqbal, "Raman Scattering Studies of the Oxygen Vibrations in La, Y-Ba, Sr-Cu Oxides," IUPAC Symposium on the Chemistry of Superconductors, C. N. R. Rao (Ed.), Blackwell, Oxford.

Iqba6   Z. Iqbal, F. Reidinger, A. Bose, J. C. Barry, and B. L. Ramakrishna, Allied Signal Preprint, 1988.

Iqba7   Z. Iqbal, H. Eckhardt, F. Reidinger, A. Bose, J. C. Barry, and B. L. Ramakrishna, Phys. Rev. B *38*, 859 (1988).

Iwazu   T. Iwazumi, R. Yoshizaki, H. Sawada, H. Uwe, T. Sakudo, and E. Matsuura, Jpn. J. Appl. Phys. *26*, L386 (1987).

Iwaz1   T. Iwazumi, R. Yoshizaki, H. Sawada, H. Hayashi, H. Ikeda, and E. Matsuura, Jpn. J. Appl. Phys. *26*, L383 (1987).

Iwaz2   T. Iwazumi, R. Yoshizaki, H. Sawada, Y. Saito, H. Ikeda, Y. Abe, and E. Matsuura, Jpn. J. Appl. Phys. *26*, 1065 (1987), Suppl. 26-3.

Iwaz3   T. Iwazumi, R. Yoshizaki, M. Inoue, H. Sawada, H. Hayashi, H. Ikeda, and E. Matsuura, Jpn. J. Appl. Phys. *26*, L621 (1987).

Jakle   R. C. Jaklevic, J. Lambe, J. E. Mercereau, and A. H. Silver, Phys. Rev. Lett. *12*, 159 (1964); Phys. Rev. *140*A, 1628 (1965).

Jayar   B. Jayaram, S. K. Agarwal, A. Gupta, and A. V. Narlikar, Sol. St. Comm. *63*, 713 (1987).

Jeanz   Y. C. Jean, S. J. Wang, H. Nakanishi, W. N. Hardy, M. E. Hayden, R. F. Kiefl, R. L. Meng, H. P. Hor, J. Z. Huang, and C. W. Chu, Phys. Rev. B *36*, 3994 (1987).

Jean1   Y. C. Jean, J. Kyle, H. Nakanishi, P. E. A. Turchi, R. H. Howell, A. L. Wachs, M. J. Fluss, R. L. Meng, H. P. Hor, J. Z. Huang, and C. W. Chu, Phys. Rev. Lett. *60*, 1069 (1988).

Jenny   H. Jenny, B. Walz, G. Leeman, V. Geiser, S. Jost, T. Frey, and H. J. Güntherodt, J. Mater. Res. *2*, 775 (1987).

Jeonz   Y. Jeon, F. Lu, H. Jhans, S. A. Shaheen, G. Liang, M. Croft, P. H. Ansari, K. V. Ramanujachary, E. A. Hayri, S. M. Fine, S. Li, X. H. Feng, M. Greenblatt, L. H. Greene, and J. M. Tarascon, Phys. Rev. B *36*, 3891 (1987).

Jiang   C. Jiang, Y. Mei, S. M. Green, H. L. Luo, and C. Politis, Z. Phys. B Cond. Mat. *68*, 15 (1987).

Jinzz   S. Jin, R. C. Sherwood, R. B. van Dover, T. H. Tiefel, and D. W. Johnson, Jr., Appl. Phys. Lett. *51*, 203 (1987).

Jinz1   S. Jin, R. C. Sherwood, R. B. van Dover, T. H. Tiefel, and D. W. Johnson, Jr., MRS Anaheim Symp., 219 (1987); see also MRS Boston Symp., 773 (1987) and Phys. Rev. B *37*, 7850 (1988).

Jinz2   S. Jin, R. C. Sherwood, T. H. Tiefel, R. B. van Dover, D. W. Johnson, Jr., and G. S. Grader, Appl. Phys. Lett. *51*, 855 (1987).

Jinz3   S. Jin, T. H. Tiefel, R. C. Sherwood, G. W. Kammlott, and S. M. Zahurak, Appl. Phys. Lett. *51*, 943 (1987).

Johns   P. D. Johnson, S. L. Qiu, L. Jiang, M. W. Ruckman, M. Strongin, S. L. Hulbert, R. F. Garrett, B. Sinković, N. V. Smith, R. J. Cava, C. S. Jee, D. Nichols, E. Kaczanowicz, R. E. Salomon, and J. E. Crow, Phys. Rev. B *35*, 8811 (1987); see also Novel SC, 755 (1987).

John1   D. W. Johnson, Jr., E. M. Gyorgy, W. W. Rhodes, R. J. Cava, L. C. Feldman, and R. B. van Dover, MRS Anaheim Symp., 193 (1987).

John2   D. C. Johnston, H. Prakash, W. H. Zachariasen, and R. Viswanathan, Mat. Res. Bull. *8*, 777 (1973).

John3   D. C. Johnston, J. P. Stokes, D. P. Goshorn, and J. T. Lewandowski, Phys. Rev. B *36*, 4007 (1987).

John4   D. C. Johnston, A. J. Jacobson, J. M. Newsam, T. J. Lewandowski, D. P. Goshorn, D. Xie, and W. B. Yelon, ACS Symp., 136 (1987).

Jones   T. E. Jones and W. C. McGinnis, MRS Anaheim Symp., 235 (1987).

Jone1   R. Jones, M. F. Ashby, A. M. Campbell, P. P. Edwards, M. R. Harrison, A. D. Hibbs, D. A. Jefferson, A. I. Kirkland, T. Thanyasiri, and E. Sinn, ACS Symp., 313 (1987).

Jorge   J. D. Jorgensen, M. A. Beno, D. G. Hinks, L. Soderholm, K. J. Volin, R. L. Hitterman, J. D. Grace, I. K. Schuller, C. U. Segre, K. Zhang, and M. S. Kleefisch, Phys. Rev. B *36*, 3608 (1987); see also Schu1.

Jorg1   J. D. Jorgensen, B. W. Veal, W. K. Kwok, G. W. Crabtree, A. Umezawa, L. J. Nowicki, and A. P. Paulikas, Phys. Rev. B *36*, 5731 (1987).

Jorg2   J. D. Jorgensen, Jpn. J. Appl. Phys. *26*, Suppl. 26-3 (1987).

Jorg3   J. D. Jorgensen, H. B. Schttler, D. G. Hinks, D. W. Capone, II, K. Zang, M. B. Brodsky, and D. J. Scalapino, Phys. Rev. Lett. B *58*, 1024 (1987).

Junod   A. Junod, A. Bezinge, T. Graf, J. L. Jorda, J. Muller, L. Antognazza, D. Cattani, J. Cors, M. Decroux, O. Fischer, M. Banovski, P. Genoud, L. Hoffmann, A. A. Manuel, M. Peter, E. Walker, M. Francois, and K. Yvon, Europhys. Lett. *4*, 247 (1987); erratum 637 (1987).

Juno1   A. Junod, A. Bezinge, and J. Muller, Physica C *152*, 50 (1988).

Kadow   K. Kadowaki, Y. K. Huang, M. van Sprang, and A. A. Menovsky, Physica *145*B, 1 (1987); see also MRS Anaheim Symp., 89 (1987).

Kado1   K. Kadowaki, H. P. van der Meulen, J. C. P. Klaasse, M. van Sprang, J. Q. A. Koster, L. W. Roeland, F. R. de Boer, Y. K. Huang, A. A. Menovsky, and J. J. M. Franse, Physica B *145*, 260 (1987).

Kago1   S. Kagoshima, K. I. Koga, H. Yasuoka, Y. Nogami, K. Kubo, and S. Hikami, Jpn. J. Appl. Phys. *26*, L355 (1987).

Kagos   S. Kagoshima, S. Hikami, Y. Nogami, T. Hirai, and K. Kubo, Jpn. J. Appl. Phys. *26*, L318 (1987).

Kaise   D. L. Kaiser, F. Holtzberg, B. A. Scott, and T. R. McGuire, Appl. Phys. Lett. *51*, 1040 (1987).

Kais1   D. K. Kaiser, F. Holtzberg, M. F. Chisholm, and T. K. Worthington, J. Cryst. Growth *85* (1987).

Kamar   K. Kamarás, C. D. Porter, M. G. Doss, S. L. Herr, D. B. Tanner, D. A. Bonn, J. E. Greedan, A. H. O'Reilly, C. V. Stager, and T. Timusk, Phys. Rev. Lett. *59*, 919 (1987).

Kangz   W. Kang, G. Collin, M. Ribault, J. Friedel, D. Jérôme, J. M. Bassat, J. P. Coutures, and Ph. Odier, J. Phys. *48*, 1181 (1987).

Kasow   R. V. Kasowski, W. Y. Hsu, and F. Herman, MRS Anaheim Symp., 41 (1987).

Kaso1   R. V. Kasowski, W. Y. Hsu, and F. Herman, Phys. Rev. B *36*, 7248 (1987).

Kastn   M. A. Kastner, R. J. Birgeneau, C. Y. Chen, Y. M. Chiang, D. R. Gabbe, H. P. Jenssen, T. Junk, C. J. Peters, P. J. Picone, T. Thio, T. R. Thurston, and H. L. Tulter, Phys. Rev. B *37*, 111 (1988).

Katay   H. Katayama-Yoshida, T. Hirooka, A. J. Mascarenhas, Y. Okabe, T. Takahashi, T. Sasaki, A. Ochiai, T. Suzuki, J. J. Pankove, T. Ciszek, and S. K. Deb, Jpn. J. Appl. Phys. *26*, L2085 (1987).

Kawab   U. Kawabe, H. Hasegawa, T. Nishino, Y. Ito, K. Miyauchi, K. Takagi, and F. Nagata, MRS Anaheim Symp., 253 (1987).

Kawas   M. Kawasawi, M. Funabashi, S. Nagata, K. Fueki, and H. Koinuma, Jpn. J. Appl. Phys. *26*, L388 (1987).

Kawa1   U. Kawabe, H. Hasegawa, T. Aita, and T. Ishiba, MRS Anaheim Symp., 251 (1987).

Kedve   F. J. Kedves, S. Mészáros, K. Vad, G. Halász, D. Keszei, and L. Mihály, Sol. St. Comm. *63*, 991 (1987).

Kelle   S. W. Keller, K. J. Leary, T. A. Faltens, J. M. Michaels, and A. M. Stacy, ACS Symp., 114 (1987).

Kello   G. L. Kellogg and S. S. Brenner, Appl. Phys. Lett. *51*, 1851 (1987).

Khach   K. Khachaturyan, E. R. Weber, P. Tejedor, A. M. Stacy, and A. M. Portis, MRS Boston Symp., 383 (1987).

Khac1   A. G. Khachaturyan and J. W. Morris, Phys. Rev. Lett. *59*, 2776 (1987).

Khimz   Z. G. Khim, S. C. Lee, J. H. Lee, B. J. Suh, Y. W. Park, C. Park, I. S. Yu, and J. C. Park, Phys. Rev. B *36*, 2305 (1987).

Khura   A. Khurana, Phys. Today, April 1987, p. 17.

Khur1   A. Khurana, Phys. Today, April 1988, p. 21.

Khlys   I. N. Khlyustikov and A. I. Buzdin, Adv. Phys. *36*, 271 (1987).

Kilco   S. H. Kilcoyne and R. Cywinski, J. Phys. D: Appl. Phys. *20*, 1327 (1987).

Kimba   C. W. Kimball, J. L. Matykiewicz, J. Giapintzakis, H. Lee, B. D. Dunlap, M. Slaski, F. Y. Fradin, C. Segre, and J. D. Joregensen, MRS Boston, 107 (1987).

Kimzz   Y. H. Kim, A. J. Heeger, L. Acedo, G. Stucky, and F. Wudl, Phys. Rev. B *36*, 7252 (1987).

Kirkz   M. D. Kirk, D. P. E. Smith, D. B. Mitzi, J. Z. Sun, D. J. Webb, K. Char, M. R. Hahn, M. Naito, B. Oh, M. R. Beasley, T. H. Geballe, R. H. Hammond, A. Kapitulnik, and C. F. Quate, Phys. Rev. B *35*, 8850 (1987).

Kirsc   I. Kirschner, J. Bánkuti, M. Gál, K. Torkos, K. G. Sólymos, and G. Horváth, Europhys. Lett. *3*, 1309 (1987).

Kirs1   I. Kirschner, J. Bánkuti, M. Gál, K. Torkos, K. G. Sólymos, and G. Horváth, Europhys. Lett. *4*, 371 (1987).

Kirs2   I. Kirschner, J. Bánkuti, M. Gál, and K. Torkos, Phys. Rev. B, *36*, 2313 (1987).

Kirt1   J. R. Kirtley, C. C. Tsuei, S. I. Park, C. C. Chi, J. Rozen, and M. W. Shafer, Phys. Rev. B *35*, 7216 (1987).

Kirt2   J. R. Kirtley, R. M. Feenstra, A. P. Fein, S. I. Raider, W. J. Gallagher, R. Sandstrom, T. Dinger, M. W. Shafer, R. Koch, R. Laibowitz, and B.

Bumble, "Studies of Superconductors Using a Low-Temperature, High-Field Scanning Tunneling Microscope," IBM, T. J. Watson preprint (1987).

Kirt3 J. R. Kirtley, T. R. Collins, Z. Schlesinger, W. J. Gallagher, R. L. Sandstrom, T. R. Dinger, and D. A. Chance, Phys. Rev. B *35*, 8846 (1987).

Kishi K. Kishio, N. Sugii, K. Kitazawa, and K. Fueki, Jpn. J. Appl. Phys. *26*, L466 (1987).

Kish1 K. Kishio, K. Kitazawa, T. Hasegawa, M. Aoki, K. Fueki, S. I. Uchida, and S. Tanaka, Jpn. J. Appl. Phys. *26*, L391 (1987).

Kish2 K. Kishio, K. Kitazawa, S. Kanbe, I. Yasuda, N. Sugii, H. Takagi, S. I. Uchida, K. Fueki, and S. Tanaka, Chem. Lett. (Japan), 429 (1987).

Kiste T. J. Kistenmacher, Phys. Rev. B *36*, 7197 (1987).

Kitan Y. Kitano, K. Kifune, I. Mukouda, H. Kamimura, J. Sakurai, Y. Komura, K. Hoshino, M. Suzuki, A. Minami, Y. Maeno, M. Kato, and T. Fujita, Jpn. J. Appl. Phys. *26*, L394 (1987).

Kitao Y. Kitaoka, S. Hiramatsu, T. Kohara, K. Asayama, K. Oh-ishi, M. Kikuchi, and N. Koyabashi, Jpn. J. Appl. Phys. *26*, L397 (1987).

Kitaz K. Kitazawa, K. Kishio, H. Takagi, T. Hasegawa, S. Kanbe, S. I. Uchida, S. Tanaka, and K. Fueki, Jpn. J. Appl. Phys. *26*, L339 (1987).

Kita1 K. Kitazawa, M. Sakai, S. I. Uchida, H. Takagi, K. Kishio, S. Kanbe, S. Tanaka, and K. Fueki, Jpn. J. Appl. Phys. *26*, L342 (1987).

Kita2 Y. Kitaoka, S. Hiramatsu, K. Ishida, T. Kohara, and K. Asayama, J. Phys. Soc. Jpn. *56*, 3024 (1987).

Kive1 S. A. Kivelson, D. S. Rokhsar, and J. P. Sethna, Phys. Rev. B *35*, 8865 (1987); ibid. B *36*, 7237 (1987).

Kive2 S. Kivelson, Phys. Rev. B *36*, 7237 (1987).

Klein U. Klein, J. Low Temp. Phys. *69*, 1 (1987).

Kobay N. Kobayashi, T. Sasaoka, K. Oh-Ishi, T. Sasaki, M. Kikuchi, A. Endo, K. Matsuzaki, A. Inque, K. Noto, Y. Syono, Y. Saito, T. Masumoto, and Y. Muto, Jpn. J. Appl. Phys. *26*, L358 (1987).

Koba1 N. Kobayashi, K. Oh-Ishi, T. Sasaoka, M. Kikuchi, T. Sasaki, S. Murase, K. Noto, Y. Syono, and Y. Muto, J. Phys. Soc. Jpn. *56*, 1309 (1987).

Kobes R. L. Kobes and J. P. Whitehead, Phys. Rev. B *36*, 121 (1987).

Kochz H. Koch, R. Cantor, J. F. March, H. Eickenbusch, and R. Schölhorn, Phys. Rev. B *36*, 722 (1987).

Koch1 R. H. Koch, C. P. Umbach, G. J. Clark, P. Chaudhari, and R. B. Laibowitz, Appl. Phys. Lett. *51*, 200 (1987).

Koch2 R. H. Koch, R. B. Laibowitz, P. Chaudhari, R. J. Gambino, G. J. Clark, A. D. Marwick, and C. P. Umbach, MRS Anaheim Symp., 81 (1987).

Kohik S. Kohiki, T. Hamada, and T. Wada, Phys. Rev. B *36*, 2290 (1987).

Koinu H. Koinuma, T. Hashimoto, M. Kawasaki, and K. Fueki, Jpn. J. Appl. Phys. *26*, L399 (1987).

Koin1 H. Koinuma, M. Kawasaki, M. Funabashi, T. Hasegawa, K. Kishio, K. Kitazawa, K. Fueki, and S. Nagata, J. Appl. Phys. *62*, 1524 (1987).

Koss1   W. J. Kossler, J. R. Kempton, X. H. Yu, H. E. Schone, Y. J. Uemura, A. R. Moodenbaugh, M. Suenaga, and C. E. Stronach, Phys. Rev. B *35*, 7133 (1987).

Koss2   W. J. Kossler, J. R. Kempton, A. R. Moodenbaugh, D. Opie, H. Schone, C. E. Stronach, M. Suegnaga, Y. J. Uemura, and X. H. Yu, Novel SC, 757 (1987).

Kosty   T. Kostyrko, Phys. Stat. Sol. *143*, 149 (1987).

Kouro   G. A. Kourouklis, A. Jayaraman, W. Weber, J. P. Remeika, G. P. Espinosa, A. S. Cooper, and R. G. Maines, Sr., Phys. Rev. B *36*, 7218 (1987).

Koyam   Y. Koyama and Y. Hasebe, Phys. Rev. B *36*, 7256 (1987).

Kraka   H. Krakauer, W. E. Pickett, D. A. Papaconstantopoulos, and L. L. Boyer, Jpn. J. Appl. Phys. *26*, Suppl. 26-3 (1987).

Krak1   H. Krakauer and W. E. Pickett, Novel SC, 501 (1987).

Krak2   H. Krakauer and W. E. Pickett, Phys. Rev. Lett. *60*, 1665 (1988).

Krame   G. J. Kramer, H. B. Brom, J. van den Berg, P. H. Kes, and D. J. W. Ijdo, Sol. St. Comm. *64*, 705 (1987).

Kram1   S. Kramer, K. Wu, and G. Kordas, MRS Boston Symp., 323 (1987).

Kresi   V. Z. Kresin, MRS Anaheim Symp. 1987, p. 19; see also V. Z. Kresin and H. Morawitz, Novel SC, 445 (1987) and V. Z. Kresin, J. Mater. Res. *2*, 793 (1987).

Kres1   V. Z. Kresin, Sol. St. Comm. *63*, 725 (1987); Phys. Rev. B *35*, 8716 (1987).

Kres2   V. Z. Kresin and S. A. Wolf, Novel SC, 287 (1987).

Kres3   V. Z. Kresin, Novel SC, 309, 1111 (1987).

Krolz   D. M. Krol, M. Stavola, L. F. Schneemeyer, J. V. Waszczak, and W. Weber, MRS Boston Symp., 781 (1987).

Krusi   L. Krusin-Elbaum, A. P. Malozemoff, and Y. Yeshurun, MRS Boston Symp., 221 (1987).

Kuboz   Y. Kubo, J. Tabuchi, T. Yoshitake, Y. Nakabayashi, A. Ochi, K. Utsumi, H. Igarashi, M. Yonezawa, and T. Satoh, MRS Anaheim Symp., 265 (1987).

Kumak   H. Kumakura, M. Uehara, and K. Togano, Appl. Phys. Lett. *51*, 1557 (1987).

Kuma1   H. Kumakura, M. Uehara, Y. Yoshida, and K. Togano, Phys. Lett. A *124*, 367 (1987).

Kungz   J. H. Kung, H. H. Yen, Y. C. Chen, C. M. Wang, and P. T. Wu, MRS Boston Symp., 785 (1987); see also Jpn. J. Appl. Phys. *26*, L657, L832 (1987).

Kuram   Y. Kuramoto and T. Watanabe, Sol. St. Comm., *63*, 821 (1987).

Kurih   S. Kurihara, S. Tsurumi, M. Hikita, T. Iwata, K. Semba, and J. Noda, MRS Anaheim Symp., 145 (1987).

Kuris   M. Kurisu, H. Kadomatsu, H. Fujiwara, Y. Maeno, and T. Fujita, Jpn. J. Appl. Phys. *26*, L361 (1987).

Kurtz   R. L. Kurtz, R. L. Stockbauer, D. Mueller, A. Shih, L. E. Toth, M. Osofsky, and S. A. Wolf, Phys. Rev. B *35*, 8818 (1987).

Kuzni   J. Kuzník, M. Odehnal, S. Safrata, and J. Endal, J. Low Temp. Phys. *69*, 313 (1987).

Kuzzz   H. C. Ku, H. D. Yang, R. W. McCallum, M. A. Noack, P. Klavins, R. N. Shelton, and A. R. Moodenbaugh, MRS Anaheim Symp., 177 (1987).

Kwokz   W. K. Kwok, G. W. Crabtree, D. G. Hinks, D. W. Capone, J. D. Jorgensen, and K. Zhang, Phys. Rev. B *35*, 5343 (1987).

Kwok1   W. K. Kwok, G. W. Crabtree, A. Umezawa, B. W. Veal, J. D. Jorgensen, S. K. Malik, L. J. Nowicki, A. P. Paulikas, and L. Nunez, Phys. Rev. B *37*, 106 (1988).

Kwok2   W. K. Kwok, G. W. Crabtree, A. Umezawa, E. E. Alp, L. Morss, and L. Soderholm, Jpn. J. Appl. Phys. *26*, Suppl. 26-3 (1987).

Kwozz   J. Kwo, M. Hong, R. M. Fleming, T. C. Hseiah, S. H. Liou, and B. A. Davidson, Novel SC, 699 (1987).

Kwoz1   J. Kwo, T. C. Hsieh, R. M. Fleming, M. Hong, S. H. Liou, B. A. Davidson, and L. C. Feldman, Phys. Rev. B. *36*, 4039 (1987).

Labbe   J. Labbé and J. Bok, Europhys. Lett. *3*, 1225 (1987).

Laegr   T. Laegreid, K. Fossheim, S. Sathish, F. Vassenden, O. Traetteberg, E. Sandvold, and T. Bye, Phys. Sripta (1988) in press.

Laibo   R. B. Laibowitz, R. H. Koch, P. Chaudhari, and R. J. Gambino, Phys. Rev. B *35*, 8821 (1987).

Laizz   W. Lai and P. N. Butcher, Sol. St. Comm. *64*, 317 (1987).

Larba   D. C. Larbalestier, M. Daeumling, P. J. Lee, T. F. Kelly, J. Seuntjens, C. Meingast, X. Cai, J. McKinnell, R. D. Ray, R. G. Dillenburg, and E. E. Hellstrom, Cryogenics *27*, 411 (1987).

Larb1   D. C. Larbalestier, M. Daeumling, X. Cai, J. Seuntjens, J. McKinnell, D. Hampshire, P. Lee, C. Meingast, T. Willis, H. Muller, R. D. Ray, R. G. Dillenburg, E. E. Hellstrom, and R. Joynt, J. Appl. Phys. *62*, 3308 (1987); see also MRS Anaheim Symp., 91 (1987).

Larb2   D. C. Larbalestier, CERN Courier, May, 3 (1987).

Lathr   D. K. Lathrop, S. E. Russek, and R. A. Buhrman, Appl. Phys. Lett. *51*, 1554 (1987).

Leary   K. J. Leary, H. C. zur Loye, S. W. Keller, T. A. Faltens, W. K. Ham, J. N. Michaels, and A. M. Stacy, Phys. Rev. Lett. *59*, 1236 (1987).

Ledbe   H. M. Ledbetter, M. W. Austin, S. A. Kim, T. Datta, and C. E. Violet, J. Mater. Res. *2*, 790 (1987).

Ledb1   H. M. Ledbetter, J. Metals *40*, 24 (1988).

Ledb2   H. M. Ledbetter, M. W. Austin, S. A. Kim, and M. Lei, J. Mater. Res. *2*, 796 (1987).

Ledb3   H. M. Ledbetter, S. A. Kim, M. W. Austin, T. Datta, J. Estrada and C. E. Violet, NBS preprint (1987).

Ledb4   H. M. Ledbetter, S. A. Kim, and A. M. Hermann, NBS preprint (1987).

Leder   P. Lederer, G. Montambaux, and D. Poilblanc, J. Phys. *48*, 1613 (1987).

Leezz   D. H. Lee and J. Ihm, "Two-Band Model for High T_c Superconductivity in $La_{2-x}(Ba,Sr)_xCuO_4$," preprint (1987); see also Novel SC, 451 (1987).

Leez1   P. A. Lee and N. Read, Phys. Rev. Lett. *58*, 2691 (1987).

Leez2   S. I. Lee, J. P. Golben, Y. Song, S. Y. Lee, T. W. Noh, X. D. Chen, J. Testa, J. R. Gaines, and R. T. Tettenhorst, Appl. Phys. Lett. *51*, 282 (1987).

Leez3   S. I. Lee, J. P. Golben, S. Y. Lee, X. D. Chen, Y. Song, T. W. Noh, R. D. McMichael, J. R. Gaines, D. L. Cox, and B. R. Patton, Phys. Rev. *36*, 2417 (1987); see also MRS Anaheim, 53 (1987).

Leez4   M. Lee, M. Yudkowsky, W. P. Halperin, J. Thiel, S. J. Hwu, and K. R. Poeppelmeier, Phys. Rev. B, *36*, 2378 (1987).

Leez5   S. J. Lee, E. D. Rippert, B. Y. Jin, S. N. Song, S. J. Hwu, K. Poeppelmeier, and J. B. Ketterson, Appl. Phys. Lett. *51*, 1194 (1987).

Leez6   B. W. Lee, J. M. Ferreira, Y. Dalichaouch, M. S. Torikachvili, K. N. Yang, and M. B. Maple, Phys. Rev. B *37*, 2368 (1988).

Leez7   S. I. Lee, J. P. Golben, Y. Song, X. D. Chen, R.D. McMichael, and J. R. Gaines, ACS Symp. (1987), p. 272.

Leez8   T. D. Lee, Nature *330*, 460 (1987).

Leide   P. Leiderer, R. Feile, B. Renker, and D. Ewert, Z. Phys. B-Cond. Mat. *67*, 25 (1987).

LePag   Y. LePage, W. R. McKinnon, J. M. Tarascon, L. H. Greene, G. W. Hull, and D. M. Hwang, Phys. Rev. B *35*, 7245 (1987).

LePa1   Y. Le Page, T. Siegrist, S. A. Sunshine, L. F. Schneemeyer, D. W. Murphy, S. M. Zahurak, J. V. Waszczak, W. R. McKinnon, J. M. Tarascon, G. W. Hull, and L. H. Greene, Phys. Rev. B *36*, 3617 (1987).

Liang   N. T. Liang, K. H. Lii, Y. C. Chou, M. F. Tai, and T. T. Chen, Sol. St. Comm. *64*, 761 (1987).

Lian1   J. K. Liang, X. T. Xu, G. H. Rao, S. S. Xie, X. Y. Shao, and Z. G. Duan, J. Phys. D: Appl. Phys. *20*, 1324 (1987).

Liizz   K. H. Lii, M. F. Tai, H. C. Ku, and S. L. Wang, Sol. St. Comm. *64*, 339 (1987).

Linzz   C. Lin, G. Lu, Z. X. Liu, Y. X. Sun, J. Lan, S. Q. Feng, C. D. Wei, X. Zhu, G. C. Li, Z. H. Shen, Z. Z. Gan, F. X. Chen, J. W. Chen, N. Li, and J. S. Liu, Sol. St. Comm. *64*, 691 (1987).

Litt1   W. A. Little, J. P. Collman, and J. T. McDevitt, MRS Anaheim Symp., 37 (1987); see also Litt2.

Litt2   W. A. Little, Novel SC, 341 (1987).

Liuzz   J. Z. Liu, G. W. Crabtree, A. Umeszawa, and Li Zongquan, Phys. Rev. Lett. A *121*, 305 (1987).

Liuz1   R. Liu, R. Merlin, M. Cardona, Hj. Mattausch, W. Bauhofer, A. Simon, F. García-Alvarado, E. Moran, M. Vallet, J. M. Gonazáles–Calbet, and M. A. Alario, Sol. St. Comm. *63*, 839 (1987).

Lizzz   F. Li, Q. Li, G. Lu, K. Wu, Y. Zhou, C. Li, and D. Yin, Sol. St. Comm. *64*, 209 (1987).

Lobbz   C. J. Lobb, Phys. Rev. B *36*, 3930 (1987).

Londo   F. London, *Superfluids*, Vol. 1, Dover, NY, 1961, p. 34.

Longo   J. M. Longo and P. M. Raccah, J. Sol. State Chem. *6*, 526 (1973).

Luozz   Y. L. Luo, A. H. Morrish, Q. A. Pankhurst, G. H. Pelletier, G. J. Roy, D. Y. Zhang, and X. Z. Zhou, Can. J. Phys. *65*, 438 (1987).

Lutge   H. Lütgemeier and M. W. Pieper, Sol. St. Comm. *64*, 267 (1987).

Lynnz   J. W. Lynn, W-H. Li, Q. Li, H. C. Ku, H. D. Yang, and R. N. Shelton, Phys. Rev. B *36*, 2374 (1987).

Lynto   E. A. Lynton, Superconductivity, Methuen, London, 1969.

Lyons   K. B. Lyons, S. H. Liou, M. Hong, H. S. Chen, J. Kwo, and T. J. Negran, Phys. Rev. B *36*, 5592 (1987).

Macfa   R. M. Macfarlane, H. Rosen, and H. Seki, Sol. St. Comm. *63*, 831 (1987).

Macf1   J. C. Macfarlane, R. Driver, and R. B. Roberts, Appl. Phys. Lett. *51*, 1038 (1987).

Machi   K. Machida and M. Kato, Phys. Rev. B *36*, 854 (1987).

Madak   P. Madakson, J. J. Cuomo, D. S. Yee, R. A. Roy, and G. Scilla, J. Appl. Phys. *63*, 2046 (1988).

Maeda   H. Maeda, Y. Tanaka, M. Fukutomi, and T. Asano, Jpn. J. Appl. Phys. Lett. *27*, L209 (1988).

Maeka   S. Maekawa, H. Ebisawa, and Y. Isawa, Jpn. J. Appl. Phys. *26*, L468 (1987); Novel SC, 411 (1987).

Maeno   Y. Maeno, Y. Aoki, H. Kamimura, J. Sakurai, and T. Fujita, Jpn. J. Appl. Phys. *26*, L402 (1987).

Maen1   Y. Maeno, M. Kato, and T. Fujita, Jpn. J. Appl. Phys. *26*, L329 (1987).

Maen2   Y. Maeno, T. Tomita, M. Kyogoku, S. Awaji, Y. Aoki, K. Hoshino, A. Minami, and T. Fujita, Nature *328*, 512 (1987).

Malet   H. Maletta, R. L. Greene, T. S. Plaskett, J. G. Bednorz, and K. A. Müller, Jpn. J. Appl. Phys. *26*, Suppl. 26-3 (1987).

Male1   H. Maletta, A. P. Malozemoff, D. C. Cronemeyer, C. C. Tsuei, R. L. Greene, J. G. Bednorz, and K. A. Müller, Sol. St. Comm. *62*, 323 (1987).

Malik   S. K. Malik, C. V. Tomy, A. M. Umarji, D. T. Adroja, R. Prasad, N. C. Soni, A. Mohan, and C. K. Gupta, J. Phys. C: Sol. St. Phys. *20*, L417 (1987).

Maliz   M. Mali, D. Brinkmann, L. Pauli, J. Roos, H. Zimmermann, and J. Hulliger, Phys. Lett. A *124*, 112 (1987).

Mali1   S. K. Malik, A. M. Umarji, D. T. Adroja, C. V. Tomy, R. Prasad, N. C. Soni, A. Mohan, and C. K. Gupta, J. Phys. C: Sol. St. Phys. *20*, L347 (1987).

Maloz   A. P. Malozemoff and P. M. Grant, "High Temperature Superconductivity Research at the IBM Thomas J. Watson and Almaden Research Centers," IBM preprint (1987).

Malo1   A. P. Malozemoff, W. J. Gallagher, and R. E. Schwall, ACS Symp., 280 (1987).

Manda   P. Mandal, A. Poddar, P. Choudhury, A. N. Das, and B. Ghosh, J. Phys. C: Sol. St. Phys. *20*, L553 (1987).

Manki   P. M. Mankiewich, J. H. Scofield, W. J. Skocpol, R. E. Howard, A. H. Dayem, and E. Good, Appl. Phys. Lett. *51*, 1753 (1987).

Mansf   J. F. Mansfield, S. Chevacharoenkul, and A. I. Kingon, Appl. Phys. Lett. *51*, 1035 (1987).

Manth   A. Manthiram and J. B. Goodenough, Nature *329*, 701 (1987).

Maple   M. B. Maple, K. N. Yang, M. S. Torikachvili, J. M. Ferreira, J. J. Neumeier, H. Zhou, Y. Dalichaouch, and B. W. Lee, Sol. St. Comm. *63*, 635 (1987).

Mapl1   M. B. Maple, Y. Dalichaouch, J. M. Ferreira, R. R. Hake, S. E. Lambert, B. W. Lee, J. J. Neumeier, M. S. Torikachvili, K. N. Yang, H. Zhou, Z. Fisk, M. W. McElfresh, and J. L. Smith, Novel SC, 839 (1987).

Mapl2   M. B. Maple, Y. Dalichaouch, J. M. Ferreira, R. R. Hake, B. W. Lee, J. J. Neumeier, M. S. Torikachvili, K. N. Yang, H. Zhou, R. P. Guertin, and M. V. Kuric, Physica B: Proc. Yamada Conf. XVIII, Superconductivity in Highly Correlated Fermion Systems, Sendai, Japan, August 31, 1987.

Marcu   J. Marcus, C. Escribe-Filippini, C. Schlenker, R. Buder, J. Devenyi, and P. L. Reydet, Sol. St. Comm. *63*, 129 (1987).

Marke   J. T. Markert, T. W. Noh, S. E. Russek, and R. M. Cotts, Sol. St. Comm., *63*, 847 (1987).

Marki   R. S. Markiewicz, MRS Boston Symp. (1987), abstract AA4.55.

Marsi   F. Marsiglio, and J. P. Carbotte, Sol. St. Comm., *63*, 419 (1987).

Mars1   F. Marsiglio, R. Akis, and J. P. Carbotte, Sol. St. Comm. *64*, 905 (1987).

Mars2   F. Marsiglio, and J. P. Carbotte, Phys. Rev. B *36*, 3937 (1987).

Mars3   F. Marsiglio, R. Akis, and J. P. Carbotte, Phys. Rev. B *36*, 5245 (1987).

Maruc   J. F. Marucco, C. Noguera, P. Garoche, and G. Collin, J. Mater. Res. *2*, 757 (1987).

Masak   A. Masaki, H. Sato, S. I. Uchida, K. Kitazawa, S. Tanaka, and K. Inoue, Jpn. J. Appl. Phys. *26*, 405 (1987).

Masca   A. Mascarenhas, H. Katayama-Yoshida, S. Geller, J. I. Pankove and S. Debe, MRS Boston Symp. (1987), p. 415.

Masum   T. Masumi, H. Schimada, and H. Minami, J. Phys. Soc. Jpn. *56*, 3009 (1987).

Masu1   T. Masumi, H. Minami, and H. Shimada, J. Phys. Soc. Jpn. *56*, 3013 (1987).

Matac   F. C. Matacotta, G. Nobile, G. Serrini, M. D. Giardina, and A. E. Merlini, MRS Boston Symp., 561 (1987).

Mathi   H. Mathias, W. Moulton, H. K. Ng, S. J. Pan, K. K. Pan, L. H. Peirce, L. R. Testardi, R. J. Kennedy, Phys. Rev. B *36*, 2411 (1987).

Matsu   K. Matsuzaki, A. Inoue, H. Kimura, K. Moroishi, and T. Masumoto, Jpn. J. Appl. Phys. *26*, L334 (1987).

Mats1   A. Matsushita, T. Hatano, T. Matsumoto, H. Aoki, Y. Asada, K. Nakamura, K. Honda, T. Oguchi, and K. Ogawa, Jpn. J. Appl. Phys. *26*, L332 (1987).

Mats2   T. Matsuura and K. Miyake, Jpn. J. Appl. Phys. *26*, L407 (1987).

Matth   L. F. Mattheiss and D. R. Hamann, Phys. Rev. B *28*, 4227 (1983).

Matti   D. C. Mattis, Phys. Rev. B *36*, 3933 (1987).

Matt1   B. T. Matthias, in "Superconductivity in d and f Band Metals," AIP Conference Proceedings, No. 4, D. H. Douglass (Ed.), New York, 1972.

Matt2   B. T. Matthias, Phys. Rev. *92*, 874 (1953).

Matt3   D. N. Matthews, A. Bailey, R. A. Vaile, G. J. Russell, and K. N. R. Taylor, Nature *328*, 786 (1987).

Matt4   L. F. Mattheiss and D. R. Hamann, Sol. St. Comm. *63*, 395 (1987).

Matt5   L. F. Mattheiss, Phys. Rev. Lett. *58*, 1028 (1987).

Matt6   L. F. Mattheiss, MRS Anaheim Symp. (1987), p. 23.

Matt7   L. F. Mattheiss and D. R. Hamann, Phys. Rev. B *28*, 4227 (1983).

Matt8   D. C. Mattis and M. P. Mattis, Phys. Rev. Lett. *59*, 2780 (1987).

Matyk   J. M. Matykiewicz, C. W. Kimball, J. Giapintzakis, A. E. Dwight, M. B. Brodsky, B. D. Dunlap, M. Slaski, and F. Y. Fradin, Phys. Lett. A *124*, 453 (1987).

Mawds   A. Mawdsley, H. J. Trodahl, J. Tallon, J. Sarfati, and A. B. Kaiser, Nature *328*, 233 (1987).

Maxwe   E. Maxwell, Phys. Rev. *78*, 477 (1950).

Mazum   S. Mazumdar, Phys. Rev. B *36*, 7190 (1987); Phys. Rev. Lett. *59*, 2617 (1987).

McAnd   T. P. McAndrew, K. G. Frase, and R. R. Shaw, A.V.S. Topical Conference on Thin Film Processing and Characteristics of High Temp. Supcond., Anaheim, California, Nov. 1987; AIP #165.

McGra   W. R. McGrath, H. K. Olsson, T. Claeson, S. Eriksson, and L. G. Johansson, Europhys. Lett. *4*, 357 (1987).

McGui   T. R. McGuire, T. R. Dinger, P. J. P. Freitas, W. J. Gallagher, T. S. Plaskett, R. L. Sandstrom, and T. M. Shaw, Phys. Rev. B *36*, 4032 (1987).

McHen   M. E. McHenry, J. McKittrick, S. Sasayama, V. Kwapong, R. C. O'Handley, and G. Kalonji, Phys. Rev. B *37*, 623 (1988).

McKin   W. R. McKinnon, J. M. Tarascon, L. H. Greene, and G. W. Hull, MRS Anaheim Symp. (1987), p. 185.

McKi1   W. R. McKinnon, J. R. Morton, and G. Pleizier, Sol. St. Comm. *66*, 1093 (1988).

McKi2   W. R. McKinnon, J. R. Morton, K. F. Preston, and L. S. Selwyn, Sol. St. Comm. *65*, 855 (1988).

McMil   M. L. McMillan, Phys. Rev. *167*, 331 (1968).

Mehra   F. Mehran, S. E. Barnes, T. R. McGuire, W. J. Gallagher, R. L. Sandstrom, T. R. Dinger, and D. A. Chance, Phys. Rev. B *36*, 740 (1987).

Mehr1   F. Mehran, S. E. Barnes, T. R. McGuire, T. R. Dinger, D. L. Kaiser, and F. Holtzberg, Sol. St. Comm. *66*, 299 (1988).

Mehr2   F. Mehran, S. E. Barnes, C. C. Tsuei, and T. R. McGuire, Phys. Rev. B *36*, 7266 (1987).

Mengz   X. F. Meng, Y. D. Dai, H. M. Jiang, X. M. Ren, Y. Zhang, M. X. Yan, Y. C. Du, J. C. Mao, X. W. Wu, and G. J. Cui, Sol. St. Comm. *63*, 853 (1987).

Meser   R. Meservey and B. B. Schwartz, Chapter 3 in Parks.

Messm   R. P. Messmer and R. D. Murphy, ACS Symp. (1987), p. 13.

Meyer   H. M. Meyer, III, T. J. Wagener, D. M. Hill, Y. Gao, S. G. Anderson, S. D. Krahn, J. H. Weaver, B. Flandermeyer, and D. W. Capone, II, Appl. Phys. Lett. *51*, 1118 (1987).

Meye1   H. M. Meyer, III, D. M. Hill, S. G. Anderson, J. H. Weaver, and D. W. Capone, II, Appl. Phys. Lett. *51*, 1750 (1987).

Miche   C. Michel and B. Raveau, Chim. Min. Miner. *21*, 407 (1984).

Mich1   C. Michel, M. Hervieu, M. M. Borel, A. Grandin, F. Deslandes, J. Provost, and B. Raveau, Z. Phys. B Cond. Matt. *68*, 421 (1987).

Mich2   C. Michel and B. Raveau, J. Sol. State Chem. *43*, 73 (1982).

Micna   R. Micnas, J. Ranninger, and S. Robaszkiewicz, Phys. Rev. B *36*, 4051 (1987).

Migli   A. Migliori, T. Chen, B. Alavi, and G. Grüner, Sol. St. Comm. *63*, 827 (1987).

Migl1   A. Migliori, D. W. Reagor, D. E. Peterson, J. O. Willis, Z. Fisk, and R. C. Smith, Los Alamos preprint (1987).

Mihai   D. Mihailović, M. Zgonik, M. Copic, and M. Hrovat, Phys. Rev. B *36*, 3997 (1987).

Miha1   L. Mihály, G. Hutiray, S. Pekker, G. Kriza, M. Prester, L. Forró, N. Brnicevik, and A. Hamzic, Sol. St. Comm. *63*, 133 (1987).

Miha2   L. Mihály, L. Rosta, G. Coddens, F. Mezei, G. Hutiray, G. Kriza, and
        B. Keszei, Phys. Rev. B *36*, 7137 (1987).

Minam   H. Minami, T. Masumi, and S. Hikami, Jpn. J. Appl. Phys. *26*, L345 (1987).

Mingr   J. Mingrong, H. Zhenghui, W. Jianxin, Z. Han, P. Guoqiang, C. Zhuyao,
        Q. Yitai, Z. Yong, H. Liping, X. Jiansen, and Z. Qirui, Sol. St. Comm. *63*,
        511 (1987).

Mingu   Z. Min-Guang, Z. Xiao-Lan, Q. You-Ping, Y. Jun, and Z. Xiao-Ning, J. Phys.
        C: Sol. St. Phys. *20*, L917 (1987).

Mitra   N. Mitra, J. Trefny, M. Young, and B. Yarar, Phys. Rev. B *36*, 5581 (1987).

Mitsu   S. Mitsu, G. Shirane, S. K. Sinha, D. C. Johnson, M. S. Alvarez, D. Vaknin,
        and D. E. Moneton, Phys. Rev. B *36*, 822 (1987).

Mohan   M. M. Mohan, and N. Kumar, J. Phys. C: Sol. St. Phys. *20*, L527 (1987).

Momin   A. C. Momin, M. D. Mathews, V. S. Jakkal, I. K. Gopalakrishnan, J. V.
        Yakhmi, and R. M. Iyer, Sol. St. Comm. *64*, 329 (1987).

Monec   J. Monecke, Phys. Stat. Sol. *143*, K 43 (1987).

Monie   H. Monien, K. Scharnberg, and D. Walker, Sol. St. Comm. *63*, 263 (1987).

Moode   A. R. Moodenbaugh, M. Suenaga, T. Asano, R. N. Shelton, H. C. Ku, R. W.
        McCallum, and P. Klavins, Phys. Rev. Lett. *58*, 1885 (1987).

Mood1   A. R. Moodenbaugh, J. J. Hurst, Jr., R. H. Jones, and M. Suenaga, MRS
        Anaheim Symp. (1987), p. 101.

Mood2   A. R. Moodenbaugh, J. J. Hurst, T. Asano, R. L. Sabatini, and M. Suenaga,
        Novel SC (1987), p. 767.

Moogz   E. R. Moog, S. D. Bader, A. J. Arko, and B. K. Flandenmeyer, Phys. Rev. B
        36, 5583 (1987).

Moorj   K. Moorjani, J. Bohandy, F. J. Adrian, B. F. Kim, R. D. Shull, C. K. Chiang,
        L. J. Swartzendruber, and L. H. Bennett, Phys. Rev. B *36*, 4036 (1987).

Moret   R. Moret, J. P. Pouget, R. Comes, and G. Collin, MRS Anaheim Symp.
        (1987), p. 206.

More1   J. Moreland, A. F. Clark, L. F. Goodrich, H. C. Ku, and R. N. Shelton, Phys.
        Rev. B *35*, 8711 (1987).

More2   J. Moreland, J. W. Ekin, L. F. Goodrich, T. E. Capobianco, and A. F. Clark,
        MRS Anaheim Symp., 73, 273 (1987).

More3   J. Moreland, L. F. Goodrich, J. W. Ekin, T. E. Capobianco, and A. F. Clark,
        Jpn. J. Appl. Phys. *26*, Suppl 263 (1987).

More4   J. Moreland, A. F. Clark, H. C. Ku, and R. N. Shelton, Cryogenics *27*, 227
        (1987).

More5   J. Moreland, J. W. Ekin, L. F. Goodrich, T. E. Capobianco, A. F. Clark,
        J. Kwo, M. Hong, and S. H. Liou, Phys. Rev. B *35*, 8856 (1987).

More6   D. T. Morelli, J. Heremans, and D. E. Swets, Phys. Rev. B *36*, 3917 (1987).

More7   R. Moret and G. Collin, MRS Boston Symp., 497 (1987).

More8   R. Moret, J. P. Pouget, and G. Collin, Europhys. Lett. *4*, 365 (1987).

Moriw   K. Moriwaki, Y. Enomoto, and T. Murakami, Jpn. J. Appl. Phys. *26*, 521
        (1987).

Mori1   K. Moriwaki, M. Suzuki, Y. Enomoto, and T. Murakami, MRS Anaheim
        Symp. (1987), p. 85.

Morri   D. E. Morris, U. M. Scheven, L. C. Bourne, M. L. Cohen, U. F. Crommie, and A. Zettl, MRS Anaheim Symp. (1987), p. 209.

Mossz   S. C. Moss, K. Forster, J. D. Axe, H. You, D. Hohlwein, D. E. Cox, P. H. Hor, R. L. Meng, and C. W. Chu, Phys. Rev. B 35, 7195 (1987).

Motaz   A. C. Mota, A. Pollini, P. Visani, K. A. Müller, and J. G. Bednorz, Phys. Rev. B 36, 4011 (1987).

Mulle   K. A. Müller, M. Takashige, and J. G. Bednorz, Phys. Rev. Lett. 58, 1143 (1987); see Preje for comment.

Mull1   K. A. Müller and J. G. Bednorz, "High-Temperature Superconductivity," American Physical Society Meeting, New York, March (1987).

Mull2   K. A. Müller and J. G. Bednorz, Science 237, 1133 (1987).

Murat   K. Murata, H. Ihara, M. Tokumoto, M. Hirabayashi, N. Terada, K. Senzaki, and Y. Kimura, Jpn. J. Appl. Phys. 26, L471 (1987).

Mura1   K. Murata, H. Ihara, M. Tokumoto, M. Hirabayashi, N. Terada, K. Senzaki, and Y. Kimura, Jpn. J. Appl. Phys. 26, L473 (1987).

Murph   D. W. Murphy, S. A. Sunshine, R. B. van Dover, R. J. Cava, B. Batlogg, S. M. Zahurak, and L. F. Schneemeyer, Phys. Rev. Lett. 58, 1888 (1987).

Murp1   D. W. Murphy, S. A. Sunshine, P. K. Gallagher, H. M. O'Bryan, R. J. Cava, B. Batlogg, R. B. van Dover, L. F. Schneemeyer, and S. M. Zahurak, ACS Symp. (1987), p. 181.

Murrz   L. E. Murr, A. W. Hare, and N. G. Eror, Nature 329, 37 (1987).

Murr1   L. E. Murr, A. W. Hare, and N. G. Eror, MRS Boston Symp. (1987), abstract AA6.5.

Mydos   J. A. Mydosh, Z. Phys. B Cond. Mat. 68, 1 (1987).

Mzoug   T. Mzoughi, M. Mesa, E. Quagliata, H. A. Farach, C. P. Poole, Jr., R. Creswick, T. Datta, Z. Z. Sheng, and A. M. Hermann, "Microwave Penetration in a TlBaCaCuO Superconductor," University of South Carolina preprint (1988).

Nagas   K. Nagasaka, M. Sato, H. Ihara, M. Tokumoto, M. Hirabayashi, N. Terada, K. Senzaki, and Y. Kimura, Jpn. J. Appl. Phys. 26, L479 (1987).

Nagat   S. Nagata, M. Kawasaki, M. Funabashi, K. Fueki, and H. Koinuma, Jpn. J. Appl. Phys. 26, L410 (1987).

Naito   M. Naito, D. P. E. Smith, M. D. Kirk, B. Oh, M. R. Hahn, K. Char, D. B. Mitzi, J. Z. Sun, D. J. Webb, M. R. Beasley, O. Fischer, T. H. Geballe, R. H. Hammond, A. Kapitulnik, and C. F. Quate, Phys. Rev. B 35, 7228 (1987), see also J. Mater. Res. 2, 713 (1987).

Nakah   S. Nakahara, G. J. Fisanick, M. F. Yan, R. B. Van Dover, T. Boone, and M. Moore, MRS Boston Symp. (1987), p. 1575.

Nakai   I. Nakai, K. Imai, T. Kawashima, and R. Yoshizaki, Jpn. J. Appl. Phys. 26, L1244 (1987).

Nakaj   S. Nakajima and Y. Kurihara, J. Phys. Soc. Jpn. 56, 3021 (1987).

Nakam   K. Nakamura, T. Hatano, A. Matsushita, T. Oguchi, H. Aoki, Y. Asada, S. Ikeda, T. Matsumoto, and K. Ogawa, MRS Anaheim Symp. (1987), p. 239.

Nakao   K. Nakao, N. Miura, S. I. Uchida, H. Takagi, S. Tanaka, K. Kishio, J. I. Shimoyama, K. Kitazawa, and K. Fueki, Jpn. J. Appl. Phys. 26, L413 (1987).

Namzz   S. B. Nam, MRS Anaheim Symp. (1987), p. 115; S. B. Nam, S. W. Nam, and J. O. Nam, Novel SC (1987), p. 993.

Naray   J. Narayan, V. N. Shukla, S. J. Lukasiewicz, N. Biunno, R. Singh, A. F. Schreiner, and S. J. Pennycook, Appl. Phys. Lett. *51*, 940 (1987).

Nara1   J. Narayan, N. Biunno, R. Singh, O. W. Holland, and O. Auciello, Appl. Phys. Lett. *51*, 1845 (1987).

Nasta   M. Nastasi, P. N. Arendt, J. R. Tesmer, C. J. Maggiore, R. C. Cordi, D. L. Bish, J. D. Thompson, S. W. Cheong, N. Bordes, J. F. Smith, and I. D. Raistrick, J. Mater. Res. *2*, 726 (1987).

Nauen   M. Nauenberg, Phys. Rev. B *36*, 7207 (1987).

Nelso   D. L. Nelson, M. S. Whittingham, and T. F. George, ACS Symp. (1987), p. 308.

Neume   J. J. Neumeier, Y. Dalichaouch, J. M. Ferreira, R. R. Hake, B. W. Lee, M. B. Maple, M. S. Torikachvili, K. N. Yang, and H. Zhou, Appl. Phys. Lett. *51*, 371 (1987).

Nevit   M. V. Nevitt, G. W. Crabtree, and T. E. Klippert, Phys. Rev. B *36*, 2398 (1987).

Newho   V. L. Newhouse, Chapter 22 in Parks.

Newns   D. M. Newns, Phys. Rev. B *36*, 5595 (1987).

Newn1   D. M. Newns, Novel SC, 515 (1987).

Nieme   J. Niemeyer, M. R. Dietrich, C. Politis, Z. Phys. B Cond. Mat. *67*, 155 (1987).

Ningz   C. Ning, D. Zhanguo, S. Xiuyu, Z. Jiaqi, R. Qize, L. Jinxiang, C. Yinchuan, H. Desen, F. Hui, C. Xichen, and G. Weiyan, Sol. St. Comm. *63*, 965 (1987).

Noelz   H. Noel, P. Gougeon, J. Padiou, J. C. Levet, M. Potel, O. Laborde, and P. Monceau, Sol. St. Comm. *63*, 915 (1987).

Norto   M. L. Norton, ACS Symp. (1987), p. 56.

Nucke   N. Nücker, J. Fink, B. Renker, D. Ewert, C. Politis, P. J. W. Weijs, and J. C. Fuggle, Z. Phys. B. Cond. Mat. *67*, 9 (1987).

Obrad   X. Obradors, A. Labarta, J. Tejada, F. García–Alvarado, E. Morán, M. Vallet, J. M. González-Calvet, and M. A. Alario, Sol. St. Comm. *64*, 707 (1987).

Odazz   Y. Oda, I. Nakada, T. Kohara, H. Fujita, T. Kaneko, H. Toyoda, E. Sakagami, and K. Asayama, Jpn. J. Appl. Phys. *26*, L481 (1987).

Ogale   S. B. Ogale, D. Dijkkamp, T. Venkatesan, X. D. Wu, and A. Inam, Phys. Rev. B *36*, 7210 (1987).

Ogita   N. Ogita, K. Ohbayashi, M. Udagawa, Y. Aoki, Y. Maeno, and T. Fujita, Jpn. J. Appl. Phys. *26*, L415 (1987).

Oguch   T. Oguchi, Jpn. J. Appl. Phys. *26*, L417 (1987).

Ohana   I. Ohana, Y. C. Liu, M. S. Dresselhaus, G. Dresselhaus, A. J. Strauss, H. J. Zeiger, P. J. Picone, H. P. Jenssen, and D. R. Gabbe, MRS Boston Symp. (1987), p. 439.

Ohbay   K. Ohbayashi, N. Ogita, M. Udagawa, Y. Aoki, Y. Maeno, and T. Fujita, Jpn. J. Appl. Phys. *26*, L420 (1987).

Ohba1   K. Ohbayashi, N. Ogita, M. Udagawa, Y. Aoki, Y. Maeno, and T. Fujita, Jpn. J. Appl. Phys. *26*, L423 (1987).

Ohish   K. Oh-Ishi, M. Kikuchi, Y. Syono, K. Hiraga, and Y. Moroika, Jpn. J. Appl. Phys. *26*, L484 (1987).

Ohkaw   F. J. Ohkawa, J. Phys. Soc. Jpn. *56*, 2615 (1987).

Ohka1   F. J. Ohkawa, J. Phys. Soc. Jpn. *56*, 2623 (1987).

Ohka2   F. J. Ohkawa, J. Phys. Soc. Jpn. *56*, 3017 (1987).

Ohzzz   B. Oh, M. Naito, S. Arnason, P. Rosenthal, R. Barton, M. R. Beasley, T. H. Geballe, R. H. Hammond, and A. Kapitulnik, Appl. Phys. Lett. *51*, 852 (1987).

Okabe   Y. Okabe, Y. Suzumura, T. Sasaki, and H. Katayama–Yoshida, Sol. St. Comm. *64*, 483 (1987).

Onell   M. Onellion, Y. Chang, D. W. Niles, R. Joynt, G. Margaritondo, N. G. Stoffel, and J. M. Tarascon, Phys. Rev. B *36*, 819 (1987).

Ongzz   N. P. Ong, Z. Z. Wang, J. Clayhold, J. M. Tarascon, L. H. Greene, and W. R. McKinnon, Phys. Rev. B *35*, 8807 (1987).

Ongz1   N. P. Ong, Z. Z. Wang, J. Clayhold, J. M. Tarascon, L. H. Greene, and W. R. McKinnon, Novel SC, 1061 (1987).

Onoda   M. Onoda, S. Shamoto, M. Sato, and S. Hosoya, Jpn. J. Appl. Phys. *26*, L363 (1987); Novel SC (1987), p. 919.

Orbac   R. Orbach, private communication.

Orens   J. Orenstein, G. A. Thomas, D. H. Rapkine, C. G. Bethea, B. F. Levine, R. J. Cava, E. A. Rietman, and D. W. Johnson, Jr., Phys. Rev. B *36*, 729 (1987).

Orlan   T. P. Orlando, K. A. Delin, S. Foner, E. J. McNiff, Jr., J. M. Tarascon, L. H. Greene, W. R. McKinnon, and G. W. Hull, Phys. Rev. B *36*, 2394 (1987).

Orla1   T. P. Orlando, K. A. Delin, S. Foner, E. J. McNiff, Jr., J. M. Tarascon, L. H. Greene, W. R. McKinnon, and G. W. Hull, Phys. Rev. B *35*, 7249 (1987).

Orla2   T. P. Orlando, K. A. Delin, S. Foner, E. J. McNiff, Jr., J. M. Tarascon, L. H. Greene, W. R. McKinnon, and G. W. Hull, Phys. Rev. B *35*, 5347 (1987).

Orla3   T. P. Orlando, K. A. Delin, S. Foner, E. J. McNiff, Jr., J. M. Tarascon, L. H. Greene, W. R. McKinnon, and G. W. Hull, MRS Anaheim Symp. (1987), p. 257.

Ortiz   E. Orti, P. Lambin, J. L. Brédas, J. P. Vigneron, E. G. Derouane, A. A. Lucas, and J. M. André, Sol. St. Comm. *64*, 313 (1987).

Osero   S. B. Oseroff, D. C. Vier, J. F. Smyth, C. T. Salling, S. Schultz, Y. Dalichaouch, B. W. Lee, M. B. Maple, Z. Fisk, J. D. Thompson, J. L. Smith, and E. Zirngiebl, Novel SC (1987), p. 679.

Oser1   S. B. Oseroff, D. C. Vier, J. F. Smyth, C. T. Salling, B. Schultz, Y. Dalichaouch, B. W. Lee, M. B. Maple, Z. Fisk, J. S. Thompson, J. L. Smith, and E. Zirngiebl, Sol. St. Comm. *64*, 241 (1987).

Osofs   M. Osofsky, L. E. Toth, S. Lawrence, S. B. Qadri, A. Shih, D. Mueller, R. A. Hein, W. W. Fuller, F. J. Rachford, E. F. Skelton, T. Elam, D. U. Gubser, S. A. Wolf, J. A. Gotaas, J. J. Rhyne, R. Kurtz, and R. Stockbauer, MRS Anaheim Symp. (1987), p. 173.

Otter   F. A. Otter, J. I. Budnick, B. R. Weinberger, L. Lynds, D. P. Yang, S. F. Galasso, M. Filipkowski, W. A. Hines, and D. M. Potrepka, MRS Boston Symp. (1987), p. 443.

Ottzz   H. R. Ott, Novel SC, 187 (1987).

Ottz1   H. R. Ott, H. Rudigier, Z. Fisk, and J. L. Smith, Phys. Rev. Lett. *56*, 1595 (1983).

Ourma   A. Ourmazd, J. C. H. Spence, M. O'Keefe, R. J. Graham, D. W. Johnson, Jr., J. A. Rentschler, and W. W. Rhodes, MRS Anaheim Symp. (1987), p. 153.

Ourm1   A. Ourmazd, J. A. Rentschler, W. J. Skocpol, and D. W. Johnson, Jr., Phys. Rev. B *36*, 8914 (1987).

Ourm2   A. Ourmazd and J. C. H. Spence, Nature *329*, 425 (1987).

Ousse   M. Oussena, S. Senoussi, and G. Collin, Europhys. Lett. *4*, 625 (1987).

Ouss1   M. Oussena, S. Senoussi, G. Collin, J. M. Broto, H. Rakoto, S. Askenazy, and J. C. Ousset, Phys. Rev. B *36*, 4014 (1987).

Ouss2   J. C. Ousset, M. F. Ravet, M. Maurer, J. Durand, J. P. Ulmet, H. Rakoto, and S. Askenazy, Europhys. Lett. *4*, 743 (1987).

Ovshi   S. R. Ovshinsky, R. T. Young, D. D. Allred, G. DeMaggio, and G. A. Van der Leeden, Phys. Rev. Lett. *58*, 2579 (1987).

Owens   F. Owens, B. L. Ramakrishna, and Z. Iqbal, EPR in $YBa_2Cu_3O_{7-\delta}$ and $YBa_2Cu_3O_{6+x}$, preprint (1988).

Oyana   H. Oyanagi, H. Ihara, T. Matsushita, M. Tokumoto, M. Hirabayashi, N. Terada, K. Senzaki, Y. Kimura, and T. Yao, Jpn. J. Appl. Phys. *26*, L488 (1987).

Palca   J. Palca, Nature *330*, 511 (1987).

Pande   C. S. Pande, A. K. Singh, L. Toth, D. U. Gubser, and S. Wolf, Phys. Rev. B *36*, 5669 (1987).

Panso   A. J. Panson, A. I. Braginski, J. R. Gavaler, J. K. Hulm, M. A. Janocko, H. C. Pohl, A. M. Stewart, J. Talvacchio, and G. R. Wagner, Phys. Rev. B *35*, 8774 (1987).

Panzz   S. Pan, K. W. Ng, A. L. de Lozanne, J. M. Tarascon, and L. H. Greene, Phys. Rev. B *35*, 7220 (1987).

Panz1   K. K. Pan, H. Mathias, C. M. Rey, W. G. Moulton, H. K. Ng, L. R. Testardi, and Y. L. Wang, Phys. Lett. A *125*, 147 (1987).

Papac   D. A. Papaconstantopoulos, W. E. Pickett, H. Krakauer, and L. L. Boyer, Jpn. J. Appl. Phys. Suppl. 26-3, 1091 (1987).

Papa1   D. A. Papaconstantopoulos and L. L. Boyer, Novel SC, 493 (1987).

Param   R. H. Parmenter, Phys. Rev. Lett. *59*, 923 (1987).

Parki   S. S. P. Parkin, V. Y. Lee, and E. M. Engler, "Magnetic Properties and Critical Fields of $RBa_2Cu_3O_{7-x}$ (R = Y, Pr, Eu, Gd, Dy, Ho)," IBM Almaden preprint (1987).

Parks   R. D. Parks (Ed.), *Superconductivity*, Vols. 1 and 2, Marcel Dekker, New York, 1969.

Park1   S. S. P. Parkin, E. M. Engler, V. Y. Lee, and R. B. Beyers, Phys. Rev. B *37*, 131 (1987).

Park2   S. S. P. Parkin, V. Y. Lee, E. M. Engler, A. I. Nazzal, T. C. Huang, G. Gorman, R. Savoy, and R. Beyers, IBM Almaden preprint, (1988).

Parmi   F. Parmigiani, G. Chiarello, N. Ripamonti, H. Goretzki, and U. Roll, Phys. Rev. B *36*, 7148 (1987).

Pauli   L. Pauling, Phys. Rev. Lett. *59*, 225 (1987).

Paulo   P. L. Paulose, V. Nagarajan, A. K. Grover, S. K. Dhar, and E. V. Sampathkumaran, J. Phys. F. Met. Phys. *17*, L91 (1987).

Paulz   D. McK. Paul, G. Balakrishnan, N. R. Bernhoeft, W. I. F. David, and W. T. A. Harrison, Phys. Rev. Lett. *58*, 1976 (1987).

Paul1   D. McK. Paul, H. A. Mook, A. W. Hewat, B. C. Sales, L. A. Boatner, J. R. Thompson, and M. Mostoller, Phys. Rev. B *37*, 2341 (1988).

Pegru   C. M. Pegrum, G. B. Donaldson, A. H. Carr, and A. Hendry, Appl. Phys. Lett. *51*, 1364 (1987).

Penne   T. Penney, M. W. Shafer, B. L. Olson, and T. S. Plaskett, Adv. Ceram. Mater. *2*, 577 (1987).

Perko   S. Perkowitz, G. L. Carr, B. Lou, S. S. Yom, R. Sudharsanan, and D. S. Ginley, Sol. St. Comm. *64*, 721 (1987).

Phata   G. M. Phatak, A. M. Umarji, J. V. Yakhmi, L. C. Gupta, K. Gangadnaran, R. M. Iyer, and R. Vijayaragnavan, Sol. St. Comm., *63*, 905 (1987).

Phil1   J. C. Phillips, Phys. Rev. B *36*, 861 (1987).

Phil2   J. C. Phillips, Phys. Rev. Lett. *59*, 1856 (1987).

Phil3   N. E. Phillips, R. A. Fisher, S. E. Lacy, G. Marcenat, J. A. Olsen, W. K. Ham, and A. M. Stacy, Novel SC, 739 (1987).

Picke   W. E. Pickett, H. Krakauer, D. A. Papaconstantopoulos, L. L. Boyer, and R. E. Cohen, MRS Anaheim Symp. (1987), p. 31.

Pick1   W. E. Pickett, H. Krakauer, D. A. Papaconstantopoulos, and L. L. Boyer, Phys. Rev. B *35*, 7252 (1987).

Podda   A. Poddar, P. Mandal, P. Choudhury, A. N. Das, and B. Ghosh, J. Phys. C: Sol. St. Phys. *20*, L669 (1987).

Poepp   R. B. Poeppel, B. K. Flandermeyer, J. T. Dusek, and L. D. Bloom, ACS Symp., 261 (1987).

Poiri   M. Poirier, G. Quirion, K. R. Poeppelmeier, and J. P. Thiel, Phys. Rev. B *36*, 3906 (1987).

Polit   C. Politis, J. Geerk, M. Dietrich, and B. Obst, Z. Phys. B: Cond. Matt. *66*, 141 (1987).

Poli1   C. Politis, M. R. Dietrich, G. M. Friedman, J. Geerk, S. M. Green, R. Hu, S. Hüfner, C. Jiang, W. Krauss, H. Küpper, H. Leitz, G. Linker, H. L. Luo, Y. Mei, O. Meyer, B. Obst, P. Steiner, and H. Wühl, MRS Anaheim Symp. (1987), p. 141.

Polle   E. Pollert, J. Hejtmánek, and D. Zemanová, Czech. J. Phys. B *37*, 655 (1987).

Poltu   E. Polturak and B. Fisher, Phys. Rev. B *36*, 5586 (1987).

Poole   C. P. Poole, Jr., *Electron Spin Resonance*, Wiley, New York, 1983.

Pool1   C. P. Poole, Jr. and H. A. Farach, *Theory of Magnetic Resonance*, Wiley, New York, 1987.

Pool2   C. P. Poole, Jr. and H. A. Farach, *Relaxation in Magnetic Resonance*, Academic Press, New York, 1971.

Pool3   C. P. Poole, Jr., C. Almasan, J. Estrada, T. Datta, and H. A. Farach, MRS Boston Symp. (1987).

Pool4   R. Poole, Science *240*, 146 (1988).

260   REFERENCES

Pool5   C. P. Poole, Jr., T. Datta, and H. A. Farach, Structural Commonalities of High Temperature Superconductors, University of South Carolina preprint (1988).

Porti   A. M. Portis, K. W. Blazey, K. A. Müller, and J. G. Bednorz, Europhys. Lett. 5, 467 (1988).

Preje   J. J. Prejean and J. Souletic, Phys. Rev. Lett. 60, 1884 (1988); comment on Mulle.

Przys   P. Przyslipski, J. Igalson, J. Rauluszkiewicz, and T. Skośkiewicz, Phys. Rev. B 36, 743 (1987).

Przy1   P. Przyslupski, M. Baran, J. Igalson, W. Dobrowolski, T. Skośkiewicz, and J. Rauluszkiewicz, Phys. Lett. A 124, 460 (1987).

Pureu   P. Pureur and J. Schaf, J. Magn. Magn. Mat. 69, L215 (1987).

Qadri   S. B. Qadri, L. E. Toth, M. Osofsky, S. Lawrence, D. U. Gubser, and S. A. Wolf, Phys. Rev. B 35, 7235 (1987).

Qirui   Z. Qi-rui, C. Lie-zhao, Q. Yi-tai, C. Zu-yao, G. Wei-yan, Z. Yong, P. Guo-gang, Z. Han, X. Jian-sheng, Z. Ming-Jian, Y. Dao-qi, H. Zheng-Hui, S. Shi-fang, F. Ming-hu, and Z. Tao, Sol. St. Comm. 63, 535 (1987).

Qiru1   Z. Qi-rui, Q. Yi-tai, C. Zu-yao, G. Wei-yan, Z. Yong, Z. Han, C. Lie-zhao, X. Jian-sheng, P. Guo-gang, Z. Ming-jian, H. Zheng-hui, Y. Dao-qi, S. Shi-fang, Z. Tao, F. Ming-hu, Y. Zhi-ping, Sol. St. Comm. 63, 497 (1987).

Qiru2   Z. Qi-rui, Q. Yi-tai, C. Zu-yao, G. Wei-yan, Z. Yong, X. Jian-sheng, C. Lie-zhao, P. Guo-qang, Z. Han, Y. Dao-qi, H. Zheng-hui, Z. Ming-jian, S. Shi-fang, F. Min-hu, and Z. Tao, Sol. St. Comm. 63, 415 (1987).

Qiru3   Z. Qi-rui, Q. Yi-tai, C. Zu-yao, G. Wei-yan, Z. Yong, Z. Hang, C. Lie-zhao, X. Jian-sheng, P. Guo-giang, Z. Min-jian, S. Shi-fang, Y. Dao-qi, H. Zheng-hui, F. Min-hu, and Z. Tao, Sol. St. Comm., 63, 961 (1987).

Qiuzz   Z. Q. Qiu, Y. W. Du, H. Tang, J. C. Walker, W. A. Bryden, and K. Moorjani, J. Magn. Magn. Mat. 69, L221 (1987).

Quagl   E. Quagliata, T. Mzoughi, M. Mesa, H. A. Farach, C. P. Poole, Jr., and R. Creswick, "Monitoring the Field Inside Superconductors by Free Radical Markers," University of South Carolina preprint (1988).

Ramak   D. E. Ramaker, N. H. Turner, J. S. Murday, L. E. Toth, M. Osofsky, and F. L. Hutson, Phys. Rev. B 36, 5672 (1987).

Rama1   B. L. Ramakrishna, E. W. Ong, and Z. Iqbal, MRS Boston Symp. (1987).

Ramir   A. P. Ramirez, B. Batlogg, G. Aeppli, R. J. Cava, E. Rietman, A. Goldman, and G. Shirane, Phys. Rev. B 35, 8833 (1987).

Rami1   A. P. Ramirez, L. F. Schneemeyer, and J. V. Waszczak, Phys. Rev. B 36, 7145 (1987).

Ramme   J. Rammer, Phys. Rev. B 36, 5665 (1987).

Raozz   C. N. R. Rao, P. Ganguly, A. K. Raychaudhuri, R. A. Mohan Ram, and K. Sreedhar, Nature 326, 856 (1987).

Raoz1   K. V. Rao, D-X. Chen, J. Nogues, C. Politis, C. Gallo, and J. A. Gerber, MRS Anaheim Symp. (1987), p. 133.

Ravea   B. Raveau and C. Michel, Novel SC, 599 (1987).

Ravel   B. Raveau, C. Michel, and M. Hervieu, ACS Symp. (1987), p. 122.

Raych   A. K. Raychaudhuri, K. Sreedhar, K. P. Rajeev, R. A. Mohan Ram, P. Ganguly, and C. N. R. Rao, Phil. Mag. Lett. *56*, 29 (1987).

Razav   F. S. Razavi, F. P. Koffyberg, and B. Mitrović, Phys. Rev. B *35*, 5323 (1987).

Reago   D. W. Reagor, R. C. Smith, A. Migliori, K. Wilson, Z. Fisk, D. E. Peterson, and J. L. Smith, Los Alamos preprint (1987).

Redin   J. Redinger, A. J. Freeman, J. Yu, and S. Massidda, Phys. Lett. A *124*, 469 (1987).

Redi1   J. Redinger, J. Yu, A. J. Freeman, and P. Weinberger, Phys. Lett. A *124*, 463 (1987).

Reeve   M. E. Reeves, T. A. Friedmann, and D. M. Ginsberg, Phys. Rev. B *35*, 7207 (1987); erratum *36*, 2349 (1987).

Reev1   M. E. Reeves, D. S. Citrin, B. G. Pazol, T. A. Friedmann, and D. M. Ginsberg, Phys. Rev. B *36*, 6915 (1987).

Reihl   B. Reihl, T. Riesterer, J. G. Bednorz, and K. A. Müller, Phys. Rev. B *35*, 8804 (1987).

Reill   J. J. Reilly, M. Suenaga, J. R. Johnson, P. Thompson, and A. R. Moodenbaugh, Phys. Rev. *36*, 5694 (1987).

Relle   A. Reller, J. G. Bednorz, and K. A. Müller, "Alternate Structure for $Ba_2Y-Cu_3O_7$," IBM Zürich preprint (1987).

Renau   A. Renault, G. J. McIntyre, G. Collin, J. P. Pouget, and R. Comes, J. Phys. *48*, 1407 (1987).

Renke   B. Renker, I. Apfelstedt, H. Küpfer, C. Politis, H. Rietschel, W. Schauer, H. Wühl, U. Gottwick, H. Kneissel, U. Rauchschwalbe, H. Spille, and F. Steglich, Z. Phys. B Cond. Mat. *67*, 1 (1987).

Renk1   B. Renker, F. Gompf, E. Gering, N. Nücker, D. Ewert, W. Reichardt, and H. Rietschel, Z. Phys. B Cond. Mat. *67*, 15 (1987).

Retto   C. Rettori, D. Davidov, I. Belaish, and I. Felner, Phys. Rev. B *36*, 4028 (1987).

Rhyne   J. J. Rhyne, D. A. Neuman, J. A. Gotaas, F. Beech, L. Toth, S. Lawrence, S. Wolf, M. Osofsky, and D. U. Gubser, Phys. Rev. B *36*, 2294 (1987).

Ricez   T. M. Rice, Z. Phys. B Cond. Mat. *67*, 141 (1987).

Rice1   C. E. Rice, R. B. van Dover, and G. J. Fisanick, Appl. Phys. Lett. *51*, 1842 (1987).

Riese   H. Riesemeir, C. Grabow, E. W. Scheidt, V. Müller, K. Lüders, and D. Riegel, Sol. St. Comm. *64*, 309 (1987).

Riest   T. Riesterer, J. G. Bednorz, K. A. Müller, and B. Reihl, Appl. Phys. A *44*, 81 (1987).

Rigne   M. M. Rigney, C. P. Poole, Jr., and H. A. Farach, J. Phys. Chem. Solids *49*, (1988), in press.

Robas   S. Robaszkiewicz, R. Micnas, and J. Ranninger, Phys. Rev. B *36*, 180 (1987).

Robin   A. L. Robinson, Science *235*, 531 (1987).

Robi1   A. L. Robinson, Science *237*, 1115 (1987).

Rodri   J. P. Rodriguez, Phys. Rev. B *36*, 168 (1987).

Ronay   M. Ronay, Phys. Rev. B *36*, 8860 (1987).

Rosei   A. C. Rose Innes and E. H. Rhoderick, *Introduction to Superconductivity*, Pergamon, Oxford, 1978.

Rosen   H. Rosen, E. M. Engler, T. C. Strand, V. Y. Lee, and D. Bethune, Phys. Rev. B *36*, 726 (1987).

Rossa   J. Rossat-Mignod, P. Burlet, M. J. G. M. Jurgens, J. Y. Henry, and C. Vettier, Physica C *152*, 19 (1988).

Rossz   N. L. Ross, R. J. Angel, L. W. Finger, R. M. Hazen, and C. T. Prewitt, ACS Symp.(1987), p. 164.

Rucke   A. E. Ruckenstein, P. J. Hirschfeld, and J. Appel, Phys. Rev. B *36*, 857 (1987).

Ruval   J. Ruvalds, Phys. Rev. B *35*, 8869 (1987); see also Novel SC (1987), p. 455.

Ruzic   J. Ruzicka, T. Tethal, J. Pracharová, V. Gregor, and S. Safrata, Czech. J. Phys. B *37*, 653 (1987).

Sagee   G. S. Grader, P. K. Gallagher, and E. M. Gyorgy, Appl. Phys. Lett. *51*, 1115 (1987).

Saito   Y. Saito, T. Noji, A. Endo, N. Matsuzaki, M. Katsumata, and N. Higuchi, Jpn. J. Appl. Phys. *26*, L491 (1987).

Sait1   Y. Saito, T. Noji, A. Endo, N. Matsuzaki, M. Katsumata, and N. Higuchi, Jpn. J. App. Phys. *26*, L366 (1987).

Sait2   Y. Saito, T. Nakamura, and T. Atake, MRS Boston Symp. (1987), p. 475.

Sait3   Y. Saito, R. Yoshizaki, H. Sawada, T. Iwazumi, Y. Abe, and E. Matsuura, Jpn. J. Appl. Phys. *26*, 1011 (1987), Suppl. 26-3.

Salam   M. B. Salamon and J. Bardeen, Phys. Rev. Lett. *59*, 2615 (1987); comment on Caval responded to by Batl3.

Salom   E. Salomons, H. Hemmes, J. J. Scholtz, N. Koeman, R. Brouwer, A. Driessen, D. G. De Groot, and R. Griessen, Physica *145B*, 253 (1987).

Sampa   E. V. Sampathkumaran, P. L. Paulose, A. K. Grover, V. Nagarajan, and S. K. Dhar, J. Phys. F: Met. Phys. *17*, L87 (1987).

Sanju   J. A. Sanjurjo, E. López Cruz, R. S. Katiyar, I. Torriani, C. Rettori, D. Davidov, and I. Felner, Sol. St. Comm. *64*, 505 (1987).

Sarik   M. Sarikaya, B. L. Thiel, I. A. Aksay, W. J. Weber, and W. S. Frydrych, J. Mater. Res. *2*, 736 (1987).

Sarma   D. D. Sarma, K. Sreedhar, P. Ganguly, and C. N. R. Rao, Phys. Rev. B, *36*, 2371 (1987).

Sarm1   D. D. Sarma and C. N. R. Rao, J. Phys. C: Sol. St. Phys. *20*, L659 (1987).

Sastr   M. D. Sastry, A. G. I. Dalvi, Y. Babu, R. M. Kadam, J. V. Yakhmi, and R. M. Iyer, Nature *330*, 49 (1987).

Satoz   M. Sato, S. Shamoto, M. Onoda, M. Sera, K. Fukuda, S. Hosoya, J. Akimitsu, T. Ekino, and K. Imaeda, Novel SC (1987), p. 927.

Satpa   S. Satpathy and R. M. Martin, Phys. Rev. B *36*, 7269 (1987).

Sawad   H. Sawada, Y. Saito, T. Iwazumi, R. Yoshizaki, Y. Abe, and E. Matsuura, Jpn. J. App. Phys. *26*, L426 (1987).

Sawan   Y. Sawan, M. Abu-Zeid, and Y. A. Yousef, MRS Anaheim Symp. (1987), p. 269.

Sawa1   H. Sawada, T. Iwazumi, Y. Saito, Y. Abe, H. Ikeda, and R. Yoshizaki, Jpn. J. Appl. Phys. *26*, L1054 (1987).

Scala   D. J. Scalapino, MRS Anaheim Symp. (1987), p. 35;

Scal1   D. J. Scalapino, R. T. Scaleter, and N. E. Bickers, Novel SC (1987), p. 475.

Schen   A. Schenstrom, M. F. Xu, H. P. Baum, B. K. Sarma, M. Levy, K. J. Sun, L. E. Toth, M. Osovsky, S. A. Wolf, and D. U. Gubser, MRS Boston Symp. (1987), abstract AA4.50.

Scheu   M. Scheuermann, C. C. Chi, C. C. Tsuei, D. S. Yee, J. J. Cuomo, R. B. Laibowitz, R. H. Koch, B. Braren, R. Srinivasan, and M. M. Plechaty, Appl. Phys. Lett. *51,* 1951 (1987).

Schir   J. E. Schirber, D. S. Ginley, E. L. Venturini, and B. Morosin, Phys. Rev. B *35,* 8709 (1987).

Schi1   J. E. Schirber, E. L. Venturini, J. F. Kwak, D. S. Ginly, and B. Morosin, J. Mater. Res. *2,* 421 (1987).

Schle   Z. Schlesinger, R. T. Collins, M. W. Shafer, and E. M. Engler, Phys. Rev. B *36,* 5275 (1987).

Schl1   Z. Schlesinger, R. L. Greene, J. G. Bednorz, and K. A. Muller, Phys. Rev. B *35,* 5334 (1987).

Schl2   Z. Schlesinger, R. T. Collins, D. L. Kaiser, and F. Holtzberg, Phys. Rev. Lett. *59,* 1958 (1987).

Schl3   Z. Schlesinger, R. T. Collins, and M. W. Shafer, Phys. Rev. B *35,* 7232 (1987).

Schne   J. W. Schneider, H. Baumeler, H. Keller, W. Odermatt, B. D. Patterson, K. A. Müller, J. G. Bednorz, K. W. Blazey, I. Morgenstern, and I. M. Savić, Phys. Lett. A *124,* 107 (1987).

Schn1   L. F. Schneemeyer, J. V. Waszczak, T. Siegrist, R. B. van Dover, L. W. Rupp, B. Batlogg, R. J. Cava, and D. W. Murphy, Nature, *328* 601 (1987).

Schn2   L. F. Schneemeyer, R. B. van Dover, S. H. Glarum, S. A. Sunshine, R. M. Fleming, B. Batlogg, T. Silgrist, J. H. Marshall, J. V. Waszczach, and L. W. Rupp, Nature *332,* 422 (1988).

Schos   M. Schossmann, F. Marsiglio, and J. P. Carbotte, Phys. Rev. B *36,* 3627 (1987).

Schri   J. R. Schrieffer, *Theory of Superconductivity,* Benjamin, New York, 1964.

Schro   A. G. Schrott, S. I. Park, and C. C. Tsuei, unpublished.

Schr1   A. G. Schrott, S. L. Cohen, T. R. Dinger, F. J. Himpsel, J. A. Yarmoff, K. G. Frase, S. I. Park, and R. Purtell, A.V.S. Topical Conf. on Thin Film Processing and Characterization of High Temp. Supercond., Anaheim, California, November, 1987; AIP #160.

Schr2   J. R. Schrieffer, X. G. Wen, and S. C. Zhang, Phys. Rev. Lett. *60,* 944 (1988).

Schut   H. B. Schüttler, M. Jarrell, and D. J. Scalapino, J. Low Temp. Phys. *69,* 159 (1987).

Schu1   I. K. Schuller, D. G. Hinks, M. A. Beno, D. W. Capone, II, L. Soderholm, J. P. Locquet, Y. Bruynseraede, C. U. Segre, and K. Zhang, Sol. St. Comm. *63,* 385 (1987).

Schu2   I. K. Schuller, D. G. Hinks, J. D. Jorgensen, L. Soderholm, M. Beno, K. Zhang, C. U. Segre, Y. Bruynseraede, and J. P. Lacquet, Novel SC (1987), p. 647.

Schu3   H. B. Schüttler, M. Jarrell, and D. J. Scalapino, Novel SC (1987), p. 481.

Schwa   B. B. Schwartz, and S. Foner (Eds.), *Superconductor Applications, Squids and Machines,* Plenum, New York, 1976.

Sebek   J. Sebek, J. Stehno, S. Safrata, J. Sramek, L. Havela, V. Sechovský, Z. Sme-tana, P. Svoboda, and V. Valvoda, Czech. J. Phys. B *37*, 664 (1987).

Segre   C. U. Segre, B. Dabrowski, D. G. Hinks, K. Zhang, J. D. Jorgensen, M. A. Beno, and I. K. Schuller, Nature *329*, 227 (1987).

Semba   K. Semba, S. Tsurumi, M. Hikita, T. Iwata, J. Noda, and S. Kurihara, Jpn. J. App. Phys. *26*, L429 (1987).

Senou   S. Senoussi, M. Oussena, M. Ribault, and G. Collin, Phys. Rev. B *36*, 4003 (1987).

Shafe   M. W. Shafer, T. Penney, and B. L. Olson, Phys. Rev. B *36*, 4047 (1987).

Shaf1   M. W. Shafer, T. Penney, and B. L. Olson, Novel SC (1987), p. 771.

Shahe   S. A. Shaheen, N. Jisrawi, Y. H. Lee, Y. Z. Zhang, M. Croft, W. L. McLean, H. Zhen, L. Rebelsky, and S. Horn, Phys. Rev. B *36*, 7214 (1987).

Shalt   D. Shaltiel, J. Genossar, A. Grayevsky, Z. H. Kalman, B. Fisher, and N. Kaplan, Sol. St. Comm. *63*, 987 (1987).

Shamo   S. I. Shamoto, S. Hosoya, M. Onoda, and M. Sato, Jpn. J. Appl. Phys. *26*, 493 (1987).

Shapi   B. Ya. Shapiro and L. V. Yefimova, Sol. St. Comm. *62*, 253 (1987).

Shap1   B. Ya. Shapiro and L. V. Yefimova, J. Low Temp. Phys. *69*, 167 (1987).

Shelt   R. N. Shelton, T. J. Folkerts, P. Klavins, and H. C. Ku, MRS Anaheim Symp. (1987), p. 49.

Sheng   Z. Z. Sheng, A. M. Hermann, A. El Ali, C. Almasan, J. Estrada, T. Datta, and R. J. Matson, Phys. Rev. Lett. *60*, 937 (1988).

Shen1   Z. Z. Sheng and A. M. Hermann, Nature *332*, 55 (1988).

Shen2   Z. Z. Sheng, W. Kiehl, J. Bennett, A. El Ali, D. Marsh, G. D. Mooney, F. Arammash, J. Smith, D. Viar, and A. M. Hermann, Appl. Phys. Lett. *52*, 1738 (1988).

Shira   G. Shirane, Y. Endoh, R. J. Birgeneau, M. A. Kastner, Y. Hidaka, M. Oda, M. Suzuki, and T. Murakami, Phys. Rev. Lett. *59*, 1613 (1987).

Shizz   Y. H. Shi, H. S. Wang, Y. G. Wang, Y. Lu, B. R. Zhao, Y. Y. Zhao, and L. Li, Sol. St. Comm., *63*, 641 (1987).

Shriv   K. N. Shrivastava, J. Phys. C: Sol. St. Phys. *20*, L789 (1987).

Siegr   T. Siegrist, S. Sunshine, D. W. Murphy, R. J. Cava, and S. M. Zahurak, Phys. Rev. B *35*, 7137 (1987).

Sigri   M. Sigrist and T. M. Rice, Z. Phys. B Cond. Mat. *68*, 9 (1987).

Simiz   S. Simizu, S. A. Friedberg, E. A. Hayri, and M. Greenblatt, Phys. Rev. B *36*, 7129 (1987).

Sishe   X. Sishen, Y. Cuiying, W. Xiaojing, C. Guangcan, F. Hanjie, C. Wei, Z. Yuq-ing, Z. Zhongxian, Y. Qiansheng, C. Genghua, L. Jingkui, and L. Fanghua, Phys. Rev. B *36*, 2311 (1987).

Skelt   E. F. Skelton, W. T. Elam, D. U. Gubser, S. H. Lawrence, M. S. Osofsky, L. E. Toth, and S. A. Wolf, Phys. Rev. B *35*, 7140 (1987).

Skel1   E. F. Skelton, S. B. Qadri, B. A. Bender, A. S. Edelstein, W. T. Elam, T. L. Francavilla, D. U. Gubser, R. L. Hotlz, S. H. Lawrence, M. S. Osofsky, L. E. Toth, and S. A. Wolf, MRS Anaheim, Symp. (1987), p. 161.

Skel2   E. F. Skelton, W. T. Elam, D. U. Gubser, V. Letourneau, M. S. Osofsky, S. B. Qadri, L. E. Toth, and S. A. Wolf, Phys. Rev. B *36*, 5713 (1987).

Skoln   M. S. Skolnick, M. K. Saker, D. S. Robertson, J. S. Satchell, L. J. Reed, J. Singleton, and D. M. S. Bagguley, J. Phys. C: Sol. St. Phys. *20*, L435 (1987).

Sleig   A. W. Sleight, J. L. Gillson, and P. E. Bierstedt, Sol. St. Comm. *17*, 27 (1975).

Slei1   A. W. Sleight, ACS Symp. (1987), p. 2.

Smeds   L. C. Smedskjaer, J. L. Routbort, B. K. Flandermeyer, S. J. Rothman, D. G. Legnini, and J. E. Baker, Phys. Rev. B *36*, 3903 (1987).

Smitz   H. H. A. Smit, M. W. Dirken, R. C. Thiel, and L. J. de Jongh, Sol. St. Comm. *64*, 695 (1987).

Smrck   O. Smrckova, D. Sýkorová, J. Dominec, K. Jurek, and L. Smrcka, Phys. Stat. Sol. *103*, K33 (1987).

Soder   L. Soderholm, K. Zhang, D. G. Hinks, M. A. Beno, J. D. Jorgensen, C. U. Segre, and I. K. Schuller, Nature *328*, 604 (1987).

Sokol   J. B. Sokoloff, Sol. St. Comm. *64*, 915 (1987).

Somek   R. E. Somekh, M. G. Blamire, Z. H. Barber, K. Butler, J. H. James, G. W. Morris, E. J. Tomlinson, A. P. Schwarzenberger, W. M. Stobbs, and J. E. Evetts, Nature *326*, 857 (1987).

Sonde   D. Sondericker, Z. Fu, D. C. Johnston, and W. Eberhardt, Phys. Rev. B *36*, 3983 (1987).

Songz   S. N. Song, Q. Robinson, S. J. Hwu, D. L. Johnson, K. R. Poeppelmeier, and J. B. Ketterson, Appl. Phys. Lett. *51*, 1376 (1987).

Soule   R. J. Soulen, Jr. and D. Van Vechten, Phys. Rev. B *36*, 239 (1987).

Sreed   K. Sreedhar, T. V. Ramakrishnan, and C. N. R. Rao, Sol. St. Comm. *63*, 835 (1987).

Sridh   S. Sridhar, C. A. Shiffman, and H. Hamdeh, Phys. Rev. B, *36*, 2301 (1987).

Stacy   A. M. Stacy, W. K. Ham, S. W. Keller, K. J. Leary, J. N. Michaels, and H. C. zur Loye, MRS Boston Symp. (1987), abstract AA5.3.

Stank   J. Stankowski, P. K. Kahol, N. S. Dalal, and J. S. Moodera, Phys. Rev. B *36*, 7126 (1987).

Stavo   M. Stavola, R. J. Cava, and E. A. Rietman, Phys. Rev. Lett. *58*, 1571 (1987).

Stav1   M. Stavola, D. M. Krol, W. Weber, S. A. Sunshine, A. Jayaraman, G. A. Kourouklis, R. J. Cava, and E. A. Rietman, Phys. Rev. B *36*, 850 (1987).

Stegl   F. Steglich, J. Arts, C. D. Brendl, W. Lieke, D. Meschede, W. Franz, and J. Schaefer, Phys. Rev. Lett. *43*, 1892 (1979).

Stein   P. Steiner, V. Kinsinger, I. Sander, B. Siegwart, S. Hüfner, and C. Politis, Z. Phys. B Cond. Mat. *67*, 19 (1987).

Stei1   H. Steinfink, J. S. Swinnea, A. Manthiram, Z. T. Sui, and J. B. Goodenough, Novel SC (1987), p. 1067.

Stern   H. Stern, Phys. Rev. *B8*, 5109 (1973); *B12*, 951 (1975).

Stewa   S. R. Stewart, Rev. Mod. Phys. *56*, 755 (1984).

Stew1   S. R. Stewart, Z. Fisk, J. O. Willis, and T. J. Smith, Phys. Rev. Lett. B *52*, 679 (1984).

Stoff   N. G. Stoffel, J. M. Tarascon, Y. Chang, M. Onellion, D. W. Niles, and G. Margaritondo, Phys. Rev. B *36*, 3986 (1987).

Stof1   N. G. Stoffel, W. A. Bonner, P. A. Morris, and B. J. Wilkens, MRS Boston Symp. (1987), p. 507.

Stras   M. Strasik and N. G. Eror, MRS Boston Symp. (1987), abstract AA4.1.

Strob   P. Strobel, J. J. Capponi, M. Marezio, and P. Monod, Sol. St. Comm. *64,* 513 (1987).

Stron   M. Strongin, D. O. Welch, and V. J. Emery, Nature *326,* 540 (1987).

Subra   M. A. Subramahian, J. C. Calabrese, C. C. Torardi, J. Gopalakrishnan, T. R. Askew, R. B. Flippen, K. R. Morrissey, U. Chowdhry, and A. W. Sleight, Nature *332,* 420 (1988).

Suena   M. Suenaga, A. Ghosh, T. Asano, R. L. Sabatini, and A. R. Moodenbaugh, MRS Anaheim Symp. (1987), p. 247; see also Novel SC (1987), p. 767.

Sugah   M. Sugahara, M. Kojima, N. Yoshikawa, T. Akeyoshi, and N. Haneji, Phys. Lett. A *125,* 429 (1987).

Sugai   S. Sugai, M. Sato, and S. Hosoya, Jpn. J. Appl. Phys. *26,* L495 (1987).

Suga1   S. Sugai, Phys. Rev. B *36,* 7133 (1987).

Sulew   P. E. Sulewski, T. W. Noh, J. T. McWhirter, A. J. Sievers, S. E. Russek, R. A. Buhrman, C. S. Jee, J. E. Crow, R. E. Salomon, and G. Myer, Phys. Rev. B *36,* 2357 (1987).

Sule1   P. E. Sulewski, A. J. Sievers, R. A. Buhrman, J. M. Tarascon, L. H. Greene, and W. A. Curtin, Phys. Rev. B *35,* 8829 (1987).

Sule2   P. E. Sulewski, T. W. Noh, J. T. McWhirter, and A. J. Sievers, Phys. Rev. B *36,* 5735 (1987).

Sunzz   J. Z. Sun, D. J. Webb, M. Naito, K. Char, M. R. Hahn, J. W. P. Hsu, A. D. Kent, D. B. Mitzi, B. Oh, M. R. Beasley, T. H. Geballe, R. H. Hammond, and A. Kapitulnik, Phys. Rev. Lett. *58,* 1574 (1987).

Sutto   C. Sutton, New Sci., Jan. 29 (1987), p. 33.

Suzuk   M. Suzuki and T. Murakami, Jpn. J. App. Phys. *26,* L524 (1987).

Suzu1   M. Suzuki, Y. Enemoto, T. Murakami, and T. Inamura, Proc. 3rd Meeting Ferroelectric Materials and their Applications, Kyoto, 1981; Jpn. J. Appl. Phys. *20,* Suppl. 20-4, 13 (1981).

Suzu2   M. Suzuki, T. Murakami, Y. Enomoto, and T. Inamura, Jpn. J. Appl. Phys. *21,* L437 (1982).

Svozi   K. Svozil, Phys. Rev. B *36,* 715 (1987).

Swinb   D. Swinbanks, Nature *328,* 750 (1987).

Swinn   J. S. Swinnea and H. Steinfink, J. Mater. Res. *2,* 424 (1987).

Syono   Y. Syono, M. Kikuchi, K. Oh-Ishi, K. Hiraga, H. Arai, Y. Matsui, N. Kobayashi, T. Sasaoka, and Y. Muto, Jpn. J. Appl. Phys. *26,* L498 (1987).

Suzu2   M. Suzuki, T. Murakami, Y. Enomoto, and T. Inamura, Jpn. J. Appl. Phys. *21,* L437 (1982).

Tachi   K. Tachikawa, M. Sugimoto, N. Sadakata, and O. Kohno, MRS Boston Symp. (1987), p. 727.

Tajim   S. Tajima, S. I. Uchida, S. Tanaka, S. Kanbe, K. Kitazawa, and K. Fueki, Jpn. J. Appl. Phys. *26,* L432 (1987).

Taji1   Y. Tajima, M. Hikita, T. Ishii, H. Fuke, K. Sugiyama, M. Date, A. Yamagishi, A. Katsui, Y. Hidaka, T. Iwata, and S. Tsurumi, Phys. Rev. B *37,* 7956 (1988).

Takab   T. Takabatake, H. Takeya, Y. Nakazawa, and M. Ishikawa, Jpn. J. Appl. Phys. *26,* L502 (1987).

Takag   H. Takagi, S. I. Uchida, H. Obara, K. Kishio, K. Kitazawa, K. Fueki, and
         S. Tanaka, Jpn. J. App. Phys. *26,* L434 (1987).

Takah   H. Takahashi, C. Murayama, S. Yomo, N. Mori, K. Kishio, K. Kitazawa,
         and K. Fueki, Jpn. J. Appl. Phys. *26,* L504 (1987).

Takay   E. Takayama Muromachi, Y. Uchida, Y. Matsui, and K. Kato, Jpn. J. App.
         Phys. *26,* 476 (1987).

Taka1   H. Takagi, S. I. Uchida, K. Kishio, K. Kitazawa, K. Fueki, and S. Tanaka,
         Jpn. J. Appl. Phys. *26,* L320 (1987).

Taka2   T. Takahashi, F. Maeda, S. Hosoya, and M. Sato, Jpn. J. Appl. Phys. *26,*
         L349 (1987).

Taka3   T. Takahashi, F. Maeda, H. Arai, H. Katayama-Yoshida, Y. Okabe, T. Su-
         zuki, S. Hosoya, A. Fujimori, T. Shidara, T. Koide, T. Miyahara, M.
         Onoda, S. Shamoto, and M. Sato, Phys. Rev. B *36,* 5686 (1987).

Taka4   K. Takagi, M. Hirao, M. Hiratani, H. Kakibayashi, T. Aida, and S. Takay-
         ama, MRS Boston Symp. (1987) p. 647.

Taka5   H. Takagi, S. I. Uchida, K. Kitazawa, and S. Tanaka, Jpn. J. Appl. Phys. *26,*
         L1 (1987).

Takeg   K. Takegahara, Jpn. J. Appl. Phys. *26,* L437 (1987).

Take1   K. Takegahara, H. Harima, and A. Yanase, Jpn. J. App. Phys. *26,* L352
         (1987).

Takit   K. Takita, T. Ipposhi, T. Uchino, T. Gochou, and K. Masuda, Jpn. J. Appl.
         Phys. *26,* L506 (1987).

Talia   C. Taliani, R. Zamboni, and F. Licci, Sol. St. Comm. *64,* 911 (1987).

Tana1   S. Tanaka, S. I. Uchida, H. Takagi, K. Kitazawa, K. Kishio, S. Tajima, and
         K. Fueki, MRS Anaheim Symp. (1987), p. 5.

Tangz   H. Tang, Z. Q. Qiu, Y. W. Du, G. Xiao, C. L. Chien, and J. C. Walker, Phys.
         Rev. B *36,* 4018 (1987).

Tanig   S. Tanigawa, Y. Mizuhara, Y. Hidaka, M. Oda, M. Suzuki, and T. Mura-
         kami, MRS Boston Symp. (1987), p. 57.

Taras   J. M. Tarascon, L. H. Greene, W. R. McKinnon, and G. W. Hull, Phys. Rev.
         B *35,* 7115 (1987).

Tara1   J. M. Tarascon, L. H. Greene, W. R. McKinnon, G. W. Hull, and T. H.
         Geballe, Science *235,* 1373 (1987).

Tara2   J. M. Tarascon, W. R. McKinnon, L. H. Greene, G. W. Hull, B. G. Bagley,
         E. M. Vogel, and Y. LePage, "Processing and Superconducting Properties
         of Perovskite Oxides," Bellcore preprint (1987).

Tara3   J. M. Tarascon, L. H. Greene, B. G. Bagley, W. R. McKinnon, P. Barboux,
         and G. W. Hull, Novel SC (1987), p. 705.

Tara4   J. M. Tarascon, W. R. McKinnon, L. H. Greene, G. W. Hull, and E. M.
         Vogel, Phys. Rev. B *36,* 226 (1987).

Tara5   J. M. Tarascon, L. H. Greene, W. R. McKinnon, and G. W. Hull, Sol. St.
         Comm. *63,* 499 (1987).

Tara6   J. M. Tarascon, L. H. Greene, P. Barboux, W. R. McKinnon, G. W. Hull,
         T. P. Orlando, K. A. Delin, S. Foner, and E. J. McNiff, Jr., Phys. Rev. B
         36, 8393 (1987).

Tara7   J. M. Tarascon, W. R. McKinnon, L. H. Greene, G. W. Hull, B. G. Bagley,
         E. M. Vogel, and Y. LePage, MRS Anaheim Symp. (1987), p. 65.

Tara8   J. M. Tarascon, P. Barboux, B. G. Bagley, L. H. Greene, W. R. McKinnon, and G. W. Hull, ACS Symp. (1987), p. 198.

Tara9   J. M. Tarascon, Y. LePage, P. Barboux, B. G. Bagley, L. H. Greene, W. R. McKinnon, G. W. Hull, M. Giroud, and D. M. Hwang, Phys. Rev. B *37*, 9382 (1988).

Temme   W. M. Temmerman, G. M. Stocks, P. J. Durham, and P. A. Sterne, J. Phys. F: Met. Phys. *17*, L135 (1987).

Tengz   M. K. Teng, D. X. Shen, L. Chen, C. Y. Yi, and G. H. Wang, Phys. Lett. A *124*, 363 (1987).

Terad   N. Terada, H. Ihara, M. Hirabayashi, K. Senzaki, Y. Kimura, K. Murata, M. Tokumoto, O. Shimomura, and T. Kikegawa, Jpn. J. Appl. Phys. *26*, L510 (1987).

Terak   K. Terakura, H. Ishida, K. T. Park, A. Yanase, and N. Hamada, Jpn. J. Appl. Phys. *26*, L512 (1987).

Tera1   N. Terada, H. Ihara, M. Hirabayashi, K. Senzaki, Y. Kimura, K. Murata, and M. Tokumoto, Jpn. J. Appl. Phys. *26*, L508 (1987).

Tesan   Z. Tesanović, Phys. Rev. B *36*, 2364 (1987).

Tesme   J. R. Tesmer, C. J. Maggiore, M. Nastasi, S. W. Cheong, and C. M. Dick, MRS Boston Symp. (1987), p. 643.

Testa   J. A. Testa, Y. Song, X. D. Chen, J. P. Golben, R. D. McMichael, S. I. Lee, B. R. Patton, and J. R. Gaines, MRS Boston Symp. (1987), p. 357.

Thanh   T. D. Than, A. Koma, and S. Tanaka, Appl. Phys. *22*, 205 (1980).

Thiel   J. Thiel, S. Song, J. B. Ketterson, and K. Poepplemeier, ACS Symp. (1987), p. 173.

Thoma   G. A. Thomas, R. N. Bhatt, A. Millis, R. Cava, and E. Rietman, Jpn. J. Appl. Phys. *26*, Suppl. 26-3, 1001 (1987).

Thomp   J. R. Thompson, S. T. Sekula, D. K. Christen, B. C. Sales, L. A. Boatner, and Y. C. Kim, Phys. Rev. B *36*, 718 (1987); see also Thom4.

Thom1   G. A. Thomas, A. J. Millis, R. N. Bhatt, R. J. Cava, and E. A. Rietman, Phys. Rev. B *36*, 736 (1987).

Thom2   J. R. Thompson, D. K. Christen, S. T. Sekula, B. C. Sales, and L. A. Boatner, Phys. Rev. B *36*, 836 (1987).

Thom3   G. A. Thomas, H. K. Ng, A. J. Millis, R. N. Bhatt, R. J. Cava, E. A. Rietman, D. W. Johnson, Jr., G. P. Espinosa, and J. M. Vandenberg, Phys. Rev. B *36*, 846 (1987).

Thom4   J. R. Thompson, D. K. Christen, S. T. Sekula, J. Brynestad, and Y. C. Kim, J. Mater. Res. *2*, 779 (1987).

Thorn   R. J. Thorn, ACS Symp. (1987), p. 25.

Thoul   D. J. Thouless, Phys. Rev. B *36*, 7187 (1987).

Tinkh   M. Tinkham, *Introduction to Superconductivity*, Krieger, Florida, 1985.

Tjuka   E. Tjukanov, R. W. Cline, R. Krahn, M. Hayden, M. W. Reynolds, W. N. Hardy, J. F. Carolan, and R. C. Thompson, Phys. Rev. B *36*, 7244 (1987).

Togan   K. Togano, H. Kumakura, K. Fukutomi, and K. Tachikawa, Appl. Phys. Lett. *51*, 136 (1987).

Tokum   M. Tokumoto, H. Ihara, K. Murata, M. Hirabayashi, N. Terada, K. Senzaki, and Y. Kimura, Jpn. J. App. Phys. *26*, L515 (1987).

Tokur   Y. Tokura, J. B. Torrance, A. I. Nazzal, T. C. Huang, and C. Ortiz, J. Am. Chem. Soc. *109*, 7555 (1987).

Toku1   M. Tokumoto, M. Hirabayashi, H. Ihara, K. Murata, N. Terada, K. Senzaki, and Y. Kimura, Jpn. J. Appl. Phys. *26*, L517 (1987).

Toku2   M. Tokumoto, Y. Nishihara, K. Oka, and H. Unoki, Nature *330*, 48 (1987).

Tomyz   C. V. Tomy, A. M. Umarji, D. T. Adroja, S. K. Malik, R. Prasad, N. C. Soni, A. Mohan, and C. K. Gupta, Sol. St. Comm. *64*, 889 (1987).

Tonou   M. Tonouchi, Y. Fujiwara, S. Kita, T. Kobayashi, M. Takata, and T. Yamashita, Jpn. J. Appl. Phys. *26*, L519 (1987).

Torar   C. C. Torardi, E. M. McCarron, P. E. Bierstedt, A. W. Sleight, and D. E. Cox, Sol. St. Comm. *64*, 497 (1987).

Tora1   C. C. Torardi, E. M. McCarron, M. A. Subramanian, H. S. Horowitz, J. B. Michel, A. W. Sleight, and D. E. Cox, ACS Symp. (1987), p. 153.

Tora2   C. C. Torardi, M. A. Subramanian, J. C. Calabrese, J. Gopalakrishnan, K. J. Morrissey, T. R. Askew, R. B. Flippen, U. Chowdhry, and A. W. Sleight, Science *240*, 631 (1988).

Torra   J. B. Torrance, E. M. Engler, V. Y. Lee, A. I. Nazzal, Y. Tokura, M. L. Ramirez, J. E. Vazquez, R. D. Jacowitz, and P. M. Grant, ACS Symp. (1987), p. 85.

Torr1   J. B. Torrance, Y. Tokura, A. Nazzal, and S. S. P. Parkin, Phys. Rev. Lett. *60*, 542 (1988).

Tothz   L. E. Toth, M. S. Osofsky, S. A. Wolf, E. F. Skelton, S. B. Quadri, W. W. Fuller, D. U. Gubser, J. Wallace, C. S. Pende, A. K. Singh, S. Lawrence, W. T. Elam, B. Bender, and J. R. Spann, ACS Symp. (1987), p. 228.

Tozer   S. W. Tozer, A. W. Kleinsasser, T. Penney, D. Kaiser, and F. Holtzberg, Phys. Rev. Lett. *59*, 1768 (1987).

Tranq   J. M. Tranquada, S. M. Heald, A. R. Moodenbaugh, and M. Suenaga, Phys. Rev. B *35*, 7187 (1987).

Tran1   J. M. Tranquada, S. M. Heald, and A. R. Moodenbaugh, Phys. Rev. B *36*, 5263 (1987).

Trocz   R. Troc, Z. Bukowski, R. Horyn, and J. Klamut, Phys. Lett. A *125*, 222 (1987).

Tsaiz   J. S. Tsai, Y. Kubo, and J. Tabuchi, Phys. Rev. Lett. *58*, 1979 (1987).

Tsai1   J. S. Tsai, Y. Kubo, and J. Tabuchi, MRS Anaheim Symp. (1987), p. 125.

Tsaur   B. Y. Tsaur, M. S. Dilorio, and A. J. Strauss, Appl. Phys. Lett. *51*, 858 (1987).

Tsuda   T. Tsuda, H. Yasuoka, and J. P. Remeika, J. Phys. Soc. Jpn. *56*, 3032 (1987).

Tsuei   C. C. Tsuei, P. P. Freitas, J. R. Kirtley, S. I. Park, W. J. Gallagher, C. C. Chi, T. S. Plasskett, M. W. Shafer, T. R. Dinger, R. L. Sandstrom, D. A. Chance, J. R. Rozen, Z. Schlesinger, R. L. Greene, and R. T. Collins, MRS Anaheim Symp. (1987), p. 103.

Tuomi   M. Tuominen, A. M. Goldman, and M. L. Mecartney, Phys. Rev. B *37*, 548 (1988).

Tuzzz   K. N. Tu, S. I. Park, and C. C. Tsuei, MRS Boston Symp. (1987) abstract AA3.7.

Uchid   S. I. Uchida, H. Takagi, K. Kishio, K. Kitazawa, K. Fueki, and S. Tanaka, Jpn. J. Appl. Phys. *26*, L443 (1987).

Uchi1   S. I. Uchida, H. Takagi, H. Ishii, H. Eisaki, T. Yabe, S. Tajima, and S. Tanaka, Jpn. J. Appl. Phys. *26*, L440 (1987).

Uchi2   S. I. Uchida, H. Takagi, H. Yanagisawa, K. Kishio, K. Kitazawa, K. Fueki, and S. Tanaka, Jpn. J. Appl. Phys. *26*, L445 (1987).

Uemur   Y. J. Uemura, W. J. Kossler, X. H. Yu, J. R. Kempton, H. E. Schone, D. Opie, C. E. Stronach, D. C. Johnston, M. S. Alvarez, and D. P. Goshorn, Phys. Rev. Lett. *59*, 1045 (1987).

Uherz   C. Uher, J. Appl. Phys. *62*, 4636 (1987).

Uher1   C. Uher, A. B. Kaiser, E. Gmelin, and L. Walz. Phys. Rev. B *36*, 5676 (1987).

Uher2   C. Uher and A. B. Kaiser, Phys. Rev. B *36*, 5680 (1987).

Uher3   C. Uher and A. B. Kaiser, Phys. Lett. A *125*, 421 (1987).

Uher4   C. Uher and A. B. Kaiser, Phys. Rev. *37*, 127 (1988).

Umeza   A. Umezawa, G. W. Crabtree, J. Z. Liu, H. W. Weber, W. K. Kwok, L. H. Nunez, T. J. Moran, C. H. Sowers, and H. Claus, Phys. Rev. B *36*, 7151 (1987).

Vakni   D. Vaknin, S. K. Sinha, D. E. Moncton, D. C. Johnston, J. M. Newsam, C. R. Safinya, and H. E. King, Jr., Phys. Rev. Lett. *58*, 2802 (1987).

VanBe   P. J. M. van Bentum, L. E. C. van de Leemput, L. W. M. Schreurs, P. A. A. Teunissen, and H. van Kempen, Phys. Rev. B *36*, 843 (1987).

VanB1   P. J. M. van Bentum, H. van Kempen, L. E. C. van de Leemput, J. A. A. J. Perenboom, L. W. M. Schreurs, and P. A. A. Teunissen, Phys. Rev. B *36*, 5279 (1987).

Vande   J. van den Berg, C. J. van der Beek, P. H. Kes, J. A. Mydosh, G. J. Nieuwenhuys, and L. J. de Jongh, Sol. St. Comm. *64*, 699 (1987).

Vand1   J. van den Berg, C. J. van der Beek, P. H. Kes, G. J. Nieuwenhuys, J. A. Mydosh, H. W. Zandbergen, F. P. F. van Berkel, R. Steens, and D. J. W. Ijdo, Europhys. Lett. *4*, 737 (1987).

Vand2   J. van der Maas, V. A. Gasparov, and D. Pavuna, Nature *328*, 603 (1987).

Vante   G. Van Tendeloo, H. W. Zandbergen, and S. Amelinckx, Sol. St. Comm. *63*, 389 (1987).

Vant1   G. Van Tendeloo, H. W. Zandbergen, T. Okabe, and S. Amelinckx, Sol. St. Comm. *63*, 969 (1987).

Vant2   G. van Tendeloo and S. Amelinckx, Phys. Stat. Sol. *103*, K1 (1987).

Varma   C. M. Varma, S. Schmit-Rink, and E. Abrahams, Sol. St. Comm. *62*, 681 (1987); Novel SC, 355 (1987).

Vealz   B. W. Veal, W. K. Kwok, A. Umezawa, G. W. Crabtree, J. D. Jorgensen, J. W. Downey, L. J. Nowicki, A. W. Mitchell, A. P. Paulikas, and C. H. Sowers, Appl. Phys. Lett. *51*, 279 (1987).

Ventu   E. L. Venturini, D. S. Ginley, J. F. Kwak, R. J. Baughman, J. E. Schirber, and B. Morosin, MRS Anaheim Symp. (1987), p. 97.

Viege   M. P. A. Viegers, D. M. de Leeuw, C. A. H. A. Mutsaers, H. A. M. van Hal, H. C. A. Smoorenburg, J. H. T. Hengst, J. W. C. de Vries, and P. C. Zalm, J. Mater. Res. *2*, 743 (1987).

Viole   C. E. Violet, T. Datta, and H. M. Ledbetter, MRS Boston Symp. (1987), p. 375.

Vonso   S. V. Vonsovsky, Yu. A. Izyumov, and E. Z. Kurmaev, *Superconductivity in Transition Metals,* Springer Verlag, New York, 1982.

Vuong   T. H. H. Vuong, D. C. Tsui, V. J. Goldman, P. H. Hor, R. L. Meng, and C. W. Chu, Sol. St. Comm. *63,* 525 (1987).

Wagen   T. J. Wagener, Y. Gao, J. H. Weaver, A. J. Arko, B. Flandermeyer, and D. W. Capone, II, Phys. Rev. B *36,* 3899 (1987).

Wagne   G. R. Wagner, A. J. Panson, and A. I. Braginski, Phys. Rev. B *36,* 7124 (1987).

Walst   R. E. Walstedt, W. W. Warren, Jr., R. F. Bell, G. F. Brennert, G. P. Espinosa, J. P. Remeika, R. J. Cava, and E. A. Rietman, Phys. Rev. B *36,* 5727 (1987).

Wangz   Z. Z. Wang, J. Clayhold, N. P. Ong, J. M. Tarascon, L. H. Greene, W. R. McKinnon, and G. W. Hull, Phys. Rev. *36,* 7222 (1987).

Wang1   Z. Wang, N. Zou, J. Pang, and C. Gong, Sol. St. Comm. *64,* 531 (1987).

Wang2   X. Z. Wang, M. Henry, J. Livage, and I. Rosenman, Sol. St. Comm. *64,* 881 (1987).

Wappl   R. Wäppling, O. Hartmann, J. P. Senateur, R. Madar, A. Rouault, and A. Yaouanc, Phys. Lett. A *122,* 209 (1987).

Warre   W. W. Warren, Jr., R. E. Walstedt, G. F. Brennert, G. P. Espinosa, and J. P. Remeika, Phys. Rev. Lett. *59,* 1860 (1987).

Warr1   W. W. Warren, Jr., R. E. Walstedt, R. F. Bell, G. F. Brennert, R. J. Cava, G. P. Espinosa, J. P. Remeika, and E. A. Reitman, MRS Boston Symp. (1987), p. 379.

Watan   I. Watanabe, K. I. Kumagai, Y. Nakamura, T. Kimura, Y. Nakamichi, and H. Nakajima, J. Phys. Soc. Jpn. *56,* 3028 (1987).

Weave   J. H. Weaver, Y. Gao, T. J. Wagener, B. Flandermeyer, and D. W. Capone, II, Phys. Rev. B *36,* 3975 (1987).

Webbz   C. Webb, S. L. Weng, J. N. Eckstein, N. Missert, K. Char, D. G. Schlom, E. S. Hellman, M. R. Beasley, A. Kapitulnik, and J. S. Harris, Jr., Appl. Phys. Lett. *51,* 1191 (1987).

Weber   W. Weber, Phys. Rev. Lett. *58,* 1371 (1987); erratum *58* (1987).

Welch   D. O. Welch, M. Suenaga, and T. Asano, Phys. Rev. B, *36,* 2390 (1987).

Welc1   D. O. Welch, M. Suenaga, and T. Asano, Novel SC (1987), p. 764.

Wells   F. C. Wellstood, M. J. Ferrari, J. Clarke, A. M. Stacy, A. Zettl, and M. L. Cohen, Phys. Lett. A *122,* 61 (1987).

Wenge   L. E. Wenger, J. T. Chen, G. W. Hunter, and E. M. Logothetis, Phys. Rev. B *35,* 7213 (1987).

Weng1   L. E. Wenger, J. T. Chen, C. J. McEwan, and E. M. Logothetis, MRS Anaheim Symp. (1987), p. 121.

White   A. E. White, K. T. Short, D. C. Jacobson, J. M. Poate, R. C. Dynes, P. M. Mankiewich, W. J. Skocpol, R. E. Howard, M. Anzlowar, K. W. Baldwin, A. F. J. Levi, J. R. Kwo, T. Hsieh, and M. Hong, MRS Boston Symp. (1987), p. 531.

Will1   J. O. Willis, J. R. Cost, R. D. Brown, J. D. Thompson, and D. E. Peterson, MRS Boston Symp. (1987), p. 391.

Wilso   J. A. Wilson, J. Phys. C: Sol. St. Phys. *20*, L911 (1987).

Wittl   A. Wittlin, R. Liu, M. Cardona, L. Genzel, W. König, and F. García Alvarado, Sol. St. Comm. *64*, 477 (1987).

Worth   T. K. Worthington, W. J. Gallagher, T. R. Dinger, and R. L. Sandstrom, Novel SC (1987), p. 781.

Wort1   T. K. Worthington, W. J. Gallagher, and T. R. Dinger, Phys. Rev. Lett. *59*, 1160 (1987).

Wrigh   W. H. Wright, D. J. Holmgren, T. A. Friedmann, M. P. Maher, B. G. Pazol, and D. M. Ginsberg, J. Low Temp. Phys. *68*, 109 (1987).

Wrobe   J. M. Wrobel, S. Wang, S. Gygax, B. P. Clayman, and L. K. Peterson, Phys. Rev. B, *36*, 2368 (1987).

Wuhlz   H. Wuhl, I. Apfelstedt, M. Dietrich, J. Ecke, W. H. Fietz, J. Fink, R. Flukiger, E. Gering, F. Gompf, H. Küpfer, N. Nücker, B. Obst, C. Politis, W. Reichardt, B. Renker, H. Rietschel, W. Schauer, and F. Weiss, MRS Anaheim Symp. (1987), p. 189.

Wuzzz   M. K. Wu, J. R. Ashburn, C. J. Torng, P. H. Hor, R. L. Meng, L. Gao, Z. J. Huang, Y. Q. Wang, and C. W. Chu, Phys. Rev. Lett. *58*, 908 (1987).

Wuzz1   X. D. Wu, T. Venkatesan, D. Dijkkamp, P. Barboux, and J. M. Tarascon, MRS Boston Symp. (1987), abstract AA7.27

Wuzz2   M. K. Wu, J. R. Ashburn, C. J. Torng, G. L. E. Peng, F. R. Szofran, P. H. Hor, and C. W. Chu, MRS Anaheim Symp. (1987), p. 69.

Wuzz3   P. T. Wu, J. H. Kung, A. K. Li, C. C. Kao, C. M. Wang, L. Chang, H. H. Yen, S. C. Tsai, G. F. Chi, M. F. Tai, and H. C. Ku, MRS Anaheim Symp. (1987), p. 197.

Wuzz4   X. D. Wu, D. Dijkkamp, S. B. Ogale, A. Inam, E. W. Chase, P. F. Miceli, C. C. Chang, J. M. Tarascon, and T. Venkatesan, Appl. Phys. Lett. *51* 861 (1987).

Wuzz5   H. S. Wu, Z. Y. Weng, G. Ji, and Z. F. Zhou, J. Phys. Chem. Solids *48*, 395 (1987).

Wyck2   R. W. G. Wyckoff, *Crystal Structures*, Vol. 2, Wiley, New York, 1964.

Wyck3   R. W. G. Wyckoff, *Crystal Structures*, Vol. 3, Wiley, New York, 1965.

Wysok   K. I. Wysokiński, Sol. St. Comm. *64*, 89 (1987).

Xianr   M. Xian-Ren, R. Yan-Ru, L. Ming-Zhu, T. Qing-Yun, L. Zhen-Jin, S. Li-Hua, D. Wei-Qing, F. Min-Hua, M. Qing-Yun, Li Chang-Jiang, L. Xiu-Hai, Q. Guan-Liang, and C. Mou-Yuan, Sol. St. Comm. *64*, 325 (1987).

Xiaoz   G. Xiao, F. H. Streitz, A. Gavrin, Y. W. Du, and C. L. Chien, Phys. Rev. B, *35*, 8782 (1987).

Xiao1   G. Xiao, F. H. Streitz, A. Gavrin, and C. L. Chien, Sol. St. Comm., *63*, 817 (1987).

Xiao2   G. Xiao, F. H. Streitz, A. Gavrin, M. Z. Cieplak, J. Childress, M. Lu, A. Zwicker, and C. L. Chien, Phys. Rev. B *36*, 2382 (1987).

Xiao3   G. Xiao, M. Z. Cieplak, A. Gavrin, F. H. Streitz, A. Bakhshai, and C. L. Chien, Phys. Rev. Lett. *60*, 1446 (1988); see also MRS Boston Symp. (1987), p. 399.

Xiazz   T. K. Xia and X. C. Zeng, J. Phys. C: Sol. St. Phys. *20*, L907 (1987).

Xuzzz   J. H. Xu, T. J. Watson-Yang, J. Yu, and A. J. Freeman, Phys. Lett. A *120*, 489 (1987).

Xuzz1   X. T. Xu, J. K. Liang, S. S. Xie, G. C. Che, X. Y. Shao, Z. G. Duan, and C. G. Cui, Sol. St. Comm. *63*, 649 (1987).

Xuzz2   J. H. Xu, Sol. St. Comm. *64*, 893 (1987).

Xuzz3   M. F. Xu, H. P. Baum, A. Schenstrom, B. K. Sarma, M. Levy, K. J. Sun, L. E. Toth, M. Osofsky, S. A. Wolf, and D. U. Gubser, MRS Boston Symp. (1987), abstract AA4.50.

Yakhm   J. V. Yakhmi, I. K. Gopalakrishnan, L. C. Gupta, A. M. Umarji, R. Vijay-araghavan, and R. M. Iyer, Phys. Rev. B *35*, 7122 (1987).

Yamad   T. Yamada, K. Kinoshita, A. Matsuda, T. Watanabe, and Y. Asano, MRS Anaheim Symp. (1987), p. 119.

Yamag   Y. Yamaguchi, H. Yamauchi, M. Ohashi, H. Yamamoto, N. Shimoda, M. Kikuchi, and Y. Syono, Jpn. J. Appl. Phys. *26*, L447 (1987).

Yama1   K. Yamada, E. Kudo, Y. Endoh, Y. Hidaka, M. Oda, M. Suzuki, and T. Murakami, Sol. St. Comm. *64*, 753 (1987).

Yangz   K. N. Yang, Y. Dalichaouch, J. M. Ferreira, B. W. Lee, J. J. Neumeier, M. S. Torikachvili, H. Zhou, and M. B. Maple, Sol. St. Comm. *63*, 515 (1987).

Yang1   K. N. Yang, Y. Dalichaouch, J.M. Ferreira, R. R. Hake, B. W. Lee, M. B. Maple, J. J. Neumeier, M. S. Torikachvili, and H. Zhou, MRS Anaheim Symp. (1987), p. 77.

Yang2   K. N. Yang, Y. D. Dalichaouch, J. M. Ferreira, R. R. Hake, B. W. Lee, J. J. Neumeier, M. S. Torikachvili, H. Zhou, and M. B. Maple, Jpn. J. Appl. Phys. *26*, Suppl. 26-3 (1987).

Yang3   C. Y. Yang, S. M. Heald, J. J. Reilly, and M. Suenaga, MRS Boston Symp. (1987), p. 969.

Yang4   B. X. Yang, S. Mitsuda, G. Shirane, Y. Yamaguchi, H. Yamauchi, and Y. Syono, J. Phys. Soc. Jpn. *56*, 2283 (1987).

Yanzz   M. F. Yan, R. L. Barns, H. M. O'Bryan, Jr., P. K. Gallagher, R. C. Sher-wood, and S. Jin, Appl. Phys. Lett. *51*, 532 (1987).

Yanz1   S. Yan, P. Lu, H. Ma, Q. Jia, and X. Wang, Sol. St. Comm. *64*, 537 (1987).

Yanz2   Q. W. Yan, P. L. Zhang, L. Jin, Z. G. Shen, J. K. Zhao, Y. Ren, Y. N. Wei, T. D. Mao, C. X. Liu, T. S. Ning, K. Sun, and Q. S. Yang, Phys. Rev. B *36*, 5599 (1987).

Yaozh   R. Yaozhong, H. Xuelong, Z. Yong, Q. Yitai, C. Zuyao, W. Ruiping, and Z. Qirui, Sol. St. Comm. *64*, 467 (1987).

Yarmo   J. A. Yarmoff, D. R. Clarke, W. Drube, U. O. Karlsson, A. Taleb-Ibrahimi, and F. J. Himpsel, Phys. Rev. B *36*, 3967 (1987).

Yehzz   W. J. Yeh, L. Chen, F. Xu, B. Bi, and P. Yang, Phys. Rev. B *36*, 2414 (1987).

Yenin   W. Yening, S. Huimin, Z. Jinsong, X. Ziran, G. Min. N. Zhongmin, and Z. Zhifang, J. Phys. C: Sol. St. Phys. *20*, L665 (1987).

Yeshu   Y. Yeshurun, I. Felner, and H. Sompolinsky, Phys. Rev. B *36*, 840 (1987).

Yinhu   S. Yinhuan, Z. Bairu, Z. Yuying, W. Yonggang, W. Huishen, L. Yong, and L. Lin, Sol. St. Comm. *63*, 661 (1987).

274 REFERENCES

Yizzz   S. Yi, S. Yin-long, and F. S. Liu, Sol. St. Comm. *64*, 93 (1987).

Yongz   Z. Yong, Z. Qirui, K. Weiyan, X. Jansheng, H. Zhenhui, S. Shifang, C. Zuyao, Q. Yitai, and P. Guoqiang, Sol. St. Comm. *64*, 885 (1987).

Yoshi   R. Yoshizaki, T. Iwazumi, H. Sawada, I. Ikeda, and E. Matsuura, Jpn. J. Appl. Phys. *26*, L311 (1987).

Yosh1   R. Yoshizaki, T. Iwazumi, H. Sawada, and E. Matsuura, Jpn. J. Appl. Phys. *26*, L316 (1987).

Yosh2   R. Yoshizaki, T. Hikata, T. Han, T. Iwazumi, H. Sawada, T. Sakudo, T. Suzuki, and E. Matsuura, Jpn. J. Appl. Phys. *26*, 1129 (1987), Suppl. 26-3.

Yosh3   R. Yoshizaki, H. Sawada, T. Iwazumi, Y. Saito, Y. Abe, H. Ikeda, K. Imai, and I. Nakai, Jpn. J. Appl. Phys. *26*, L1703 (1987).

Youzz   H. You, R. K. McMullan, J. D. Axe, D. E. Cox, J. Z. Liu, G. W. Crabtree, and D. J. Lam, Sol. St. Comm. *64*, 739 (1987).

Yuech   Z. Yuechao, Z. Shuyuan, Z. Guien, W. Ziqin, C. Zuyao, Q. Yitai and Z. Qirui, Sol. St. Comm. *64*, 493 (1987).

Yushe   H. Yusheng, Z. Baiwen, L. Sihan, X. Jiong, L. Yongming, and C. Haoming, J. Phys. F: Met. Phys. *17*, L243 (1987).

Yuzzz   J. Yu, A. J. Freeman, and J. H. Xu, Phys. Rev. Lett. *58*, 1035 (1987).

Yuzz1   J. Yu, A. J. Freeman, and S. Massidda, Novel SC (1987), p. 367.

Yuzz2   J. Yu, S. Massidda, and A. J. Freeman, Northwestern University preprint (1988).

Yuzz3   R. C. Yu, M. J. Naughton, P. M. Chaikin, F. Holtzberg, R. L. Greene, J. Stuart, and P. Davies, Phys. Rev. B *37*, 7963 (1988).

Zache   R. A. Zacher, Phys. Rev. B *36*, 7115 (1987).

Zandb   H. W. Zandbergen, G. van Tendeloo, T. Okabe, and S. Amelinckx, Phys. Stat. Sol. *103*, 45 (1987).

Zandl   H. W. Zandbergen, K. Wang, and R. Gronsky, MRS Boston Symp. (1987) p. 553.

Zand2   H. W. Zandbergen, Y. K. Huang, M. J. V. Menken, J. N. Li, K. Kadowaki, A. A. Menovsky, G. Van Tendeloo, and S. Amelinckx, Nature *332*, 620 (1988).

Zelle   H. R. Zeller and I. Giaever, Phys. Rev. *181*, 789 (1969).

Zhang   Y. C. Zhang, J. H. Liu, K. Dwight, P. H. Rieger, and A. Wold, Sol. St. Comm. *63*, 765 (1987).

Zhan1   Q. Zhang, Y. Zhao, J. Xia, Z. He, S. Sun, M. Zhang, L. Cao, Z. Chen, Y. Qian, G. Pan, and M. Qian, Phys. Lett. A *124*, 457 (1987).

Zhan2   J. P. Zhang, H. Shibahara, D. J. Li, L. D. Marks, J. B. Wiley, S. J. Hwu, K. R. Poeppelmeier, S. N. Song, J. B. Ketterson, and B. Wood, Northwestern University preprint, 1988.

Zhaoj   C. Zhaojia, Z. Yong, Y. Hongshun, C. Zuyao, Z. Dongnin, Q. Yitai, W. Baimei, and Z. Qirui, Sol. St. Comm. *64*, 685 (1987).

Zhaoz   Z. Zhao, L. Chen, Q. Yang, Y. Huang, G. Chen, R. Tang, G. Liu, C. Cui, L. Chen, L. Wang, S. Guo, S. Li, and J. Bi, to be published in Kexue, Tongbao (1987).

Zhao1   B. R. Zhao, Y. H. Shi, Y. Lu, H. S. Wang, Y. Y. Zhao, and L. Li, Sol. St. Comm. *63*, 409 (1987).

Zhao2   G. L. Zhao, Y. Xu, W. Y. Ching, and K. W. Wong, Phys. Rev. B *36,* 7203 (1987).

Zhong   Y. Zhongjin, Z. Naiping, J. Ziaoping, P. Dexing, Q. Hongbo, S. Guoyue, Z. Ze, and Y. Huafeng, J. Phys. C: Sol. St. Phys. *20,* L351 (1987).

Zhon1   Y. Zhongjin, Z. Jie, and X. Yinghua, J. Phys. C: Sol. St. Phys. *20,* L843 (1987).

Zhon2   Y. Zhongjin, S. Jianxian, Z. Jingsheng, Z. Jie, and C. W. Lung, J. Phys. C: Sol. St. Phys. *20,* L923 (1987).

Zhouz   W. Zhou, J. M. Thomas, D. A. Jefferson, K. D. Mackay, T. H. Shen, I. van Damme, and W. Y. Liang, J. Phys. F: Met. Phys. *17,* L173 (1987).

Zhou1   X. Z. Zhou, A. H. Morrish, J. A. Eaton, M. Raudsepp, and Y. L. Luo, J. Phys. D: Appl. Phys. *20,* 1542 (1987).

Zhou2   X. Z. Zhou, M. Raudsepp, Q. A. Pankhurst, A. H. Morrish, Y. L. Luo, and I. Maartense, Phys. Rev. B *36,* 7230 (1987).

Zhuzz   N. Zhu, L. Zhou, Y. Zhang, T. Li, G. Qiao, and C. Shi, "Temperature Dependence of Oxygen Content and Phase Transition in $Y_1Ba_2Cu_3O_{9-x}$ Compounds," Aca. Sinica Shenyang preprint (1987).

Zimme   J. E. Zimmerman, J. A. Beall, M. W. Cromar, and R. H. Ono, Appl. Phys. Lett. *51,* 617 (1987).

Zirng   E. Zirngiebl, J. O. Willis, J. D. Thompson, C. Y. Huang, J. L. Smith, Z. Fisk, P. H. Hor, R. L. Meng, C. W. Chu, and M. K. Wu, Sol. St. Comm. *63,* 721 (1987).

Zuozz   F. Zuo, B. R. Patton, D. L. Cox, S. I. Lee, Y. Song, J. P. Golben, X. D. Chen, S. Y. Lee, Y. Cao, Y. Lu, J. R. Gaines, J. C. Garland, and A. J. Epstein, Phys. Rev. B *36,* 3603 (1987).

Zuoz1   F. Zuo, B. R. Patton, T. W. Noh, S. I. Lee, Y. Song, J. P. Golben, X. D. Chen, S. Y. Lee, J. R. Gaines, J. C. Garland, and A. J. Epstein, Sol. St. Comm. *64,* 83 (1987).

INDEX

Thermal:
  conductivity, 204, 211
  cycling, 113
  expansion, 136
Thermodynamic:
  phase diagram, 60
  state variable, 26
Thermoelectric:
  effect, 209
  power, 209
Thermogravimetric analysis (TGA), 69
Thermopower, 45, 204, 210
Thin film:
  preparation, 64
  T_c, 20
  transport, 116
Ti, 127
Time dependent effects, 163
Tl:
  energy gap ratio, 177
  resistivity, 197, 198
  superconductor, 13
$Tl_2Ba_2CaCu_2O_8$, 10, 105, 106
TlBaCaCuO, 7, 23, 164, 165
  anisotropy, 116
  atom positions, 105, 106
  density of states, 178
  structure, 105, 106
  unit cell, 106
Tm, NMR, 168
Toughness, 138
Toxicity, 59
Transition, metal to insulator, 51, 78
Transition element:
  superconductor, 11
  table, 12, 14
Transition temperature (T_c), 13, 161, 205
  definition, 23, 24
  derivative curve, 24
  formula, 45
  Hall coefficient, 207
  measurement, 23
  midpoint, 23, 25
  nonreproducible, 29
  onset, 23
  peak to peak width, 24, 27
  resistivity determination, 197
  susceptibility determination, 197
  table, 12, 14, 31, 32
  very high, 29, 221
  width, 23, 24, 27
  zero resistance, 23
Transport:
  activated, 209

current, 196, 200
hopping, 209
properties, 196
Tunneling, 212
  break junction, 214
  differential conductance, 212
  energy gap, 171, 212
  point contact, 212
  sandwich, 212
Twinning:
  benefits, 116
  boundary kinetics, 113
  definition, 115
  effect on high T_c, 55
  micro, 212
  monocrystal preparation, 67
  vortex pinning site, 116
  $YBa_2Cu_3O_7$, 116
Two-dimensional, 48
  band structure, 57
  magnetism, 159
  NMR, 168
  pressure, 131
Type I superconductivity, 33
Type II superconductivity, 33, 141, 219

UBe_{13}, 20, 21
Ultrasonic, 134, 138, 171
  attenuation, 21
Ultraviolet:
  excitons, 49
  photoemission spectroscopy (UPS), 39, 185
Unit Cell, 30, 223
  dimensions, 72
  doubling, 115
  parameters of YBaCuO, 126
  stacking, 81
UPt_3, 20

V, energy gap ratio, 177
V_3Ge electron phonon coupling constant, 45
  energy gap ratio, 177
V_3Ir, T_c, 15
V_3S:
  electron phonon coupling constant, 45
  energy gap ratio, 177
  T_c, 17
Vacancies:
  ordered, 113
  ordering, 17
Valence:
  band PES, 186, 187
  Cu, 36
  electron, 13